About Island Press

Since 1984, the nonprofit Island Press has been stimulating, shaping, and communicating the ideas that are essential for solving environmental problems worldwide. With more than 800 titles in print and some 40 new releases each year, we are the nation's leading publisher on environmental issues. We identify innovative thinkers and emerging trends in the environmental field. We work with world-renowned experts and authors to develop cross-disciplinary solutions to environmental challenges.

Island Press designs and implements coordinated book publication campaigns in order to communicate our critical messages in print, in person, and online using the latest technologies, programs, and the media. Our goal: to reach targeted audiences—scientists, policymakers, environmental advocates, the media, and concerned citizens—who can and will take action to protect the plants and animals that enrich our world, the ecosystems we need to survive, the water we drink, and the air we breathe.

Island Press gratefully acknowledges the support of its work by the Agua Fund, Inc., Annenberg Foundation, The Christensen Fund, The Nathan Cummings Foundation, The Geraldine R. Dodge Foundation, Doris Duke Charitable Foundation, The Educational Foundation of America, Betsy and Jesse Fink Foundation, The William and Flora Hewlett Foundation, The Kendeda Fund, The Forrest and Frances Lattner Foundation, The Andrew W. Mellon Foundation, The Curtis and Edith Munson Foundation, Oak Foundation, The Overbrook Foundation, the David and Lucile Packard Foundation, The Summit Fund of Washington, Trust for Architectural Easements, Wallace Global Fund, The Winslow Foundation, and other generous donors.

The opinions expressed in this book are those of the author(s) and do not necessarily reflect the views of our donors.

About SCOPE

The Scientific Committee on Problems of the Environment (SCOPE) was established by the International Council for Science (ICSU) in 1969. It brings together natural and social scientists to identify emerging or potential environmental issues and to address jointly the nature and solution of environmental problems on a global basis. Operating at an interface between the science and decision-making sectors, SCOPE's interdisciplinary and critical focus on available knowledge provides analytical and practical tools to promote further research and more sustainable management of the Earth's resources. SCOPE's members, national scientific academies and research councils, and international scientific unions, committees and societies, guide and develop its scientific program.

SCOPE 70

Watersheds, Bays, and Bounded Seas

THE SCIENTIFIC COMMITTEE ON PROBLEMS OF THE ENVIRONMENT (SCOPE)

SCOPE SERIES

SCOPE 1–59 in the series were published by John Wiley & Sons, Ltd., UK. Island Press is the publisher for SCOPE 60 as well as subsequent titles in the series.

SCOPE 60: *Resilience and the Behavior of Large-Scale Systems,* edited by Lance H. Gunderson and Lowell Pritchard Jr.

SCOPE 61: *Interactions of the Major Biogeochemical Cycles: Global Change and Human Impacts*, edited by Jerry M. Melillo, Christopher B. Field, and Bedrich Moldan

SCOPE 62: *The Global Carbon Cycle: Integrating Humans, Climate, and the Natural World,* edited by Christopher B. Field and Michael R. Raupach

SCOPE 63: *Invasive Alien Species: A New Synthesis,* edited by Harold A. Mooney, Richard N. Mack, Jeffrey A. McNeely, Laurie E. Neville, Peter Johan Schei, and Jeffrey K. Waage

SCOPE 64: *Sustaining Biodiversity and Ecosystem Services in Soils and Sediments,* edited by Diana H. Wall

SCOPE 65: *Agriculture and the Nitrogen Cycle: Assessing the Impacts of Fertilizer Use on Food Production and the Environment,* edited by Arvin R. Mosier, J. Keith Syers, and John R. Freney

SCOPE 66: *The Silicon Cycle: Human Perturbations and Impacts on Aquatic Systems,* edited by Venugopalan Ittekkot, Daniela Unger, Christoph Humborg, and Nguyen Tac An

SCOPE 67: *Sustainability Indicators: A Scientific Assessment,* edited by Tomás Hák Bedrich Moldan and Arthur Lyon Dahl

SCOPE 68: *Communicating Global Change Science to Society: An Assessment and Case Studies,* edited by Holm Tiessen, Gerhard Breulmann, Michael Brklacich, and Rômulo S. C. Menezes

SCOPE 69: *Biodiversity Change and Human Health: From Ecosystem Services to Spread of Disease,* edited by Osvaldo E. Sala, Laura A. Meyerson, and Camille Parmesan

Watersheds, Bays, and Bounded Seas

The Science and Managment of Semi-Enclosed Marine Systems

Edited by
Edward R. Urban, Jr., Bjørn Sundby, Paola Malanotte-Rizzoli, and Jerry M. Melillo

A project of SCOPE, the Scientific Committee on Problems of the Environment, of the International Council for Science

Washington • Covelo • London

Library of Congress Cataloging-in-Publication Data
Watersheds, bays, and bounded seas : the science and management of semi-enclosed marine systems / edited by Edward J. Urban, Jr. ... [et al.].
p. cm. — (SCOPE ; 70)
Includes bibliographical references and index.
ISBN-13: 978-1-59726-502-7 (cloth : alk. paper)
ISBN-10: 1-59726-502-0 (cloth : alk. paper)
1. Coastal ecology. 2. Marine ecology. 3. Marine ecosystem management.
I. Urban, Edward J.
QH541.5.C65W38 2008
577.7—dc22 2008027634

Printed on recycled, acid-free paper

Manufactured in the United States of America
10 9 8 7 6 5 4 3 2 1

Contents

Figures, Tables, and Boxes

Color plate section follows page 144.

Figures

Tables

Boxes

Plates

Foreword

This volume was developed in partnership among three organizations of the International Council for Science (ICSU): the International Association for the Physical Sciences of the Oceans (IAPSO), the Scientific Committee on Oceanic Research (SCOR), and the Scientific Committee on Problems of the Environment (SCOPE).

IAPSO (http://iapso.sweweb.net) is a constituent association of the International Union of Geodesy and Geophysics (IUGG) (http://www.iugg.org), an ICSU scientific union. IAPSO has the prime goal of promoting the study of scientific problems relating to the oceans and the interactions taking place at the seafloor, coastal, and atmospheric boundaries insofar as such research is conducted by the use of mathematics, physics, and chemistry. IAPSO participates in SCOR and interacts with UNESCO's Intergovernmental Oceanographic Commission.

SCOR (http://www.scor-int.org), the first interdisciplinary body formed by ICSU, was established in 1957 in recognition that scientific questions about the ocean often require an interdisciplinary approach. SCOR activities focus on promoting international cooperation in planning and conducting oceanographic research and on solving methodological and conceptual problems that hinder research. The SCOR secretariat is located at the College of Marine and Earth Studies at the University of Delaware (Newark).

SCOPE (http://www.icsu-scope.org), founded in 1969, is an interdisciplinary body of natural and social science expertise focused on global environmental issues, operating at the interface between scientific and decision-making instances. A worldwide network of scientists and scientific institutions develops syntheses and reviews of scientific knowledge on current or potential environmental issues. The SCOPE secretariat is located at ICSU headquarters in Paris, France.

Watersheds, Bays, and Bounded Seas is Volume 70 in the SCOPE series, now published with Island Press. This cooperative approach among three ICSU organizations to synthesize and analyze information is an excellent example of how the scientific community can help to address a complex multidisciplinary subject of global significance.

We dedicate this book to Hal Mooney, who has been a mentor to all of us and a pioneer of studies of the importance of biodiversity on human well-being.

Bernard D. Goldstein
Editor-in-Chief, SCOPE Publications

1

Introduction

Edward R. Urban, Jr., Bjørn Sundby,
Paola Malanotte-Rizzoli, and Jerry M. Melillo

Why and How This Book Was Created

Observations and images from various regions of the earth provide dramatic evidence that global change is real. Large-scale weather events that erode shorelines and flood low-lying areas make it obvious that global change affects not only the remote parts of the earth, but also the daily life of each of the planet's inhabitants. Melting ice caps and glaciers, eroding shorelines, and floods are indeed spectacular manifestations of global change. But because they are so spectacular, they may overshadow other manifestations of global change that are less visible but no less important. Regrettably, out of sight often means out of mind.

In the ocean, many effects of global change are literally out of sight because they manifest themselves below the sea surface. This also is true in the coastal ocean, the buffer zone between continents and ocean. Although global change's impacts on the hydrologic cycle will vary among regions, increasing air temperatures over the ocean are predicted (on the basis of well-known physical processes) to increase evaporation from the ocean, which may lead to increasing precipitation on the continents and increasing runoff from the continents. Increased freshwater runoff to the coastal zone—carrying with it nutrients, contaminants, and sediments—is worrisome because it can be expected to alter the local ecology; reduce the quality of the habitats of many organisms, including those useful to humans; and affect their health, growth, and ability to reproduce. The effect of global warming on runoff will not be uniformly distributed; runoff may increase in some areas and decrease in others, and regional predictions are notoriously unreliable.

Semi-enclosed marine systems (SEMSs) are important in many coastal regions; they are tightly linked with land and have restricted exchange with the open ocean. SEMSs are important to humans, who are especially numerous at the edges of the continents. Humans rely on SEMSs for often-competing services such as provision of food, protection from natural disasters, navigation and transport, disposal of waste, extraction of minerals and sand/gravel, and leisure. Society has set up institutions to manage coastal areas for the benefit of all, but these institutions often lack information, understanding, and tools to help them with their task. Thus society turns to scientists, who are trained to provide information, develop understanding, and create tools for studying the ocean. The task facing scientists is difficult, for it obliges them to venture onto the often unknown terrain where nature and human forces interact.

To address the overlapping issues of management and research in SEMSs, three organizations of the International Council for Science (ICSU)—the Scientific Committee on Problems of the Environment (SCOPE), International Association for the Physical Sciences of the Oceans (IAPSO), and Scientific Committee on Oceanic Research (SCOR)—pooled their resources and expertise to bring together a carefully designed mixture of natural scientists and social scientists. This group met for four days at the Hanse-Wissenschaftskolleg (HWK) in Delmenhorst, Germany, to deliberate on the special characteristics of SEMSs and identify management approaches and research that should be applied to these special systems. This resulting book is intended to provide information for more-effective management of SEMSs and to serve as a useful reference for coastal managers, policy makers, and scientists. It also includes original analyses of special features of SEMSs (e.g., nutrient inputs, primary production, fisheries production, and socioeconomic indicators) and directions for future research on some topics relevant to understanding SEMSs.

Definition/Description of Semi-Enclosed Marine Systems for the Purposes of This Book

A semi-enclosed marine system (SEMS) is a marginal sea, bounded by land along more than half of its periphery and separated from the open ocean by one or more of the following boundaries (Figure 1-1):

- a strait (as bounds the Baltic Sea)
- a sill and/or island chain (as bounds the East China Sea)
- a front generated by physical processes separating the coastal/shelf water from the open ocean water (as bounds the Bay of Bengal)

The SEMSs that are the focus of this book (Figure 1-2) are those that are most impacted by changes taking place on the surrounding land masses. In fact, the runoff of water with its loads of sediment and chemicals is the most important driver of the systems discussed in this book. SEMSs with positive freshwater budgets (i.e., the total of runoff and precipitation exceeds loss by evaporation) are more susceptible to influx of land-derived materials and were therefore chosen for study in this book. The systems selected are meant to provide examples of a broader set of similar systems.

The degree of openness and the efficiency of water exchange between SEMSs and the

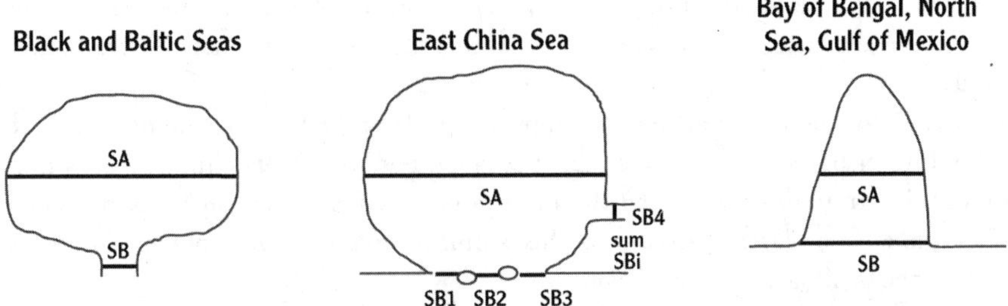

Figure 1-1. Examples of regional seas with different degrees of openness to the ocean. Provided by M. Meybeck.

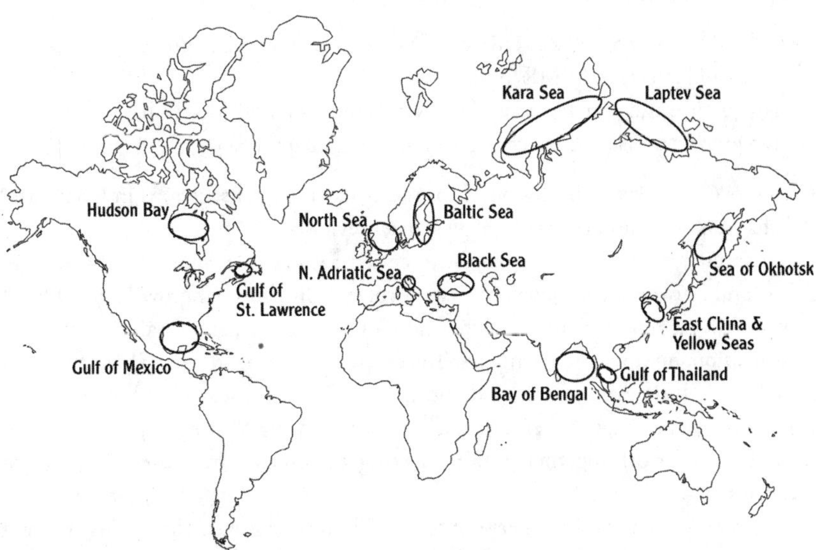

Figure 1-2. Map of the thirteen SEMSs considered in this volume (modified from Figure 10-2).

open ocean are important characteristics. The degree of openness ranges from very low (e.g., of the Black Sea) to very high (e.g., of the Bay of Bengal). As a first approximation, a measure of the degree of openness can be obtained by calculating the ratio of two cross-sectional areas: the cross-sectional area of the opening(s) to the ocean (SB in Figure 1-1) and the average cross-sectional area of the semi-enclosed body of water (SA in Figure 1-1). The thirteen systems that are the focus of this book include the Black Sea, Baltic Sea, Hudson Bay, Gulf of St. Lawrence, Northern Adriatic Sea, Gulf of Thailand, North Sea, Sea of Okhotsk, East China Sea (including the Yellow Sea), Gulf of Mexico, Laptev

Sea, Kara Sea, and the Bay of Bengal (Figure 1-2). These were selected to represent a span of latitudes and encompass different combinations of degree of openness and freshwater influence.

Because of the importance of the inputs from land, SEMSs include not only the coastal water bodies and the underlying seafloor, but also the catchment basins that drain the surrounding lands, and the adjacent ocean—hence the emphasis on system. The regional boundaries derived from this definition do not preclude potential impacts beyond the boundaries of the system.

Structure of the Book

The following chapters fall into two categories. Chapters 2–5 document the discussions that occurred at the workshop. These focused on

- vulnerability of SEMSs to environmental disturbances (Chapter 2)
- threshold effects in SEMSs (Chapter 3)
- governance and management of ecosystem services in SEMSs (Chapter 4)
- integrating tools to assess changes in SEMSs (Chapter 5)

Chapters 6–12 serve as background for the cross-cutting discussions in Chapters 2–5. All chapters and the complete book were peer reviewed.

Chapter 2 discusses the variety of forces that will cause changes to SEMSs in the future and illustrates the many processes involved. Climate change will affect SEMSs in terms of water temperatures, precipitation and ice cover, runoff, salinity, circulation, stratification, mixing, and chemistry. The chapter recommends that SEMSs be assessed as to how well they are understood and what additional research and observations are needed to enable management. Scientists, local and regional managers, and decision makers should work together in both making assessments and developing mitigation measures.

Chapter 3 discusses the concept of thresholds in ecosystems. Thresholds are environmental tipping points: When a threshold is passed, abrupt and dramatic changes, which are difficult or impossible to reverse, take place in the relationships between environmental forces and effects. Once a threshold has been surpassed, the system may not return to its initial state or it may return more slowly than expected from the speed of the initial change, even if the force or combination of forces that brought about the change is reduced to pre-threshold levels. Examples of thresholds include the oxygen level that defines hypoxia; the introduction of invasive species; the combination of light, nutrients, and mixed-layer depth that stimulates a phytoplankton bloom; the ratios of the nutrients that control the plankton species composition; the minimum population size of a fish species that ensures reproduction; and the conditions necessary for larval survival. Chapter 3 recommends that ecosystem research in SEMSs should aim to increase our understanding of thresholds and their consequences.

Chapter 4 is built around the role of the many ecosystem services that provide benefits to human society. It describes the multiscale (and often transboundary) nature of the threats to these services, discusses the vulnerability of ecosystem services in SEMSs, and describes management approaches and strategies that address loss of ecosystem services and attempt to reverse the losses. The chapter provides case studies that show the successes and failures of various governance and management strategies and discusses how past efforts to maintain ecosystem services can be used to guide future efforts in SEMSs around the world. Managing human disturbances that operate at smaller space and time scales (e.g., habitat destruction) tends to be easier than for larger-scale disturbances (e.g., ocean acidification). Habitat destruction can be handled on a local level, and preventing habitat destruction is less costly and ultimately more effective than trying to restore damaged habitats. Invasive species starting out as a local problem can spread and require national and international intervention. Point source pollution can be managed locally, but non-point source pollution entering the ocean from the air and from rivers with large watersheds and airsheds requires regional or multinational solutions. Living marine resources often travel among different political jurisdictions, and this too may require multistate or multinational management. Living marine resources pose the additional challenge that management of one species can affect other species. Most problems are associated with intermediate time and space scales.

Chapter 5 reviews the tools that are available for integrating information and knowledge across interdisciplinary boundaries and discusses how they can be used to advance our understanding of the complex dynamics and the multitude of interactions in SEMSs. It identifies the components of a conceptual integrating framework that takes into account the interactions between a SEMS and its watershed and between local people, the ecosystems, and local and international markets. Development of a comprehensive conceptual framework is a major task. Although the interactions among the various components are relatively well-known in qualitative terms, incorporating them in quantitative models remains an important scientific challenge. Integrated modeling will serve the scientific community, decision makers, and other stakeholders. The tools for integration described here should help to shape and develop policies aimed at solving the problems occurring in SEMSs.

The overall message of this book is that SEMSs are under serious pressure in most parts of the world and that the stress on a given system is often positively related to its isolation from oceanic influences. Organisms that live in SEMSs are often adapted to greater swings in environmental conditions than organisms living in less variable open coastal environments, but they are not necessarily adapted to thriving at one extreme or the other. The type of pressure and its extent differs from system to system, so it is difficult to make general statements about all the systems discussed in this book. However, a few points deserve to be highlighted.

Water Temperatures Most SEMSs will experience increasing water temperatures, which will be amplified in SEMSs that are less open to the ocean. Some organisms, notably corals, appear to be at the upper limit of their temperature tolerance, so unusually warm years can lead to widespread "coral bleaching." Systems now dominated by sea ice will be less so in a warmer world (see Chapter 6), leading to significant changes in many aspects of these ecosystems.

Ocean Acidification Tropical and subtropical SEMSs in which corals and coralline algae play a major role in ecosystem structure and function will be most affected by ocean acidification.

Runoff The changes in precipitation for the watershed of any SEMS are hard to predict, as global models are still unable to provide accurate predictions of local and regional changes in precipitation. In some systems, increased precipitation in the watershed will increase inflow of freshwater and sediments, decreasing salinity and increasing turbidity. In other systems, decreased precipitation will increase salinity and decrease turbidity. Approaches to categorizing watersheds by their hydrology and sediment fluxes (see Chapter 7) are important for improving predictions of the effects of changes of precipitation on SEMSs and for developing mitigation measures.

Pollution Increased freshwater inflow tends to increase pollution by nutrients and other chemicals. For most SEMSs considered in this book, dissolved nitrogen is more important than particulate forms of nitrogen, and dissolved phosphorus is less important (Chapters 8, 9, and 11). Dissolved nutrient inputs are enriched in nitrogen in relation to phosphorus. Nitrogen and polychlorinated biphenyls (PCBs) tend to decrease from low- to high-latitude systems. As mentioned above, different SEMSs will experience different changes in precipitation and, thus, inputs of pollutants. A major new threat on the horizon is the potential that increases in biomass production for ethanol will increase the level of inorganic nutrients entering coastal areas.

Fish Productivity Fish catch is positively related to levels of primary production in a SEMS, but not to the climate zone in which it is located (see Chapter 10). That is, tropical and high-latitude systems can be equally productive (or not). Unexpectedly, the transfer efficiency among trophic levels is not related to the size of the system or to its long-term exploitation. However, the carbon transfer per trophic level appears to increase as the degree of connectivity to the open ocean increases. This means that more-closed systems are more dependent on production generated internally and may be less resilient to overfishing and other factors that decrease their productivity, because they are less efficiently replenished and fed from the open ocean. The correlation found should be investigated further in a greater range of Large Marine Ecosystems (LMEs) to determine whether it holds as a general rule. It is imperative that climate-induced changes in marine ecosystems be taken into account in future fisheries management in SEMSs.

Management Managers should seek ways to increase ecosystem resilience to both natural changes and human pressures (see Chapter 12). (Resilience is the ability of a system to absorb temporary changes or episodic events without being shifted to an alternate state.) Ecosystem resilience seems to be positively related to biodiversity, so it is important to preserve biodiversity through marine protected areas and by moving away from single-species management and toward management of ecosystems as complex systems. Precaution in use of all ecosystem services—not utilizing the maximum services that might be available under ideal conditions—should help increase resilience and stability of natural systems. The resilience of social systems results from their ability to learn and adapt, and from stable governance.

Modeling Ecosystem management will require much greater information and more significant modeling efforts. Models will both require and help identify new indicators of ecosystem processes and status. Each SEMS should be studied and modeled separately to make it possible to institute management strategies specific to the system. Yet, scientists and managers working in one system can learn from those studying similar systems in other parts of the world. Many international marine research projects focus their research on comparable systems to gain understanding across systems; similar cooperative activities among managers working in similar systems can stimulate the sharing of successful management approaches and advancements in management practice.

Acknowledgments

The cosponsors thank the many financial contributors to the project, including ICSU (through a grant from UNESCO), the Hanse-Wissenschaftskolleg, the Center for Tropical Marine Ecology at the University of Bremen, Germany's Federal Ministry of Education and Research, the Nederlandse Organisatie voor Wetenshappelijk Onderzoek (NWO), and the Koninklijke Nederlandse Akademie van Wetenschappen (KNAW), which provided significant aid in development of the workshop and completion of the project. SCOR provided funding for travel of scientists from developing countries and countries with economies in transition with funding to SCOR from the U.S. National Science Foundation (Grant OCE-0531642). We also thank the many individuals who reviewed all or part of this book.

2

Vulnerability of Semi-Enclosed Marine Systems to Environmental Disturbances

Michael MacCracken, Elva Escobar-Briones, Denis Gilbert, Gennady Korotaev, Wajih Naqvi, Gerardo M.E. Perillo, Tim Rixen, Emil Stanev, Bjørn Sundby, Helmuth Thomas, Daniela Unger, and Edward R. Urban, Jr.

Introduction

Semi-enclosed marine systems (SEMSs) are impacted on time scales ranging from the very short (days or less) to the very long (decades and beyond) by many of the same forces and stresses as other components of the Earth system. Like the open ocean, SEMSs are linked to large-scale anthropogenic disturbances through climate change, acidification from increasing levels of atmospheric carbon dioxide (CO_2), and atmospheric deposition of contaminants. The restricted flushing of SEMSs makes them vulnerable to land-based disturbances. Locally, SEMSs are affected by agricultural runoff, urbanization, and pollution. Changes in freshwater runoff and precipitation, cloudiness, winds, and upwelling can modify sedimentation and erosion patterns and the productivity and health of these systems. Atmospheric motions, especially sea and land breezes, connect SEMSs with neighboring land areas. These motions transport gaseous and particulate matter and nutrients, buffer diurnal and seasonal temperature variations over land, and amplify them over the ocean.

While SEMSs are more vulnerable to human influences than are coastal environments that are fully connected to the open ocean, the diversity of SEMSs, in terms of location, size, depth, and coupling to adjacent land areas and the open ocean, precludes a simple, generalized description of their vulnerability. This chapter presents an overview of the most important drivers of change, the processes that are likely to be most affected, and the most important implications for SEMSs.

Significant Climatic Forcing Factors

Worldwide emissions of CO_2, other greenhouse gases, and aerosols are forcing significantly larger changes in the global climate than previous extremes and fluctuations caused by natural phenomena such as solar radiation and volcanic eruptions. Natural oscillations of the ocean–atmosphere system, such as the El Niño/Southern Oscillation, are becoming less effective in obscuring the regional manifestations of human-induced climate change. Such oscillations, which are likely to be affected by the changing climate, will continue to result in fluctuations around the changing mean climatic conditions, increasingly leading to new extremes. The Fourth Assessment Report of the Intergovernmental Panel on Climate Change (IPCC 2007) projects an increase in the global average surface temperature of about 0.2°C to 0.4°C per decade over the twenty-first century, compared with an average rate of roughly 0.06°C per decade over the twentieth century. As a result, the change from a nature-dominated to a human-dominated world will occur at an accelerating pace, amplified further as the efficiency of the ocean carbon sink decreases.

Because the ocean will warm less rapidly than continents, the land–sea contrast will increase, which will affect atmospheric circulation, cloud distributions, precipitation, and freshwater availability in coastal regions. In addition, changes in the north–south temperature gradient will alter atmospheric circulation and storm tracks on larger scales, changing precipitation patterns, river runoff, wetness of soils, and duration and extent of snow and ice cover. These globally forced changes will be augmented at the regional scale by changes in land cover, water and soil management, agriculture and forest management, urban and coastal development, energy use and development, and other factors that alter the surface albedo, soil permeability, heat capacity of the surface, and supply of freshwater and sediment to the coastal environment.

Global warming is also accelerating the rate of sea level rise, both by increasing ocean heat content, which leads to thermal expansion of ocean waters, and by increasing the rate of loss from the world's glaciers. Over the past decade, increasing deterioration of the Greenland and West Antarctic ice sheets has accelerated the rate of sea level rise. Over the next few decades, if strong controls of greenhouse gas emissions are not implemented soon, the rate of sea level rise will likely increase from the average twentieth century value of almost 20 mm per decade to 60–100 mm per decade, or even more; indeed, the recent rate is already near 40 mm per decade. Higher sea level will cause significant impact along the low-lying coastlines that surround many SEMSs.

Higher temperatures and the elevated atmospheric CO_2 concentration are also increasing the intensity of the hydrologic cycle, with higher rates of evaporation leading to a net increase in global precipitation. However, changes in the timing and amount of runoff are uncertain because evaporation from land regions will also increase and because changes in atmospheric circulation will alter precipitation patterns, causing increases in some locations (particularly in high latitudes) and decreases in others (particularly in the subtropics). More-rapid drying of land areas is likely to amplify any initial temperature increase, and if precipitation does not simultaneously increase by a sufficient amount, the result will be higher temperatures and salinities in estuaries and coastal waters. In some regions, wind changes could also amplify or diminish upwelling of colder waters.

In addition to changing mean conditions, global warming is very likely to lead to new climatic extremes. With higher atmospheric water vapor mixing ratios, heavy rain and snow events will become more common. Tropical cyclones (variously also known as hurricanes and typhoons) are projected to become more powerful, leading to significantly higher precipitation rates and wind speeds. These changes will in turn lead to higher storm surges and waves, which will greatly increase coastal damage and shore erosion, especially because of sea level rise and reduced extent and duration of sea ice.

The manifestations of large-scale changes in particular regions are likely to shift as atmospheric and oceanic circulations change and are affected by the local coastline, coastal landforms, and bathymetry. The complexity and uniqueness of the influences and driving forces in each region make it important to carefully evaluate the consequences for each SEMS.

The following sections describe the primary processes and characteristics affected by human-induced climate change and their potential influence on SEMSs; the boxes provide specific examples from around the world.

Physical Responses to Climate Forcing

Climate change will cause a wide range of changes in the physical environment. While climate change is often described in terms of the amount of warming, the largest impacts on SEMSs are likely to result from changes in the components of the hydrologic cycle that move water through the system and from changes in the supply and distribution of sediment that, along with changes in sea level and, in some regions, sea ice, will affect erosion and induce changes in the coast.

Changes in Hydrodynamics

Circulation in regional seas is dominated by regional wind patterns. These patterns are likely to shift in response to global warming as midlatitude storm tracks shift poleward. Such shifts are likely to cause changes in the location and intensity of upwelling, ice cover extent and duration, ocean stratification, regional weather, and the balance among precipitation, evaporation, and runoff. As the hydrologic cycle intensifies, the depth and strength of ocean currents will also be affected.

The temperature of the upper ocean's mixed layer is strongly affected by latent and sensible heat fluxes at the sea surface, short-wave radiation (modulated by cloudiness, atmospheric aerosol, and seawater transparency), the surface long-wave radiation budget, and the effects of water transport and mixing. Heat fluxes caused by inflowing rivers and melting and freezing of ice can have regional to global impacts on ocean ecosystems. Horizontal mixing conditioned by mesoscale processes, vertical mixing caused by winds, shear, or convective instability, and large-scale or local upwelling phenomena can affect temperature distributions.

Because the coastal ocean is significantly impacted by freshwater fluxes, warming and changes in precipitation and evaporation can alter salinity and affect important hydrologic features. These features include river plumes, salt wedges, and freshwater and thermal fronts that can alter not only vertical exchanges of heat, but also ecological and sedimentary systems. In addition, salinity in the mixed layer is directly dependent on the amount of precipitation relative to evaporation, river runoff, and sea-ice formation and melting.

Regional weather will also be affected. Changes in land and ocean surface temperatures resulting from human activities will affect the atmospheric coupling between land and ocean, intensifying the daytime sea breeze and weakening the evening land breeze. Generally, the warm season will lengthen and the cool season will become shorter. Changes in climate will generally be larger over land areas than over ocean areas, greater at night than during the day, and greater during the cold season than during the warm season, except where land areas dry out. However, the responses of particular SEMSs are likely to vary. Where the water is relatively shallow or stratified, changes are likely to be more closely related to warming over land.

The observed trend toward more-intense precipitation events is likely to continue, which is likely to make runoff more variable and even episodic, especially as warmer temperatures increase evaporation and reduce runoff between storms. In regions exposed to tropical cyclones, the warmer ocean temperatures are projected to lead to increases in the average intensity of storms (a change already evident in the Atlantic Basin), causing higher storm surges and wind-whipped waves that will penetrate farther inland. Greater duration of hurricane-level winds is likely to extend the domain and increase the frequency of exposure of various coastal regions, leading to increased inundation and erosion of coastal lands.

Changes in Freshwater Runoff and Salinity

Changes in the amount of freshwater reaching estuaries will alter temperature, salinity, and nutrient concentrations, shifting the position of the salt intrusion boundary. Decreasing runoff will cause landward displacement of organisms that are adapted to fluctuating salinities, an increase in the frequency and severity of saltwater contamination events in low-lying coastal regions, and an increase in the extent of salt transport into coastal aquifers and groundwater, impacting municipal water supplies (Nicholls and Wong 2007).

Changes in the Supply of Sediments

River input is the major source of sediment in most coastal regions (Wang et al. 1998; Chapter 7, this volume), although atmospheric transport of dust, alongshore transport of sediments, and coastal erosion can also be important. Observations of baselines and trends reveal significant regional variations in sediment supply due to natural variations of relief and erosion and to human influences such as damming, water treatment, and flood control (Restrepo and Kjerfve 2000a, 2000b; Kjerfve and Restrepo 2002; Syvitski et al. 2005a). Both flood control measures and land-use and land-cover modification have affected sediment delivery (Syvitski et al. 2005b; Syvitski and Milliman 2007). The sediment source term has increased by an estimated 2.3 ± 0.6 billion metric tons, or gigatonnes (Gt), per year (Gt y^{-1}) because of deforestation, soil mismanagement, and other causes, but about 60% of this increase (1.4 ± 0.3 Gt y^{-1}) is retained by reservoirs. An estimated 100 Gt of sediment have been sequestered in reservoirs built over the last fifty years, depriving the coastal zone of a substantial source of sediments.

Severe storms and tropical cyclones can have devastating effects on sediment transport. For example, Hurricane Agnes (1972), with its particularly strong rains, flushed out large amounts of sediment that had built up in the drainage basin over previous decades, carrying the sediment into Chesapeake Bay and wiping out the benthic ecosystem in estuaries and on the shelf (Meade and Trimble 1974; Gross et al. 1978). Observations already indicate that a larger fraction of the precipitation is coming in heavy rainfall events (Trenberth and Jones 2007), and projections are that climate change will lead to more-powerful tropical cyclones that drench coastal regions with substantially increased precipitation (Meehl and Stocker 2007). As one example, Box 2-1 describes interactions that are likely to result as the Gulf of Mexico and Caribbean Sea warm.

Increase in Sea Level and Higher Storm Surges

Thermal expansion of ocean waters, melting of mountain glaciers, and changes in snow accumulation and melting on the surfaces of the Greenland and West Antarctic ice sheets are projected to cause an increase in global sea level of 0.18 to 0.59 m by 2100 (IPCC 2007). Significant additional contributions from the Greenland and West Antarctic ice sheets are likely as a result of rapid dynamical changes in ice flow that are already becoming evident in some glacial streams. Although sea level rose about 120 m as the last glacial period ended, relative constancy of sea levels over the past several thousand years allowed development of mangroves, corals, sea grass beds, and other coastal features that have tended to stabilize the existing coastline. The dynamic state of equilibrium that came to exist is being disturbed by human activities, including by climate change, cutting of mangroves, and building of dams (Syvitski et al. 2005b). In some areas, dredging of channels for sand and to improve navigation has altered riverbeds, and waves from ship traffic have caused erosion. Increased numbers of tourists, attracted by

Box 2-1. Gulf of Mexico and Ocean-Atmosphere Interactions

The Gulf of Mexico provides an example of the interactions of regional circulation patterns with climate. The region's weather is influenced by trade winds, with differences in ocean and atmospheric temperatures resulting in cyclogenesis from June through October that, when the tropical storms become hurricanes, poses a severe threat to humans (Escobar 2006). Although these systems cause extensive damage as a result of high wind speeds and flooding, these cyclonic systems contribute vital rainfall over an extensive area of the southern and eastern United States. Warming of the Gulf of Mexico and Caribbean Sea is likely to increase regional warming and lead to additional intensification of nascent tropical cyclones, exacerbating both the positive and negative influences of these SEMSs over the adjacent land areas. For example, as a result of the powerful hurricanes that struck the central coast of the Gulf of Mexico in 2005, roughly 300 km^2 of coastal wetlands were lost (Barras 2006), increasing the exposure of urban and industrial infrastructure in the Mississippi River delta region to future hurricanes.

At the ocean interfaces of the Gulf of Mexico, climate change will also exert influences. The interaction of ocean eddies with the continental slope (Muller-Karger 2000; Toner et al. 2003) and the confluence of along-shelf currents generate cross-shelf transports nearshore and offshore (Cochrane and Kelly 1986; Zavala-Hidalgo et al. 2003). As a result, cross-shelf transports of phytoplankton-rich waters have a seasonal cycle that is largely modulated by the wind speed and direction over time; changes in cross-shelf transports of phytoplankton due to climate change will directly affect marine life.

sandy beaches, have also accelerated coastal erosion, especially in locations where tourism development has led to loss of coastal vegetation and increased construction on beaches and foredunes. With coastal resistance to the sea weakened, rising sea level is likely to significantly impact many low-lying areas, particularly those exposed to storm surges and wind-whipped waves.

Reductions in Sea-Ice Cover

Global warming scenarios for high latitudes predict amplified warming due to positive feedbacks from loss of sea ice and to allocation of the additional energy to warming rather than increased evaporation. Loss of coastal ice cover increases the exposure to surface waves, especially during winter storms, and increases the risk of accelerated coastal erosion. In areas of coastal permafrost, shoreline erosion can be several meters per year. Sea-ice melting and increased arctic river input are also likely to lead to stronger thermohaline stratification and reduced deep-water formation in high-latitude regions, reducing CO_2 uptake, decreasing the efficiency of ocean carbon sequestration, and further amplifying global warming. Reduced ice cover is also likely to disrupt reproduction and feeding of seals, which depend on ice for reproduction, and polar bears, which prey on the seals. Changes in ice cover are already disrupting the livelihood of the humans who hunt seals and depend on other marine resources.

Reduction in sea ice can affect climatic conditions in downwind regions. For example, reduced ice cover over Hudson Bay and the Arctic Ocean will lead to sharply warmer temperatures in the fall and early winter over much of eastern North America. Similar effects will be seen as a result of diminished ice cover and reduced ice thickness, and of later formation and earlier melting of seasonal sea ice in areas such as the Sea of Okhotsk, the Baltic Sea, Hudson Bay, and the Gulf of St. Lawrence (Box 2-2).

Accumulated Impacts on Erosion, Coastal Stability, and the Coastal Edge

The natural coastline is a result of dynamic processes acting over geologic time scales. These processes include changes in sea level; uplift or subsidence of continental land masses; erosion by waves, tides, and currents; and sediment supply. About 70% of the world's sandy coast, occupying about 20% of the global coastline, has been retreating over the last century; 20%–30% has been stable, and less than 10% has been advancing (Bird 1993; Syvitski et al. 2005a). Sea level rise will cause more-rapid retreat of the coastline (Leatherman 2001).

Experience has shown that attempts to protect sandy beaches from erosion by constructing barriers such as jetties and wave breakers to limit wave action and reduce littoral sediment transport have only shifted the problem to other locations, with unintended consequences, such as complete loss of beaches and destruction of property (Perillo 2005). With projected changes in winds, temperature, precipitation, frequency and intensity of extreme events, and sea level, the risk is high for dramatic changes to sedimentary systems in ways that are yet to be understood.

Biological and Chemical Responses to Climate Forcing

In addition to the many physical responses to climate forcing, significant chemical and biological responses are expected in SEMSs. Projected changes in the hydrologic cycle will certainly alter fluxes and distributions of nutrients, CO_2, and contaminants and, as a consequence, the ecology of SEMSs.

Box 2-2. Changes in Gulf of St. Lawrence Stratification and Hypoxia

Throughout the twentieth century, the hydrologic regime of the Gulf of St. Lawrence was modified by construction of major dams along the St. Lawrence, Saguenay, Manicouagan, and several smaller rivers. Changes in the seasonal patterns of river flow caused by these dams altered the seasonal cycle of surface salinity and stratification in the Gulf of St. Lawrence, although the effects have not yet been quantified. With further construction of dams planned over the next few decades, changes in freshwater flows are likely to have further impacts, especially near the mouths of the newly harnessed rivers.

Additional changes are likely as warmer winter temperatures reduce the extent and duration of the seasonal ice cover. These reductions will likely benefit marine transportation by lengthening the ice-free navigation season, but they are likely to adversely impact wildlife. Further nutrient enrichment and increased eutrophication of the St. Lawrence River and estuary over the coming decades appear likely as the human population continues to grow, as more farmers turn to row-crop cultures of corn to produce ethanol for fuel, and as the construction of new and bigger animal farms and the associated manure disposal continue to stress the nutrient-accepting capacity of agricultural land.

Global warming will also cause changes in ocean currents in the northwest Atlantic Ocean. A continuing decrease in the western transport of Labrador Current Water along the southern edge of the Grand Banks of Newfoundland is likely to cause warmer, saltier, and lower-oxygen waters from the Gulf Stream to enter the mouth of the Laurentian Channel and then propagate landward toward the heads of the three major deep channels of the Gulf of St. Lawrence (Laurentian, Anticosti, and Esquiman channels). Entrance of low-oxygen water from the open ocean would further exacerbate the oxygen deficit in the bottom waters of the Gulf of St. Lawrence (Gilbert et al. 2005).

Changes in Fluxes and Distribution of Nutrients

Increasing amounts of nitrogen and phosphorus compounds are being released to coastal waters due to the use of fertilizers, disposal of sewage, and production of manure (Cloern 2001). The phosphorus loading in runoff has been reduced in many places as a result of improved wastewater treatment and elimination of phosphate from detergents, but the dissolved inorganic nitrogen (DIN) load is still high and even increasing. Indeed, modern sewage treatment is not designed to remove nitrogen compounds. Atmospheric inputs of nitrogen to SEMSs can be important, and in densely populated regions, atmospheric nitrogen inputs can exceed inputs from rivers.

Paradoxically, the flux of dissolved silica (DSi) to the coastal zone has diminished, apparently because of removal by algal growth in the increasing numbers of reservoirs and in eutrophic rivers (Humborg et al. 1997; Garnier et al. 2002). Because DIN consumption by algae tends to be outweighed by anthropogenic DIN additions downstream, the DSi:DIN ratio in river water has been shifting from "pristine values" of about 30 to 40 (typical of the Amazon or Zaire rivers) to < 2 (typical of the heavily impacted Changjiang (Yangtze), Mississippi, and Scheldt rivers, each of which empties into a SEMS; Chou and Wollast 2006). A full review of the effects of human perturbations on the silicon cycle is provided by Ittekkot and colleagues (2006).

Organic nitrogen compounds from natural and anthropogenic sources can comprise a significant proportion of total nitrogen in coastal seas (see Chapters 8 and 9, this volume). Organic nitrogen compounds serve as nutrients for heterotrophic bacteria and phytoplankton (Seitzinger et al. 2002). Adverse shifts in plankton species composition and development of harmful algal blooms are potential consequences of an imbalance between nitrogen and phosphorus.

In addition to runoff, the ocean is a major source of nutrients (and other constituents of seawater) for SEMSs. For example, the primary production in the North Sea, among the highest in marine areas, is fueled to a large extent by nutrients from the North Atlantic Ocean and to a lesser extent by riverine or atmospheric nutrient sources. In return, the North Sea exports inorganic carbon taken up from the atmosphere to the North Atlantic Ocean via the continental shelf pump. Land-use change can also affect riverine inputs in ways that can alter the alkalinity of coastal seas, which in turn can determine the capacity of the water to take up atmospheric CO_2 (Raymond and Cole 2003).

The characteristics of SEMSs that are not strongly coupled to the open ocean are primarily controlled by river runoff and in situ forcing, which significantly increases the vulnerability of the systems to human-induced disturbances. In the Baltic Sea, for example, anoxic conditions dominate (see Plate 1) unless sporadic events deliver oxygen into anoxic deep waters and temporarily reestablish oxic conditions (Box 2-3). In the Black Sea (Box 2-4), which is only minimally ventilated vertically or by exchange with the Mediterranean Sea (Stanev et al. 2002), stagnant anoxic conditions prevail in the bottom waters (Stanev et al. 2004).

The North Sea is made up of a shallower southern region, exhibiting shelf-like characteristics, and deeper central and northern regions that have ocean-like conditions. The flushing time scale for North Atlantic Ocean water to circulate through the North Sea is on the order of one year, so North Sea conditions are reset on annual time scales. Despite deleterious impacts of eutrophication on the North Sea's northernmost ecosystem during the second half of the twentieth century, the relatively strong mixing has prevented spreading of the anoxic conditions characterizing the German Bight and other nearshore

Box 2-3. The Baltic Sea and the North Sea: Connected but Different

While the Baltic Sea and the North Sea are connected through the Skagerrak, their physical and biogeochemical characteristics differ substantially. The Baltic Sea is a brackish environment, with the river runoff into it being three times larger than the inflow of water from the North Sea. The biogeochemical characteristics of the freshwater inputs to the Baltic are controlled by its drainage area, which includes parts of the Scandinavian Peninsula and parts of continental Europe. The Baltic's deep waters are anoxic, with only occasional oxic periods induced by the inflow of oxygenated waters from the North Sea. The stagnant conditions make this sea very sensitive to eutrophication and other human impacts. Over recent decades, international efforts to limit pollution and nutrient flows have improved the health of the Baltic Sea, which has allowed reestablishment of the codfish stock after depletion by overfishing.

areas. International management efforts have led to improvement of water quality over recent decades, although pelagic and benthic fauna are still under severe pressure from exploitation. On the other hand, the relatively good ventilation makes the North Sea susceptible to disturbances originating in or experienced by the North Atlantic Ocean, including the declining pH due to the rise in atmospheric CO_2 concentration.

The recent availability of reliable computational tools for the Black Sea ecosystem (Beckers et al. 2002; Lancelot et al. 2002b) has opened the road to predictions based on the integration of up-to-date data and numerical models (Stanev 2005). Results indicate that future improvement of environmental conditions will depend on further reductions of riverine nutrient input (Black Sea Environment Programme 2002).

Climatic Effects on Vertical Stratification

By changing temperatures and freshwater fluxes, climate change will affect the stratification of the water column. Stratification controls the vertical exchange of heat and dissolved elements, hence the supply of nutrients to the photic zone. Stronger stratification, resulting from warming of the upper ocean, will reduce upward mixing of nutrients from the subsurface layers and reduce the depth of oxygen penetration from the atmosphere. Episodic cooling, for example after volcanic eruptions or from unusual weather regimes, can also affect the formation of intermediate water masses in regional seas and contribute

Box 2-4. The Black Sea: A Nearly Closed, Freshwater-Dominated Basin

The Black Sea is a deep basin with a broad shelf zone in its northwestern area. Although its maximum depth exceeds 2,200 m, it is freshwater dominated. Eighty percent of the total river discharge of about 300 $km^3 y^{-1}$ enters the northwestern shelf. The exchange with the Atlantic Ocean via the Mediterranean Sea is restricted to narrow straits, creating a nearly enclosed environment with a surface salinity that is about half that of the Mediterranean Sea. The extremely strong vertical stratification restricts ventilation and encourages accumulation of hydrogen sulfide (H_2S) in the deep layers.

A permanent feature of the upper layer circulation is the encircling Rim Current, which forms a sharp (40–80 km wide) salinity front over the continental slope and dynamically decouples the coastal and open-sea waters. Mesoscale variability of the Rim Current provides a mechanism for coastal–open sea exchange and is further supported by quasi-permanent and/or transient anticyclonic eddies confined between the Rim Current and the coast. The Black Sea's continental shelf hosts a biodiverse ecosystem that is, however, heavily affected by a massive influx of nutrients and pollutants from the surrounding coastal areas. From the 1970s to the 1990s, the delivery of nitrogen and phosphorus to the northwestern Black Sea increased by factors of 3 and 10, respectively, mostly as a result of more-intensive agriculture, while silica decreased by a factor of about 4, leading to significant modification of inorganic nutrient ratios. Eutrophication and enhanced oxygen deficiency as a result of human intervention in river flow, introduction of a ctenophore (*Mnemiopsis leidyi*) that achieved dominance, overfishing, uncontrolled sewage discharge, and dumping of wastes have all added to the Black Sea's ecological problems (e.g., see Lancelot et al. 2002a and references therein).

Environmental monitoring during the last several years has indicated a perceptible and gradual improvement in the state of some biotic components of the ecosystem in the western coastal waters (Black Sea Environment Programme 2002; Chapter 3, this volume). One of the most noticeable improvements was the reduction in the nutrient input, reducing the frequency and intensity of algal blooms (Petranu et al. 1999). During 1995 and 1996, the mesozooplankton biomass was observed to be more abundant than in previous years, which is also reflective of a gradual improvement in the ecosystem. However, the effects of high zooplankton mortality in past decades are still evident in the reduced benthos.

to transport of nutrients from deeper layers into the photic layer. This occurred in the Black Sea during the first half of the 1990s (Stanev et al. 2003).

Increased Occurrence of Phytoplankton Blooms

Increasing runoff, to the extent that it occurs, is expected to increase the likelihood and intensity of phytoplankton blooms in coastal areas (Rabalais et al. 2002). This can occur because freshwater can increase stratification and inputs of the macronutrients that lead to phytoplankton blooms, although the end result will depend on changes in the ratios of the individual nutrients (Anderson et al. 2002; GEOHAB 2005). In high-latitude regions, changes in the timing of sea-ice formation and breakup, combined with changes in timing and amount of river runoff, are also likely to impact the timing of plankton bloom formation and related events involved in the marine food web. One of the best-known examples of the effects of river runoff is in the northern Gulf of Mexico, where seasonal hypoxia and plankton blooms are sustained by nutrient inputs provided by the Mississippi River (Turner and Rabalais 1994). Mixing of the water column during hurricanes relaxes the hypoxic conditions only temporarily because large hurricanes draw additional nutrients into the water column from the sediments and because increased freshwater inflow from rivers carries with it increased amounts of nutrients (see Chapter 11, this volume).

Officer and Ryther (1980) hypothesized that if a Si:N ratio of 1:1 for diatoms was not maintained, then a phytoplankton community of nondiatoms might become competitively enabled. The alternative community would be more likely to be composed of flagellated algae such as dinoflagellates, including noxious bloom-forming algal communities. This has been observed in various SEMSs, including the East China Sea (Gong et al. 2006; Jiao et al. 2007), the Baltic Sea (Radach et al. 1990), the North Sea (Lancelot et al. 1987), and the Northern Adriatic Sea (Granéli et al. 1999).

Changes in nutrient ratios also cause the fisheries web to switch to less desirable species. This happened in the continental shelf waters near the Mississippi River delta (Turner et al. 1998); when the Si:N ratio dropped below 1:1, the copepod abundance of the zooplankton dropped from 80% to 20%. The fecal pellet production of copepods and the relative proportion of carbon carried from the upper to the lower water column via fecal pellets also declined. Because copepod fecal pellets contain many partially decomposed diatoms, they sink much more rapidly than individual phytoplankton cells, and there is relatively less decomposition enroute to the bottom. For this reason, more of the sinking organic material is respired in the bottom layer when copepods dominate the zooplankton community than when the Si:N ratio is < 1:1 and copepods are relatively scarce. These findings are affected when eutrophication occurs in the presence of higher diatom production (i.e., when Si:N > 1:1). Under these conditions, there is greater fecal pellet production, more carbon sedimentation to the bottom layer, higher respiration rates in the lower water column, and development of hypoxia throughout the stratified water column.

Altered Size Distribution of Organisms

The size distribution of phytoplankton and zooplankton can also change if the copepod density is decreased for a period longer than one generation. For these conditions, Turner (2001) projected that a reformed phytoplankton community of smaller cells would be grazed by a new community of smaller prey of different escape velocities, growth rates, aggregation potential, and palatability or by filter feeders such as salps.* A sequence of events leading to reduced productivity of different trophic levels has been documented on the northwestern shelf of the Black Sea as nitrogen and phosphorus loads increase and silica decreases (Zaitsev 1992; Lancelot et al. 2002a).

Biogeochemical Cycling of Elements

Alterations of nutrient inputs and nutrient speciation and organic matter input influence primary production and respiration processes, which in turn play a significant role in determining the system's trophic state. While organic matter inputs can play a direct role in creating anoxic conditions that kill off the marine life on which society depends, inorganic nutrients play a more complicated role because primary production supplies not only organic matter, but also oxygen, offsetting anoxia. If the water column is shallow and/or poorly ventilated, sinking or settled organic matter will be remineralized over short time scales, promoting anoxic conditions.

The balance between oxygen consumption via remineralization of organic matter and oxygen supply determines the concentration of dissolved oxygen. The supply of oxygen is controlled by advection and mixing, both within the basin and from the open ocean. The residence time of water is largely determined by the exchange with the open ocean. The residence time of water within the oxygen minimum zone (OMZ) of the central Gulf of Mexico is less than a year, resulting in a high oxygen concentration (> 100 μM). Basins that have long residence times (thousands of years), such as the Black Sea, are essentially anoxic below the pycnocline. Oxygen concentrations within the OMZ in the Bay of Bengal are as low as about 2–3 μM. This implies that minor changes in oxygen consumption or supply could make the water column of this region denitrifying (Box 2-5).

Increased Emission of Greenhouse Gases

Production and release to the atmosphere of the greenhouse gases nitrous oxide and methane (N_2O and CH_4, respectively) are expected to increase with increasing oxygen deficiency. This holds for the Bay of Bengal, where production of greenhouse gases is large. However, recent studies of N_2O in the Black Sea did not find it to be a strong

*Salps are gelatinous barrel-shaped marine animals that have a free-floating life at sea. Salps filter feed on the smallest phytoplankton fraction. Like other tunicates, salps contribute to the carbon cycle by aggregating the carbon into fecal pellets that sink to the ocean floor.

Box 2-5. Bay of Bengal: Very Open, but Very Vulnerable to Change

The Bay of Bengal, including the Andaman Sea, is strongly influenced by monsoonal seasonality and by inputs from surrounding land masses. As a result of the high monsoonal precipitation over the bay and runoff from a number of major rivers, reduced surface salinities are found over most of this large SEMS. Consequently, the upper water column here is among the most strongly stratified in the world, affecting nutrient replenishment from deeper waters and gas exchange between the ocean and atmosphere.

The basin receives enormous quantities of dissolved and particulate matter. The dissolved inorganic nitrogen (DIN) input from the Brahmaputra-Ganges river system alone is estimated to contribute about 10% of the global total (Dumont et al. 2005); nonetheless, the region is still considered nitrogen-limited. The supply of suspended sediment by this river has led to the formation of one of the world's largest deltas and is essential for stabilizing the low-lying coastline of Bangladesh. The surrounding regions are among the most densely populated on Earth, making this oceanic region highly vulnerable to human-induced changes.

In this region, disturbances from climate change, in combination with other human influences, are likely to lead to a number of ecological and socioeconomic impacts:

- Changes in monsoon precipitation and reductions in snow cover in the Himalayas, coupled with enhanced water utilization, will alter freshwater inputs to the Bay of Bengal, affecting both stratification and dissolved silica

(continued)

source (Westley et al. 2006). Methane accumulates in the Black Sea and the Cariaco Trench to about 15 μM (Reeburgh 2007), which is 4 orders of magnitude higher than in surface waters of the open ocean. CH_4 concentrations in surface layers can be supersaturated with respect to the atmosphere in spite of the presence of methane-oxidizing bacteria in the water column.

Acidification of the Ocean

The oceanic carbonate system involves a complex balance among individual carbonate species (CO_2, H_2CO_3, HCO_3^-, CO_3^{2-}). However, the overall response of the carbonate system to perturbations is predictable: Higher atmospheric CO_2 concentration will shift

Box 2-5 *(continued)*

loading. Any relaxation of the stratification is likely to bring about enhanced emission of CO_2. Addition of nutrients from subsurface waters via mixing and upwelling would enhance productivity and most probably affect local food web structures (e.g., reducing the presently high diatom contribution to primary production).

- Eutrophication caused by enhanced inputs of nutrients from both river runoff and atmospheric deposition triggers primary production and related organic matter input to the water column, altering oxygen demand. This is especially important because the Bay of Bengal's subsurface waters are presently close to suboxic. Therefore, any disturbance to the delicate biogeochemical balance among organic matter supply, regeneration, and oxygenation has the potential to bring about large changes in regional biogeochemistry and functioning of the ecosystem (e.g., onset of water column denitrification, emission of N_2O).
- A rise in sea level of a meter or more, along with associated coastal erosion, would very likely lead to a huge loss of land, with serious socioeconomic implications, especially for Bangladesh. This effect would likely be compounded by projected increases in the strength of the tropical cyclones that frequently cause enormous loss of life and property.
- Sea level rise is a severe threat to the extended mangrove forests of the Sundarbans and would add to the intense human pressure on these forests. With a pace of change that is expected to be too rapid to allow this complex ecosystem to adjust, the Sundarbans might face partial or even complete disappearance. Thus, the natural functions of the mangroves, including provision of nursery grounds for fish, stabilization of the coastline, and modulation of land–ocean fluxes, as well as related services provided to the coastal population, would be heavily impacted.

the pH of the surface ocean to lower values and lower the concentration of CO_3^{2-}. The CO_3^{2-} concentration, in turn, controls the saturation state of carbonates (Zeebe and Wolf-Gladrow 2001). The decreasing carbonate saturation level during the next hundred years is likely to have multiple impacts. Changes to the carbonate system by acidification of the ocean are likely to make carbonate precipitation more difficult for calcifying organisms (Kleypas et al. 1999) and will interfere with the biological pump, which transfers organic matter produced in the surface waters to the deep ocean (Armstrong et al. 2002; Klaas and Archer 2002). Changes in the biological pump are likely to affect the distribution of nutrients and oxygen in the water column and the burial rate of organic matter in sediments (Volk and Hoffert 1985; Jahnke 1996; Rixen et al. 2000; Rixen and Ittekkot 2005).

Box 2-6. Impacts on Coral Reefs Fringing Tropical SEMSs

Coral reefs are among the most diverse ecosystems in the world. They also function in ways that provide ecosystem services to tropical communities and tourists from around the world. Their largest threats come from overexploitation, eutrophication, and sediment loading. Global surveys of the status of reefs around the world indicate conditions are serious, and many countries have programs to protect their reefs. In addition to local stresses, however, global warming is also impacting reefs, contributing, for example, to acidification and to the extensive coral bleaching during El Niño events (e.g., 1997–1998).

Ocean acidification caused by the rising concentration of atmospheric CO_2, especially given projected levels in the future, suggests a catastrophic future. Laboratory and model results indicate that coral calcification will be reduced by 30%–40% during the twenty-first century. Algal replacement and competition will also be more evident. Dissolution of reefs will result in loss of their ability to function and provide services, making tropical coastal areas even more vulnerable to extreme events such as tropical storms and tsunamis/floods.

For SEMSs, the most important effect is likely to be on the marine species they host. The effects of the decrease in ocean pH on pelagic and benthic calcifying organisms will be larger in the colder waters of high latitudes, and the resulting impacts on the food web and fish stocks are also likely to be greatest there. In lower latitudes, corals will be disproportionately affected. In particular, the aragonite saturation state is decreasing. In the Bay of Bengal the saturation state has decreased from between 4.0 and 4.5 to between 3.5 and 4.0 since the beginning of the Industrial Revolution (Kleypas et al. 2006) and is projected to drop below 3 by the year 2100. This would threaten coral reefs and disrupt their ability to sustain the diversity of SEMSs and their ability to deliver ecosystem services (Box 2-6).

Concluding Remarks

The intention of this chapter has been to provide an overview of examples of how human-created disturbances are affecting and foreseeably could affect SEMSs. Our purpose has been to illustrate the wide variety of processes active in these highly diverse oceanic regions. The description of the physical and chemical impacts given here provides a foundation for the three following cross-cutting chapters, which, in turn, focus on thresholds and key drivers of accelerated or even abrupt change (Chapter 3), the implications for sustaining ecosystem services (Chapter 4), and the best tools for assessing the present conditions and future states of SEMSs (Chapter 5).

Climate change will affect many of the characteristics of SEMSs, including average and extreme temperatures, precipitation intensity and patterns, runoff intensity and timing, salinity distribution, circulation, stratification, mixing, and chemical properties. We recommend regional assessments to evaluate the state of understanding of SEMSs, identify important questions that need to be investigated, and take measures to increase the resilience of these systems and reduce the potential for adverse consequences. These assessments must be made jointly by the scientific community and local and regional managers and decision makers.

References

Anderson, D.M., P.M. Gilbert, and J.M. Burkholder. 2002. Harmful algal blooms and eutrophication: Nutrient sources, composition, and consequences. *Estuaries* 25:704–726.

Armstrong, R.A., C. Lee, J.I. Hedges, S. Honjo, and S.G. Wakeham. 2002. A new, mechanistic model for organic carbon fluxes in the ocean, based on the quantitative association of POC with ballast minerals. *Deep-Sea Research II* 49:219–236.

Barras, J.A. 2006. *Land Area Change in Coastal Louisiana after the 2005 Hurricanes: A Series of Three Maps.* U.S. Geological Survey Open-File Report -06–1274. http://pubs.usgs.gov/of/2006/1274/.

Beckers, J.M., M.L. Gregoire, J.C.J. Nihoul, E. Stanev, J. Staneva, and C. Lancelot. 2002. Modelling the Danube-influenced north-western continental shelf of the Black Sea. I. Hydrodynamical processes simulated by 3-D and box models. *Estuarine Coastal and Shelf Science* 54:453–472.

Bird, E.C.F. 1993. *Submerging Coasts: The Effects of Rising Sea Level on Coastal Environments.* Chichester, UK: John Wiley & Sons.

Black Sea Environment Programme. 2002. *State of the Environment of the Black Sea: Pressures and Trends 1996–2000.* http://www.blacksea-commission.org/Publications/SOE.htm.

Chou, L., and R. Wollast. 2006. Estuarine silicon dynamic. In *The Silicon Cycle: Human Perturbations and Impacts on Aquatic Systems*, edited by V. Ittekkot, D. Unger, C. Humborg, and N. Tac An, 93–120. SCOPE Series Vol. 66. Washington, DC: Island Press.

Cloern, J.E. 2001. Our evolving conceptual model of the coastal eutrophication problem. *Marine Ecology Progress Series* 210:223–253.

Cochrane, J.D., and F.J. Kelly. 1986. Low-frequency circulation on the Texas–Louisiana continental shelf. *Journal of Geophysical Research* 91 (C9):10,645–10,659.

Dumont, E., J.A. Harrison, C. Kroeze, E.J. Bakker, and S. Seitzinger. 2005. Global distribution and sources of dissolved inorganic nitrogen export to the coastal zone: Results from a spatially explicit, global model. *Global Biogeochemical Cycles* 19:GB4S02, doi: 10.1029/2005GB002488.

Escobar, E. 2006. Structure and function in the ecosystems of the Intra-Americas Sea (IAS). In Vol. 14 of *The Sea*, edited by A.R. Robinson and K.H. Brink, 225–258. Cambridge, MA: Harvard University Press.

Garnier, J., G. Billen, E. Hannon, S. Fonbonne, Y. Videnina, and M. Soulie. 2002. Modelling the transfer and retention of nutrients in the drainage network of the Danube River. *Estuarine, Coastal and Shelf Science* 54:285–308.

GEOHAB. 2005. *Global Ecology and Oceanography of Harmful Algal Blooms, GEOHAB Core Research Project: HABs in Upwelling Systems,* edited by G. Pitcher, T. Moita, V. Trainer, R. Kudela, F. Figuieras, and T. Probyn. Paris: Intergovernmental Oceanographic Commission (IOC) and Scientific Committee on Oceanic Research (SCOR).

Gilbert, D., B. Sundby, C. Gobeil, A. Mucci, and G.-H. Tremblay. 2005. A seventy-two-year record of diminishing deep-water oxygen in the St. Lawrence Estuary: The northwest Atlantic connection. *Limnology and Oceanography* 50:1,654–1,666.

Gong, G.-C., J. Chang, K.-P. Chiang, T.-M. Hsiung, C.-C. Hung, S.-W. Duan, and L. Codispoti. 2006. Reduction of primary production and changing of nutrient ratio in the East China Sea: Effect of the Three Gorges Dam. *Geophysical Research Letters* 33:L07610, doi: 10/1029/2006GL025800.

Granéli, E., P. Carlsson, P. Tester, J.T. Turner, C. Bechemin, R. Dawson, and F. Azam. 1999. Effects of N:P:Si ratios and zooplankton grazing on phytoplankton communities in the Northern Adriatic Sea. I. Nutrients, phytoplankton, biomass, and polysaccharide production. *Aquatic Microbial Ecology* 18:37–54.

Gross, M.G., M. Karweit, W.B. Cronin, and J.R. Schubel. 1978. Suspended sediment discharge of the Susquehanna River to northern Chesapeake Bay, 1966 to 1976. *Estuaries* 1:106–110, doi: 10.2307/1351599.

Humborg, C., V. Ittekkot, A. Cociasu, and B.V. Bodungen. 1997. Effect of Danube River dam on Black Sea biogeochemistry and ecosystem structure. *Nature* 386:385–388.

IPCC. 2007. *Climate Change 2007: The Physical Science Basis.* Working Group I Contribution to the Fourth Assessment Report of the IPCC (Intergovernmental Panel on Climate Change). Cambridge, UK: Cambridge University Press.

Ittekkot, V., D. Unger, C. Humborg, and N. Tac An (eds.). 2006. *The Silicon Cycle: Human Perturbations and Impacts on Aquatic Systems.* SCOPE Series Vol. 66. Washington DC: Island Press.

Jahnke, R.A. 1996. The global ocean flux of particulate organic carbon: A real distribution and magnitude. *Global Biogeochemical Cycles* 10:71–88.

Jiao, N., Y. Zhang, Y. Zeng, W.D. Gardner, A.V. Mishonov, M.J. Richardson, N. Hong, D. Pan, X.-H. Yan, Y.-H. Jo, C.-T.A. Chen, P. Wang, Y. Chen, H. Hong, Y. Bai, X. Chen, B. Huang, H. Deng, Y. Shi, and D. Yang. 2007. Ecological anomalies in the East China Sea: Impacts of the Three Gorges Dam? *Water Research* 41:1,287–1,293.

Kjerfve, B., and J.D. Restrepo. 2002. River discharge and sediment load variability in South America. In *South American Basins: LOICZ Global Change Assessment and Synthesis of River Catchment–Coastal Sea Interaction and Human Dimensions,* edited by L.D. Lacerda, H.H. Kremer, B. Kjerfve, W. Salomons, J.I.M. Crossland, and C.J. Crossland, 87–91. LOICZ (Land–Ocean Interactions in the Coastal Zone) Reports and Studies Vol. 21. Texel, The Netherlands: LOICZ International Project Office.

Klaas, C., and D.E. Archer. 2002. Association of sinking organic matter with various types of mineral ballast in the deep sea: Implications for the rain ratio. *Global Biogeochemical Cycles* 16 (4), doi: 10.1029/2001GB001765.

Kleypas, J.A., R.W. Buddemeier, D. Archer, J.-P. Gattuso, C. Langdon, and B.N. Opdyke. 1999. Geochemical consequences of increased atmospheric carbon dioxide on coral reefs. *Science* 284:118–120.

Kleypas, J.A., R.A. Feely, V.J. Fabry, C. Langdon, C.L. Sabine, and L.L. Robbins. 2006. *Impacts of Ocean Acidification on Coral Reef and Other Marine Calcifiers: A Guide for Future Research.* Sponsored by the National Science Foundation (NSF), National Oceanic and Atmospheric Administration (NOAA), and U.S. Geological Survey (USGS). Boulder, CO: University Corporation for Atmospheric Research.

Lancelot, C., G. Billen, A. Sournia, T. Weisse, F. Colijn, M.J.W. Veldhuis, A. Davies, and P. Wassmann. 1987. *Phaeocystis* blooms and nutrient enrichment in the continental coastal zones of the North Sea. *Ambio* 16:38–46.

Lancelot, C., J.M. Martin, N. Panin, and Y. Zaitsev. 2002a. The north-western Black Sea: A pilot site to understand the complex interaction between human activities and the coastal environment. *Estuarine, Coastal and Shelf Science* 54:279–283.

Lancelot, C., J. Staneva, D. van Eeckhout, J.M. Beckers, and E. Stanev. 2002b. Modeling the Danube-influenced north-western continental shelf of the Black Sea. II. Ecosystem response to changes in nutrient delivery by the Danube River after its damming in 1972. *Estuarine, Coastal and Shelf Science* 54:473–499.

Leatherman, S.P. 2001. Social and economic costs of sea level rise. In *Sea Level Rise: History and Consequences*, edited by B.C. Douglas, M.S. Kearney, and S.P. Leatherman, 181–217. International Geophysics Series Vol. 75. San Diego: Academic Press.

Meade, R.H., and S.W. Trimble. 1974. Changes in sediment loads in rivers of the Atlantic drainage of the United States since 1900. *International Association of Hydrological Publications* 113:99–104.

Meehl, S.A., and T.F. Stocker. 2007. Global climate projections. In *Climate Change 2007: The Physical Science Basis*, 747–845. Working Group I Contribution to the Fourth Assessment Report of the IPCC (Intergovernmental Panel on Climate Change). Cambridge, UK: Cambridge University Press.

Muller-Karger, F.E. 2000. The spring 1998 northeastern Gulf of Mexico (NEGOM) cold water event: Remote sensing evidence for upwelling and for eastward advection of Mississippi water (or How an errant Loop Current anticyclone took the NEGOM for a spin). *Gulf of Mexico Science* 18:55–67.

Nicholls, R.J., and P.O. Wong. 2007. Coastal systems and low-lying areas. In *Climate Change 2007: Impacts, Adaptation and Vulnerability.* Working Group II Contribution to the Fourth Assessment Report of the IPCC (Intergovernmental Panel on Climate Change). Cambridge, UK: Cambridge University Press.

Officer, C.B., and J.H. Ryther. 1980. The possible importance of silicon in marine eutrophication. *Marine Ecology Progress Series* 3:83–91.

Perillo, G.M.E. 2005. Mar del Plata: A cautionary tale. In *Coastal Fluxes in the Anthropocene*, edited by C.J. Crossland, H.H. Kremer, H.J. Lindeboom, J.I. Marshall Crossland, and M.D.A. Le Tissier, 84. Global Change—The IGBP Series. Berlin: Springer.

Petranu, A., M. Apas, N. Bodeanu, A.S. Bologa, C. Dumitrache, M. Moldoveanu, G. Radu, and V. Tiganus. 1999. Status and evolution of the Romanian Black Sea coastal ecosystem. In *Environmental Degradation of the Black Sea: Challenges and Remedies*, edited by S. Besiktepe, U. Unluata, and A.S. Bologa, 175–195. Dordrecht, The Netherlands: Kluwer Academic Publishers.

Rabalais, N.N., R.E. Turner, Q. Dortch, D. Justic, V.J. Biermann, and W.J. Wiseman. 2002. Nutrient-enhanced productivity in the northern Gulf of Mexico. *Hydrobiologia* 475/476:39–63.

Radach, G., J. Berg, and E. Hagmeier. 1990. Long-term changes of the annual cycles of meteorological, hydrographic nutrient and phytoplankton time series at Helgoland and at LV Elbe 1 in the German Bight. *Continental Shelf Research* 10:305–328.

Raymond, P.A., and J.J. Cole. 2003. Increase in the export of alkalinity from North America's largest river. *Science* 301:88–91.

Reeburgh, W.S. 2007. Oceanic methane biogeochemistry. *Chemical Reviews* 107:86–513.

Restrepo, J.D., and B. Kjerfve. 2000a. Magdalena River: Interannual variability (1975–1995) and revised water discharge and sediment load estimates. *Journal of Hydrology* 235:137–149.

Restrepo, J.D., and B. Kjerfve. 2000b. Water and sediment discharges from the western slopes of the Colombian Andes with focus on Rio San Juan. *Journal of Geology* 108 (1):17–33.

Rixen, T., and V. Ittekkot. 2005. Nitrogen deficits in the Arabian Sea, implications from a three component mixing analysis. *Deep-Sea Research II* 52:1,879–1,891.

Rixen, T., V. Ittekkot, B. Haake-Gaye, and P. Schäfer. 2000. The influence of the SW monsoon on the deep-sea organic carbon cycle in the Holocene. *Deep-Sea Research II* 47:2,629–2,651.

Seitzinger, S.P., R.W. Sanders, and R. Styles. 2002. Bioavailability of DON from natural and anthropogenic sources to estuarine plankton. *Limnology and Oceanography* 47:353–366.

Stanev, E.V. 2005. Understanding Black Sea dynamics: Overview of recent numerical modelling. *Oceanography* 18 (2):52–71.

Stanev, E.V., J.-M. Beckers, C. Lancelot, J.V. Staneva, P.Y. Le Traon, E.L. Peneva, and M. Gregoire. 2002. Coastal–open ocean exchange in the Black Sea: Observations and modelling. *Estuarine Coastal and Shelf Science* 54:601–620.

Stanev, E.V., M.J. Bowman, E.L. Peneva, and J.V. Staneva. 2003. Control of Black Sea intermediate water mass formation by dynamics and topography: Comparisons of numerical simulations, survey and satellite data. *Journal of Marine Research* 61:59–99.

Stanev, E.V., J. Staneva, J.L. Bullister, and J.W. Murray. 2004. Ventilation of the Black Sea pycnocline: Parameterization of convection, numerical simulations and validations against observed chlorofluorocarbon data. *Deep-Sea Research* 51(12):2,137–2,169.

Syvitski, J.M.P., N. Harvey, E. Wolanski, W.C. Burnett, G.M.E. Perillo, V. Gornitz, H. Bokuniewicz, M. Huettel, W.S. Moore, Y. Saito, M. Taniguchi, P. Hesp, W.W.-S. Yim, J. Salisbury, J. Campbell, M. Snoussi, S. Haida, R. Arthurton, and S. Gao. 2005a. Dynamics of the coastal zone. In *Coastal Fluxes in the Anthropocene*, edited by C.J. Crossland, H.H. Kremer, H.J. Lindeboom, J.I.M. Crossland, and M.D.A. Le Tissier, 39–94. Global Change—The IGBP Series. Berlin: Springer.

Syvitski, J.M.P., and J.D. Milliman. 2007. Geology, geography, and humans battle for dominance over the delivery of fluvial sediment to the coastal ocean. *The Journal of Geology* 115:1–19.

Syvitski, J.M.P., C.J. Vörösmarty, A.J. Kettner, and P. Green. 2005b. Impact of humans on the flux of terrestrial sediment to the global coastal ocean. *Science* 308:376–380.

Toner, M., A.D. Kirwan, A.C. Poje, L.H. Kantha, F.E. Muller-Karger, and C.K.R.T. Jones. 2003. Chlorophyll dispersal by eddy–eddy interactions in the Gulf of Mexico. *Journal of Geophysical Research* 108 (C4):3,105, doi: 10.1029/2002JC001499.

Trenberth, K.E., and P.D. Jones. 2007. Observations: Surface and atmospheric climate change. In *Climate Change 2007: The Physical Science Basis*, 235–336. Working Group I Contribution to the Fourth Assessment Report of the IPCC (Intergovernmental Panel on Climate Change). Cambridge, UK: Cambridge University Press.

Turner, R.E. 2001. Estimating the indirect effects of hydrologic change on wetland loss: If the earth is curved, then how would we know it? *Estuaries* 24:639–646.

Turner, R.E., N. Qureshi, N.N. Rabalais, Q. Dortch, D. Justic, R.F. Shaw, and J. Cope. 1998. Fluctuating silicate:nitrate ratios and coastal plankton food webs. *Proceedings of the National Academy of Science USA* 95:13,048–13,051.

Turner, R.E., and N.N. Rabalais. 1994. Coastal eutrophication near the Mississippi River delta. *Nature* 368:619–621.

Volk, T., and M.I. Hoffert. 1985. Ocean carbon pumps: Analysis of relative strengths and efficiencies in ocean-driven atmospheric CO_2 changes. In *The Carbon Cycle and Atmospheric CO_2: Natural Variations Archean to Present*, edited by E.T. Sundquist and W.S. Broecker, 99–110. Washington, DC: American Geophysical Union.

Wang, Y., M.-E. Ren, and J.P.M. Syvitski. 1998. Sediment transport and terrigenous fluxes. In Vol. 10 of *The Sea,* edited by K.H. Brink and A.R. Robinson, 253–292. New York: John Wiley & Sons.

Westley, M.B., H. Yamagishi, B.N. Popp, and N. Yoshida. 2006. Nitrous oxide cycling in the Black Sea inferred from stable isotope and isotopomer distributions. *Deep-Sea Research II* 53, doi: 10.1016/j.dsr2.2006.03.012.

Zaitsev, Y.P. 1992. Recent changes in the trophic structure of the Black Sea. *Fisheries Oceanography* 1 (2):180–189.

Zavala-Hidalgo, J., S.L. Morey, and J.J. O'Brien. 2003. Seasonal circulation on the western shelf of the Gulf of Mexico using a high-resolution numerical model. *Journal of Geophysical Research* 108 (C12):3,389, doi: 10.1029/2003JC001879.

Zeebe, R.E., and D. Wolf-Gladrow. 2001. *CO_2 in Seawater: Equilibrium, Kinetics, Isotopes.* Elsevier Oceanography Series Vol. 65. Paris: Elsevier.

3

Threshold Effects in Semi-Enclosed Marine Systems

Johan van de Koppel, Paul Tett, Wajih Naqvi, Temel Oguz, Gerardo M.E. Perillo, Nancy N. Rabalais, Maurizio Ribera d'Alcalà, Jilan Su, and Jing Zhang

Introduction

The natural history of the twentieth century documents dramatic changes in many ecological systems, mostly due to the influence of the expansion of human civilization. Semi-enclosed marine systems (SEMSs) are no exception to this. Here is an example, from Jannson and Dahlberg (1999):

> In the 1940s, the Baltic Sea was a nutrient-poor sea with low biological production, clear water, and rocky shores with dense growths of the brown seaweed bladderwrack, providing food and shelter for many species, including spawning and nursery grounds for many fish. There was sufficient oxygen in the bottom water for cod to spawn in the deep areas of the Baltic Proper, except for periods of oxygen depletion in the Gotland Deep. Top consumers like seal and sea eagle were common and people living around the Baltic Sea could eat fish without risking their health. The Baltic Sea of today is different. Eutrophication and toxic substances now affect the entire Baltic Sea ecosystem, even the offshore areas. Filamentous green and brown algae shade the bladderwrack and may even totally replace it. Increased plankton blooms and organic particle production [have] lowered light penetration by 3 m and oxygen depletion and hydrogen sulfide formation sometimes dominate as much as one third of the total bottom area. To reduce the nutrient load to the levels of the 1940s, a reduction by 65% for phosphorus and 80% for nitrogen is needed.

Box 3-1. Glossary

Threshold: A particular level of some internal or external condition at which a large change in ecosystem state is observed

Stable state: A state at which a system will persist unless perturbed, and to which it will return after a disturbance

Regime: The dynamic conditions that characterize an ecosystem state

Pressures: Human activities causing environmental problems

Impact: The effects of human pressures on an ecosystem state

Hysteresis: The condition in which the trajectory a system goes through when pressure levels are increased differs from the trajectory when pressure levels are reduced (see Figure 3-1B)

The authors describe an apparent large change in the state of the Baltic Sea as a consequence of eutrophication and the release of toxic substances into the water. They suggest that humans should aim to decrease these perturbing inputs in order to return the sea to the original condition. However, despite a significant reduction in the extent of those disturbances, little recovery has been witnessed in the Baltic Sea. It appears that a threshold has been passed and that a return from the new degraded state to the old state is very difficult, posing significant problems to management.

A large body of ecological theory has developed to explain dramatic changes in ecosystems that are hard to reverse. We mainly draw on work by Scheffer and colleagues (1998, 2001) from temperate lakes, arid ecosystems, and coral reefs (see glossary in Box 3-1 for terminology). This theory proposes that, in general terms, two types of threshold responses can be recognized (Figure 3-1). The first type refers to a continuous response of an ecosystem to an increase in external pressure, which starts or is most apparent at a specific threshold level of the external condition (Figure 3-1A). With changing pressure levels, as conditions change, the system gradually moves from an original state or regime, through an affected and altered state, to a state or regime that humans consider degraded because of deterioration in the services that the ecosystem can provide us. Upon reduction of the pressure, the system returns to its original state, tracing the same path backward. We refer to this as a nonhysteretic threshold response (vs. hysteretic; see explanation below). The gradual nature of the change implies that managers can infer from a slowly deteriorating system that there is eminent danger of reaching a critical state and can take measures to prevent this from occurring.

The second type of threshold response is characterized by a dramatic change when external forcing exceeds a particular threshold. As a consequence, the system jumps

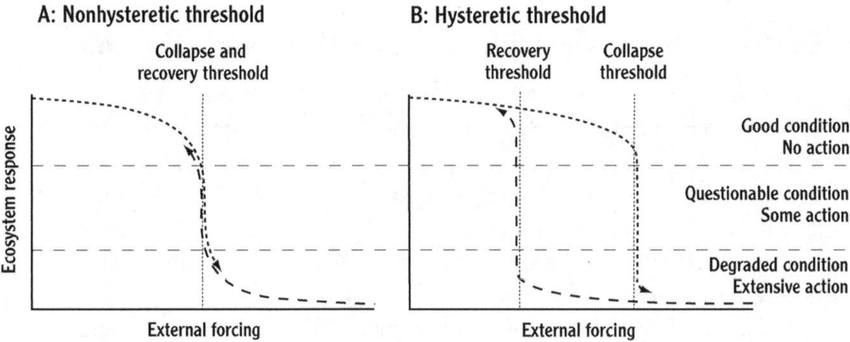

Figure 3-1. Graphical representation of two types of threshold responses of ecosystems to changing conditions. Horizontal dashed lines represent change trajectories, and vertical dotted lines represent threshold levels. Panel A represents a nonhysteretic response to changing conditions, where the dotted line is where a threshold effect is experienced. This threshold, however, is reversible. Moreover, a continuous change is experienced from a more or less pristine to a critical, highly degraded state. This change can easily be reversed. Panel B represents a hysteretic threshold, where the switch from a pristine condition to a degraded condition (short-dashed line) occurs at a different threshold level than the switch back from the degraded to the pristine condition (long-dashed line). Hence, ecosystem conditions can be reversed only by a dramatic change in the external forcing.

abruptly (in relative terms, because it may still take years in large systems) from an original regime to a degraded one with little warning. The change in ecosystem state cannot be easily reversed with a change in external forcing. External pressure has to be reduced below a second threshold at a very different level (recovery threshold in Figure 3-1B) before the ecosystem regains the old regime. Thus, we refer to this second threshold as a *hysteretic threshold*.

A hysteretic threshold may result from a self-enforcing positive feedback relation between biotic or abiotic components of a system. In Box 3-2, we explain the background behind this type of threshold, using a simplified model of algal blooms in a SEMS. In this model, positive feedback processes cause the system to have two stable states, one state in which grazers control algal growth, and another in which grazing effects on high algal stocks are minor. These two states only co-occur for a limited set of parameter values (the gray area), and hence dramatic shifts may occur when the system is pushed outside of this zone. The model presented in Box 3-2 shows that both a hysteretic and a nonhysteretic response can be found, depending on parameter values. Thus, in nature, the two responses are extremes of a continuum, and in-between situations can exist, for instance, where the hysteresis range along the external forcing axis in Figure 3-1 is very small. Nevertheless, the distinction is likely to be important for managers.

Theory on hysteretic thresholds and the associated alternative regimes has been

Box 3-2. Alternative Stable States and Hysteretic Thresholds

The emergence of alternative stable states and hysteretic thresholds in SEMSs can be explained by a simple model of, for instance, the growth of algae. We analyze an example model in this box to explain how alternative stable states are generated by nonlinear ecological interactions and how hysteresis can follow from this. Finally, we use the model to illustrate the effect of the openness of a SEMS on the potential for hysteretic thresholds.

Consider a population of algae within a SEMS (or any unenclosed water body) that is controlled by grazing by pelagic herbivores and exchange with the outside marine world. A possible description of the dynamics of such a population is given by

$$\frac{dB}{dt} = rB\left(1 - \frac{B}{K}\right) - c\frac{B^2}{B^2 + d^2} - f(B - B_0),$$

where B defines the algal biomass per unit volume, r is the intrinsic growth rate determined by nutrient availability, K is the carrying capacity set by light availability, c is the potential grazing rate, d is the half-saturation value for grazing, f is the exchange rate with the outside marine world (expressed as a fraction of the volume of the water body), and B_0 is the algal biomass concentration in the outside waters. Note that we assume, for simplicity, that population dynamics of the grazers are not controlled by algal densities but, for instance, by fisheries and that the algal concentration in the outside waters is constant. For a more detailed description in a specific setting, see Oguz (2007).

We can use the model to study the implication of increased nutrient availability, leading to an increased algal growth rate r, on the dynamics of the algae by analyzing for changes in equilibrium algal biomass. Figure A below reveals that, for low ocean exchange conditions, the

(continued)

applied extensively to enclosed systems such as small and shallow lakes as typically found in northern temperate regions (Scheffer 1998). Whether the concept can be used to explain threshold responses in large-scale systems such as SEMSs is an important topic that we will discuss in this chapter. First, we will briefly discuss two example studies in SEMSs and address whether there is evidence to categorize the thresholds as nonhysteretic or hysteretic. Then, we will discuss whether there is evidence in SEMSs for alternative regimes and associated dynamics, given their semi-open nature. The questions we address are the following:

Box 3-2 *(continued)*

relation between the equilibrium biomass and algal growth rates is strongly nonlinear. In the gray zone, two equilibria can occur side by side, separated by an unstable equilibrium (the dashed line). In the lower equilibrium, algal densities are controlled by grazing, while in the upper equilibrium, high algal growth prevents grazing control. When we move out of the gray area, a sudden shift may occur when one of the equilibria ceases to exist.

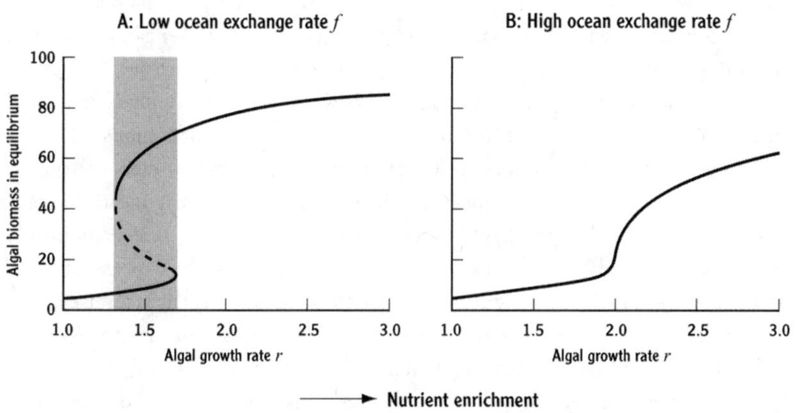

The hysteretic nature of the relation between equilibrium algal biomass and algal growth rate, as in Figure A, depends on the partially enclosed nature of the system. If we increase exchange with the open ocean, algal densities are depressed by outflow, and grazing intensity is always sufficient to control algal biomass (as in Figure B). In this instance, we find a nonhysteretic threshold. This implies that sufficient exchange with the open ocean may prevent the occurrence of hysteretic thresholds in SEMSs.

- Does the disturbing effect of, for instance, increased nutrient input increase gradually with the amount added, or is there a threshold amount beyond which a dramatic shift to a new state is observed?
- How easy is it to reverse the impact of pollution or eutrophication—does it suffice simply to reduce the load below the threshold, after which recovery will follow the same reverse trajectory?

Finally, we will discuss the implications of our findings in light of the management of coastal zones.

Nonhysteretic Thresholds—Oxygen Depletion in Arabian Sea

One of the most vivid and clear examples of a dramatic, but reversible, nonhysteretic threshold is the response of both pelagic and benthic ecosystems to declining oxygen concentrations. Low oxygen concentrations can result from natural causes, as documented for the unpolluted Scottish fjord Loch Etive during the 1970s (Edwards and Grantham 1986). However, there are many studies that report on severe hypoxia or even anoxia as a consequence of organic pollution or eutrophication from human activities. Example studies come from the western Indian shelf (Naqvi et al. 2000), Gulf of Mexico (Turner and Rabalais 1994, 2003), Black Sea, Baltic Sea, East China Sea, and many more. Oxygen depletion not only adversely affects fisheries, but also can result in shifts in ecosystem structure and biogeochemical cycling (Rabalais and Turner 2001).

Oxygen depletion occurs in subsurface layers when physical supply is restricted by stratification/stagnation, often in combination with availability of copious organic matter associated with eutrophication, which drives higher respiration rates (Turner et al. 1998). Below an O_2 threshold concentration of around 60 μM, conditions become hypoxic, under which the mobility/behavior of aerobic organisms is severely impaired, although this threshold varies from system to system. When dissolved oxygen falls below about 1 μM (~0.02 mL L^{-1} in Figure 3-2), only those organisms that can use chemical species other than O_2 for respiration, mostly microorganisms, can survive. As a consequence, microbial activity leads to a sequential reduction of chemical species such as NO_3^- and SO_4^{2-} that occurs at well-defined thresholds. Under the suboxic conditions, dramatic changes in the community structure (e.g., proliferation of bacteria) and chemical cycling (e.g., recycling of Mn at the interface, rapid turnover of N_2O) will occur.

Under even more-severe conditions when NO_3^- gets completely removed, a transition to fully anoxic conditions involving SO_4^{2-} reduction takes place. The only known area where this phenomenon occurs on a regular (seasonal) basis along an open coast is the western Indian shelf. Here low-oxygen subsurface water upwells during summer. Strong stratification and high respiration rates drive serial shifts to suboxic and anoxic conditions before a change in coastal circulation associated with the monsoon reversal restores the oxic environment. However, this essentially natural oxygen deficiency seems to have intensified in recent years, presumably due to enhanced fertilizer runoff from agriculture (Naqvi et al. 2000). This is consistent with trends observed in other coastal areas.

In SEMSs having estuarine circulation, such as the Black and Baltic seas, all three types of conditions—oxic, suboxic, and anoxic—can be found. Of these, the Baltic Sea experiences somewhat larger variability due to the inflow of saline water from the North Sea that occurs approximately at decadal scales, causing intermittent reoxygenation below the pycnocline (Walter et al. 2006). The switchovers from oxic to suboxic and anoxic conditions in both directions observed on a seasonal scale over the Indian shelf and on a decadal scale in the Baltic are indicators of reversibility of ecosystem response

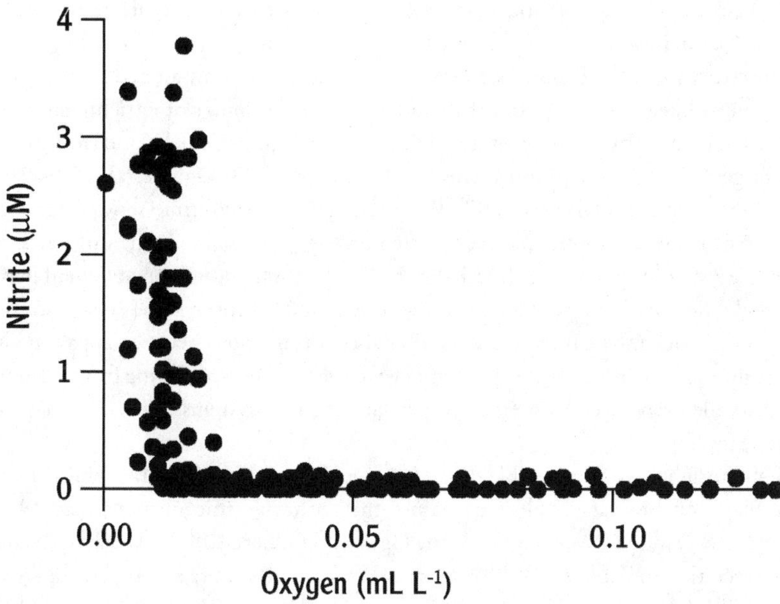

Figure 3-2. Plot of NO_2^- (an intermediate of denitrification) versus O_2 within the Arabian Sea O_2-deficient zone. Note the abrupt threshold at O_2 equal to about 1 μM for the onset of microbial NO_3^- reduction. The data were generated by S.W.A. Naqvi and L. Codispoti in October 1994 on the TN039 cruise of the USJGOFS Arabian Sea Process Study, on which O_2 was measured using the colorimetric procedure suitable for low O_2 concentrations. Figure adapted from Naqvi et al. 2003; used with permission from Blackwell Publishing/CRC Press.

to biogeochemical/physical forcing even though the system may not immediately attain exactly the same state after each cycle.

Hysteretic thresholds

Regime Shifts in the Black Sea

A number of good examples can be found of nonhysteretic thresholds, as was described in the above section. The number of case studies pointing to hysteretic thresholds with strong hysteresis effects is much more limited. Moreover, they are often prone to intense debate on the validity of the claim of hysteresis. Below, we discuss the changes observed in the Black Sea ecosystem as a potential example of a hysteretic threshold, but we also discuss alternative explanations for the observed changes. This case study highlights the difficulty in identifying potential hysteresis effects in systems that are affected by changes in multiple forcing factors.

The structure of the Black Sea ecosystem has been profoundly altered since the early 1970s, most likely resulting from the combined effects of nutrient enrichment, strong changes in temperature regime, overexploitation of pelagic fish stocks, and the introduction of gelatinous zooplankton carnivores (Daskalov 2003; Oguz et al. 2003, 2006; Oguz 2005; Oguz and Gilbert 2007). Long-term (1960–2000) time-series data suggest that the Black Sea ecosystem has reorganized through successive regime shifts affecting many ecosystem state variables (Daskalov 2003). Phytoplankton biomass, small pelagic fish stocks, Secchi disk depth, and dissolved oxygen and hydrogen sulfide concentrations close to the oxic–anoxic interface all exhibit abrupt transitions between apparent alternate regimes. These Black Sea regime shifts were abrupt events, differing from the multidecadal-scale cyclic events observed in pelagic ocean ecosystems under low-frequency climatic forcing.

Some insight into the possible drivers of this abrupt change can be obtained when phytoplankton biomass is plotted against the anthropogenic nutrient load (ANL) through the Danube River (Figure 3-3A; Oguz and Gilbert 2007). Two sharp changes can be seen that took phytoplankton from a "low biomass," more or less pristine state of the early 1970s (mean of about 3.0 g m^{-2}) to the "high biomass" state in the early 1980s (mean of about 17.5 g m^{-2}), occurring at apparent ANL thresholds of 300 and 600 kilotonnes y^{-1}. While the transition to a higher-biomass state appeared to have occurred rather abruptly, the transition back to the lower-biomass state was more gradual and followed a different trajectory. The observation that the return trajectory follows a different route from the initial trajectory indicates hysteresis, pointing to the presence of a hysteretic threshold. Hence, the data suggest the presence of two stable states: a near-pristine state characterized by a low phytoplankton density, and a second state characterized by high phytoplankton density and a drastically altered community structure.

Increase in anthropogenic nutrient load is not the only potential driver of ecosystem change in the Black Sea. There has been significant variation in the sea surface temperature (SST) during the observation period that is likely related to changes in wind regimes. When phytoplankton biomass is plotted against ANL and SST in a three-dimensional graph (Figure 3-3B), the additional likely role of climatic variations in promoting sharp changes in the phytoplankton biomass becomes evident. The high- and low-biomass states are linked to two different sea surface temperature regimes (lower than 7.8°C and greater than 8.4°C, respectively). The system preserved its poorly productive state for the entire range of ANL up to 600 kilotonnes y^{-1} even though it had accumulated appreciable nutrients in subsurface waters. These nutrients, however, were not made available by physical mechanisms into the surface layer for biological production. The Black Sea switched to a highly productive state (denoted by RS1) only after the SST became lower than 7.8°C, possibly reflecting increased mixing. Thus, it would appear that two conditions must be met before the system moves to the new high-biomass

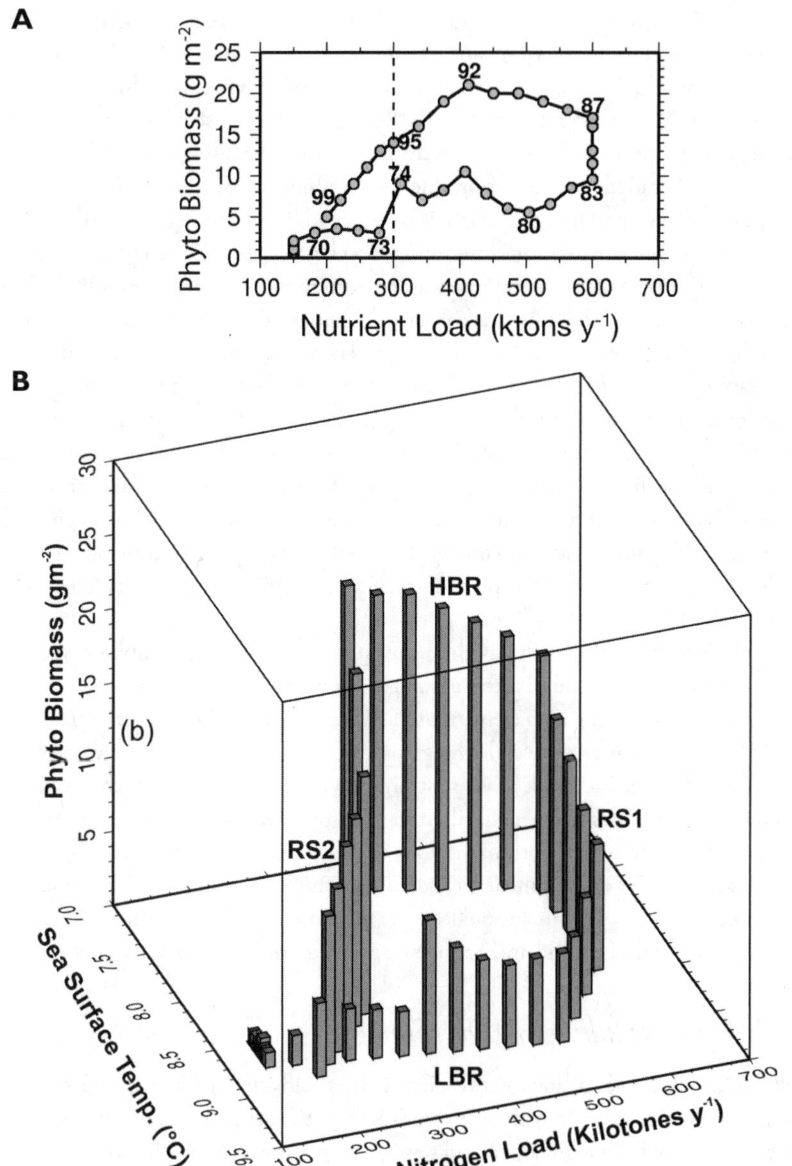

Figure 3-3. (A) Phytoplankton biomass distribution in relation to the anthropogenic nutrient load (in kilotonnes y^{-1}) showing trajectories from a low-biomass regime (LBR) to a high-biomass regime (HBR), with apparent abrupt transitions between them during 1973–1974 and 1984–1985, and a gradual recovery to the low-biomass regime. (B) Phytoplankton biomass (in g m^{-2}) as a function of both nitrogen load and sea surface temperature. Figure from Oguz and Gilbert 2007, copyright 2007; used with permission from Elsevier.

state: a sufficient annual nutrient load and a persistent change in climatic conditions leading to greater mixing between surface and subsurface waters. Once the system had moved to a new regime, it appeared resistant to decreasing ANL, to a threshold level of 300 kilotonnes y^{-1} and SST lower than 7.8°C. As the system shifted from a cold to a warm regime, the biological production declined at an increasing rate, and finally the system reverted gradually to its less productive, low-biomass state for SSTs greater than 8.4°C during the second half of the 1990s, as denoted by RS2 in Figure 3-3B.

Figure 3-3B points to an important interaction between ANL and climate variation. The ANL appears to introduce the necessary nutrients (i.e., external conditions), but these remain locked in the subsurface layers. Climate change sets in motion the necessary physical mechanisms for an abrupt change by increasing mixing between subsurface and surface layers, resulting in a flux of subsurface NO_3^- to the productive zone.

Abrupt changes may also be caused by slight changes in the monitored forcing coupled with unpredicted and frequently unmonitored forcing. More observations are needed to address the issue of whether and at what ANL level a threshold response would be observed if sea surface temperature remains unaltered. Moreover, other potential factors could also be present, such as concomitant changes in top-down control of the food web due to alterations in the top predator, resulting from excessive fisheries (Gucu 2002).

The Black Sea case shows clearly that ecosystems can display a complex response to human-induced changes and that these changes can take a long time to reverse. This is an important message for the management of SEMSs, as the costs involved in restoring the degraded ecosystem may outweigh the gains from the restored economic activity by orders of magnitude. The Black Sea case study, however, also emphasizes that a complex response trajectory does not necessarily imply the presence of alternative regimes. The multitude of ways in which both natural and anthropogenic changes affect ecosystems can generate complex developmental trajectories, which may appear as hysteresis loops when projected in a two-dimensional effect–response plot but in reality include the effects of multiple natural and anthropogenic pressures on ecosystem dynamics.

Threshold Effects Induced by Overfishing

Some classical examples of apparent hysteretic thresholds come from the impact of fishing on marine ecosystems, mostly from studies from the open ocean. Fishing is a dominant human activity in many open and semi-enclosed seas, and it is favored by the fact that continental shelves and associated coastal zones are optimal breeding grounds for many edible species, including top predators such as bluefin tuna but also for other trophic groups. A good example is the northern Atlantic cod, where following a swift collapse of the cod stocks, subsequent recovery has been limited (Hutchings 2000). Examples of stock collapse originate from both continental shelf and open-ocean fisheries. Studies on the recovery of marine stocks, however, indicate that populations that are harvested using bottom-deployed seines and trawls, typically used in SEMSs and on continental shelves on shallow benthic environments, show a more limited recovery compared with populations that are harvested using purse seines and mid-water trawls that are typically used in open-ocean fisheries (Hutchings 2000). This limited recovery results mostly from losses due to bycatches and destruction of bottom habitats, which are more difficult to avoid in bottom-deployed seines and trawls compared with midwater fishing gear. Hence, fish stocks in SEMSs may be particularly vulnerable to collapse in response to increased fishing pressure.

In many instances, fish stock collapses are not solely triggered by increased fishing efforts, but are related to recruitment failure. Recruitment fluctuates significantly from one year to another for many species, with only a few years producing a large renewal of the stock (Jennings et al. 2001). This suggests that variability in abiotic and biotic environmental processes can have a strong impact on maintenance of the stock. The need for good correspondence between environmental conditions and fish reproductive periods has been long understood (Lasker 1975, 1978; Cushing 1982). A mismatch causes higher vulnerability and larger fluctuations in recruitment, as compared with other groups of organisms, which in turn implies that the overlap of high fishing pressure, deterioration of environmental conditions due to human activities, and unfavorable recruitment years may cause population collapse. This rarely provokes extirpation of the population but generally results in long times for the recovery of the stocks, complicated by continuing fishing pressure (Hutchings 2000).

A similar collapse has been observed in the Adriatic Sea, where data on fish landings show that an abrupt change in recruitment occurred between 1977 and 1979, while the catches increased in 1980. This led to a drastic decrease of the anchovy stock over the following decade (Cingolani et al. 1996). The Northern Adriatic Sea (depth < 50 m) receives an average freshwater discharge of 3,000 $m^3 s^{-1}$ (Hopkins et al. 1999), which is transported southward by the Western Adriatic Current. Dry and cold winds facilitate the ventilation of the winter water column. During summer a weak cyclonic circulation provides renewal of the bottom layer. Increase in the residence time of low-salinity water and summer heating, coupled with sluggish circulation, cause widespread anoxic events, as in 1973, 1984, and 1988 (Stachowitsch 1991). In 1984, 90% of the sessile benthic organisms were killed, and after three years the recovery for some species was only 11%.

Stress on bivalve mollusk populations and intense dredging could be the cause of increased frequency of mucilaginous events (i.e., an anomalous accumulation of algal exudates, mostly polysaccharides, on the sea bottom and, afterward, along the coastline) in the Northern Adriatic Sea, through reduction of phytoplankton removal from the water column and accumulation of excess photosynthesized carbon in the low-nutrient summer conditions (F. Boero, Universita' Degli Studi Di Lecce, personal communication), thus producing a shift in ecosystem functioning. Even persistent anoxia may produce abrupt changes in recruitment. Eggs are vulnerable to low oxygen levels, and unfavorable coupling of physical dispersal with sinking into anoxic bottom waters may severely hamper recruitment (Voss 2007). These losses have profound consequences for bottom-feeding fish.

Regime Shifts in Other SEMSs

The case studies discussed above are not isolated incidents but provide some well-described examples of threshold effects that have been found in a wide range of SEMSs. There is an extensive body of literature reporting on threshold effects in other marine systems. Strong effects of eutrophication are found in the Gulf of Mexico. An increase in nutrient release to the Mississippi River and its tributaries resulted in enhanced surface-water primary production, and an increased accumulation in continental shelf sediments of marine organic carbon, adjacent to the Mississippi River discharge (Turner and Rabalais 1994; Rabalais et al. 2001, 2002, 2007). Seasonal development of hypoxic conditions was first noticed in the 1970s, followed by an abrupt increase in the early 1980s in the size and severity of the hypoxia (verified by changes in paleoindicators, hydrologic data, and modeled hindcasts of oxygen conditions; Scavia et al. 2003; Justic et al. 2005; Turner et al. 2006; Rabalais et al. 2007). Whether these abrupt changes in the Gulf of Mexico are reversible is not known because there is no concerted effort to reduce nutrient loads in these coastal waters.

A potential regime shift, resulting from the combined effects of eutrophication and increased fisheries, seems eminent in the SEMS to the east of China. The observed symptoms of eutrophication in this system (Li et al. 2007) lead to the diagnosis that this ecosystem is approaching a crisis. Inputs of inorganic nitrogen from large rivers have increased continuously during the last fifty years, while the supply of light-absorbing sediments has been decreased due to the construction of dams (cf. Zhang et al. 1999; Yang et al. 2006). These changes in inputs have led to eutrophication in coastal waters of the East China Sea (Zhang 2002). The phytoplankton are changing, with more frequent algal blooms and a higher ratio of dinoflagellates to diatoms in these blooms (Zhou et al. 2001).

Overfishing has removed the large fish that eat small fish, and so these smaller fish have become more abundant in commercial catches. Although this should allow a more efficient transfer of production into the fishery, the overall fish catch per unit of effort has dropped. According to our best understanding of ecosystem theory, increase in small fish abundance results in a decrease in the abundance of grazing planktonic invertebrates and hence makes more likely a switch to an ecosystem dominated by harmful algae and jellyfish, with increasing episodes of bottom-water hypoxia killing benthic animals. The outcome of the crisis will be a collapsed ecosystem—collapsed in its ability to provide the services needed to support human life in the adjacent land, as will be discussed in Chapter 12. What we do not know is how well or quickly the system will recover from such a collapse and what the regime will look like when recovery is final. Our mathematical analysis in Box 3-2 suggests that the openness of the sea may be important in the ability of a system to recover; the more rapid the exchange of water with an undisturbed ocean, the greater the reversibility (in speed and extent) of change.

In most of these studies, data series are insufficiently long to distinguish between continuous versus discontinuous change. Moreover, in many systems, the return trajectory is unknown, as the drivers of the observed regime shift have not been reversed (e.g., the northern Gulf of Mexico). Finally, in many systems, human activity has combined with climate variability to yield multiple pressures upon ecosystems, and it is unknown which one is most important in explaining the observed change. Hence, although evidence for threshold effects has been found in many rocky intertidal systems (Petraitis and Dudgeon 2004; but see Bertness et al. 2002 for a counterview) and benthic communities (van Nes et al. 2007), we are forced to conclude that for large-scale SEMSs, there is only limited and often indecisive evidence for hysteretic thresholds, and there is little understanding of the functioning of internal feedbacks in many marine ecosystems. Yet, the evidence that dramatic shifts occur is real for many SEMSs, emphasizing that the potential for nonlinear responses of marine ecosystems is of clear importance to both scientists and managers of these vulnerable ecosystems.

Management Implications

The issues raised in this chapter are of clear importance to managers, as they deal with abrupt and unexpected changes in ecosystems that can occur when a threshold is exceeded. We have attempted to distinguish two types of threshold. One kind of threshold is like a warning of a banana skin on a footpath: If you ignore the warning, you may slip and fall. However, with a bit of luck, you will be able to pick yourself up and return to the path. This *nonhysteretic* threshold is a warning that an ecosystem is about to experience a disturbance due to human activities, from which, however, it will recover as soon as the human pressure is decreased. The second kind of threshold is like a warning of a cliff ahead. If you ignore the warning, you will fall over the cliff and experience a severe change of state, likely a very permanent one, or otherwise spend a long time and much effort to get back to the top of the cliff. This is the *hysteretic* threshold.

Of course, the cliff is a metaphor. It is not always easy to see that ecological dysfunction is imminent. Thus, managing the environment is not like walking along a footpath. It is more akin to wandering on rough ground in a fog, and therefore scientists need to provide managers with *indicators* to warn when a cliff is near. Some indicators were reviewed by Tett and colleagues (2007). As an example, we will consider the measurement of dissolved oxygen. This is especially useful in the case of ecosystems that are sensitive to the undesirable disturbance associated with eutrophication, which corresponds to a catastrophic collapse in ecosystem health. Tett and colleagues (2007) pointed out that some ecosystems are more sensitive than others to nutrient enrichment (as defined by Cloern 2001). Clear examples are coastal systems that (1) become layered due to warming or freshwater supply to the surface, and (2) exchange their waters only weakly with those in the nearby ocean (see Chapter 6, this volume). As discussed in this chapter, the Baltic Sea, the Black Sea, and the Gulf of Mexico are examples of such sensitive systems. In each case, increased nutrient input (together with changed nutrient element ratios) has led to increased primary production and hence to increased sinking of organic matter. The decay of that organic matter consumes oxygen in deep water. The annual minimum of the concentration of oxygen in these waters provides a good indicator of the impact of nutrients on the ecosystem. Environmental management agencies in many countries have already set minimum allowable levels between 2 and 4 mg L^{-1} dissolved oxygen, because seabed animals begin to die when concentrations fall below this level. In some cases, this is a nonhysteretic threshold: When dissolved oxygen concentrations are restored, the benthic community recovers quickly. In other cases, such as that documented for the Baltic Sea by Laine and colleagues (1997), the extent of benthic community destruction is so widespread that recovery is a slow process, suggesting a possible hysteretic threshold. In these systems, it is wise to closely monitor dissolved oxygen. If the threshold is approached, it is a warning that there is a risk of a benthic catastrophe—that the edge of the cliff is approaching—and hence that remedial action is necessary.

As we will discuss in the final section, the ideas about nonhysteretic and hysteretic threshold changes that have been presented in this chapter are hypotheses. We have tried to explore these hypotheses with recent observations, and we conclude that mechanisms other than hysteretic thresholds might explain the changes observed in many SEMSs with equal likelihood. Nevertheless, the data themselves show that there can be abrupt changes in the state of ecosystems in SEMSs, even if we cannot fully explain these changes. Taking account of the precautionary principle, managers should act now to protect SEMSs against threshold changes. They should do this by reducing human pressures such as those of nutrient enrichment and overfishing. We urge this especially for systems that appear, from already documented changes or because of low flushing, to be sensitive to possible hysteretic threshold changes. It will also be necessary to keep in mind the effect of global warming as an uncontrollable pressure on these ecosystems.

Scientific Conclusions

We have presented observational evidence that SEMSs can change abruptly to a new regime that provides fewer ecosystem services to humans. We have attempted to interpret these changes as being nonhysteretic, and thus reversible, or hysteretic and hard to reverse. Some changes, such as those associated with hypoxia in deep waters, appear to be reversible, although restoration of an original micro-biogeochemical regime is faster than restoration of macrobenthic communities. In other cases, time series of data are not long enough to distinguish between a reversible versus irreversible change. Even in the case of the Black Sea, our best example, a complex trajectory in the driver-response phase plane (Figure 3-3) may indicate hysteresis and the presence of alternative regimes, but it may also be a two-dimensional projection of a multidimensional trajectory involving several forcing factors. The theory of alternate stable states (or regimes) that has been offered to explain hysteretic threshold dynamics has its origins in the study of relatively small and enclosed temperate lakes. SEMSs are, however, typically larger, more energetic, and more heterogeneous than such lake systems and, in addition, more likely to be subject to multiple external pressures. Hence, their response to pressure will possibly be even more nonlinear than that observed in shallow lakes.

These considerations imply that the concepts presented in Figure 3-1 provide an overly simplified picture of the complex relations between external pressures and ecosystem response. We therefore need better conceptual and mathematical models of ecosystem change in SEMSs to take account of food web complexity, exchange with the ocean, and replacement of locally damaged populations. The second research need is for sustained study of those SEMSs in which humans are, somewhat blindly, carrying out perturbation experiments. A third research need is one that we have already mentioned: the need for indicators to monitor against hysteretic threshold change in SEMSs. We need indicators of the impact of pressures on ecosystems that will provide reliable warning of the closeness of a threshold. In addition, we need indicators of the sensitivity of ecosystems, including for SEMSs, to pressures that might cause irreversible shifts. Finding these indicators will require a deep understanding of the structure and function of marine coastal ecosystems.

References

Bertness, M.D., G.C. Trussell, P.J. Ewanchuk, and B.R. Silliman. 2002. Do alternate stable community states exist in the Gulf of Maine rocky intertidal zone? *Ecology* 83:3,434–3,448.

Cingolani, N., G. Giannetti, and E. Arneri. 1996. Anchovy fisheries in the Adriatic Sea. *Scientia Marina* 60:269–277.

Cloern, J.E. 2001. Our evolving conceptual model of the coastal eutrophication problem. *Marine Ecology Progress Series* 210:223–253.

Cushing, D.H. 1982. *Climate and Fisheries*. Oxford: Academic Press/Elsevier.

Daskalov, G.M. 2003. Long-term changes in fish abundance and environmental indices in the Black Sea. *Marine Ecology Progress Series* 255:259–270.

Edwards, A., and B.E. Grantham. 1986. Inorganic nutrient regeneration in Loch Etive bottom water. In *The Role of Freshwater Outflow in Coastal Marine Ecosystems*, edited by S. Skreslet, 195–204. Berlin: Springer.

Gucu, A.C. 2002. Can overfishing be responsible for the successful establishment of *Mnemiopsis leidyi* in the Black Sea? *Estuarine, Coastal and Shelf Science* 54:439–451.

Hopkins, T.S., A. Artegiani, C. Kinder, and R. Pariente. 1999. A discussion of the northern Adriatic circulation and flushing as determined from the ELNA hydrography. In *The Adriatic Sea*, edited by T.S. Hopkins, 85–106. Ecosystem Research Report of the European Commission. Luxembourg: Office for Official Publications of the European Community.

Hutchings, J.A. 2000. Collapse and recovery of marine fishes. *Nature* 406:882–885.

Jansson, B.-O., and K. Dahlberg. 1999. The environmental status of the Baltic Sea in the 1940s, today and in the future. *Ambio* 28:312–319.

Jennings, S., M. Kaiser, and J. Reynolds. 2001. *Marine Fisheries Ecology*. Oxford: Blackwell.

Justic, D., N.N. Rabalais, and R.E. Turner. 2005. Coupling between climate variability and coastal eutrophication: Evidence and outlook for the northern Gulf of Mexico. *Journal of Sea Research* 54:25–35.

Laine, A.O., H. Sandler, A.B. Andersin, and J. Stigzelius. 1997. Long-term changes of macrozoobenthos in the Eastern Gotland Basin and the Gulf of Finland (Baltic Sea) in relation to the hydrographical regime. *Journal of Sea Research* 38:135–159.

Lasker, R. 1975. Field criteria for survival of anchovy larvae: The relation between inshore chlorophyll maximum layers and successful first feeding. *Fishery Bulletin* 73:453–462.

Lasker, R. 1978. The relation between oceanographic conditions and larval anchovy food in the California Current: Identification of factors contributing to recruitment failure. *Rapports et Procès-Verbaux des Réunions* 173:212–230.

Li, M., K. Xu, M. Watanabe, and Z. Chen. 2007. Long-term variations in dissolved silicate, nitrogen, and phosphorus flux from the Yangtze River into the East China Sea and impacts on estuarine ecosystem. *Estuarine Coastal and Shelf Science* 71:3–12.

Naqvi, S.W.A., D.A. Jayakumar, P.V. Narvekar, H. Naik, V. Sarma, W. D'Souza, S. Joseph, and M.D. George. 2000. Increased marine production of N_2O due to intensifying anoxia on the Indian continental shelf. *Nature* 408:346–349.

Naqvi, S.W.A., H. Naik, and P.V. Narvekar. 2003. The Arabian Sea. In *Biogeochemistry of Marine Systems*, edited by K.D. Black and G.B. Shimmield, 157–207. Oxford: Blackwell/CRC Press.

Oguz, T. 2005. Hydraulic adjustments of the Bosphorus exchange flow. *Geophysical Research Letters* 32:L06604, doi: 10.1029/2005GL022353.

Oguz, T. 2007. Nonlinear response of Black Sea pelagic fish stocks to over-exploitation. *Marine Ecology Progress Series* 345:211–228.

Oguz, T., T. Cokacar, P. Malanotte-Rizzoli, and H.W. Ducklow. 2003. Climatic warming and accompanying changes in the ecological regime of the Black Sea during 1990s. *Global Biogeochemical Cycles* 17 (3):1,088, doi: 10.1029/2003GB002031, 2003.

Oguz, T., J.W. Dippner, and Z. Kaymaz. 2006. Climatic regulation of the Black Sea hydrometeorological and ecological properties at interannual-to-decadal time scales. *Journal of Marine Systems* 60:235–254.

Oguz, T., and D. Gilbert. 2007. Abrupt transitions of the top-down controlled Black Sea pelagic ecosystem during 1960–2000: Evidence for regime-shifts under strong fishery exploitation and nutrient enrichment modulated by climate-induced variations. *Deep-Sea Research I* 54:220–242.

Petraitis, P.S., and S.R. Dudgeon. 2004. Detection of alternative stable states in marine communities. *Journal of Experimental Marine Biology and Ecology* 300:343–371.

Rabalais, N.N., and R.E. Turner (eds.). 2001. *Coastal Hypoxia: Consequences for Living Resources and Ecosystems*. Coastal and Estuarine Studies Vol. 58. Washington, DC: American Geophysical Union.

Rabalais, N.N., R.E. Turner, B.K. Sen Gupta, E. Platon, and M.L. Parsons. 2007. Sediments tell the history of eutrophication and hypoxia in the northern Gulf of Mexico. *Ecological Applications* 17:S129–S143.

Rabalais, N.N., R.E. Turner, and W.J. Wiseman. 2001. Hypoxia in the Gulf of Mexico. *Journal of Environmental Quality* 30:320–329.

Rabalais, N.N., R.E. Turner, and W.J. Wiseman. 2002. Gulf of Mexico hypoxia, aka "the dead zone." *Annual Review of Ecology and Systematics* 33:235–263.

Scavia, D., N.N. Rabalais, R.E. Turner, D. Justic, and W.J. Wiseman. 2003. Predicting the response of Gulf of Mexico hypoxia to variations in Mississippi River nitrogen load. *Limnology and Oceanography* 48:951–956.

Scheffer, M. 1998. *Ecology of Shallow Lakes.* 1st ed. London: Chapman & Hall.

Scheffer, M., S. Carpenter, J.A. Foley, C. Folke, and B. Walker. 2001. Catastrophic shifts in ecosystems. *Nature* 413:591–596.

Stachowitsch, M. 1991. Anoxia in the Northern Adriatic Sea: Rapid death, slow recovery. In *Modern and Ancient Continental Shelf Anoxia*, edited by R.V. Tyson and T.H. Pearson, 119–129. Special Publication 58. London: Geological Society.

Tett, P., R. Gowen, D. Mills, T. Fernandes, L. Gilpin, M. Huxham, K. Kennington, P. Read, M. Service, M. Wilkinson, and S. Malcolm. 2007. Defining and detecting undesirable disturbance in the context of eutrophication. *Marine Pollution Bulletin* 53:282–297.

Turner, R.E., N. Qureshi, N.N. Rabalais, Q. Dortch, D. Justic, R.F. Shaw, and J. Cope. 1998. Fluctuating silicate:nitrate ratios and coastal plankton food webs. *Proceedings of the National Academy of Science USA* 95:13,048–13,051.

Turner, R.E., and N.N. Rabalais. 1994. Coastal eutrophication near the Mississippi River delta. *Nature* 368:619–621.

Turner, R.E., and N.N. Rabalais. 2003. Linking landscape and water quality in the Mississippi River basin for 200 years. *Bioscience* 53:563–572.

Turner, R.E., N.N. Rabalais, and D. Justic. 2006. Predicting summer hypoxia in the northern Gulf of Mexico: Riverine N, P, and Si loading. *Marine Pollution Bulletin* 52:139–148.

Van Nes, E.H., T. Amaro, M. Scheffer, and G.C.A. Duineveld. 2007. Possible mechanisms for a marine benthic regime shift in the North Sea. *Marine Ecology Progress Series* 330:39–47.

Voss, R. 2007. Recruitment processes in the larval phase: The influence of varying transport on cod and sprat larval survival, p. 134. PhD thesis, Christian-Albrechts-Universität zu Kiel.

Walter, S., U. Breitenbach, H.W. Bange, G. Nausch, and D.W.R. Wallace. 2006. Distribution of N_2O in the Baltic Sea during transition from anoxic to oxic conditions. *Biogeosciences* 3:557–570.

Yang, S.L., M. Li, S.B. Dai, Z. Liu, J. Zhang, and P.X. Ding. 2006. Drastic decrease in sediment supply from the Yangtze River and its challenge to coastal wetland management. *Geophysical Research Letters* 33:L06408, doi: 10.1029/2005GL025507.

Zhang, J. 2002. Biogeochemistry of Chinese estuarine and coastal waters: Nutrients, trace metals and biomarkers. *Regional Environmental Change* 3:65–76.

Zhang, J., Z.F. Zhang, S.M. Liu, Y. Wu, H. Xiong, and H.T. Chen. 1999. Human impacts on the large world rivers: Would the Changjiang (Yangtze River) be an illustration? *Global Biogeochemical Cycles* 13:1,099–1,105.

Zhou, M.J., M.Y. Zhu, and J. Zhang. 2001. Status of harmful algal blooms and related research activities in China. *Chinese Bulletin of Life Science* 13:54–59.

4

Governance and Management of Ecosystem Services in Semi-Enclosed Marine Systems

Paul V.R. Snelgrove, Michael Flitner, Edward R. Urban, Jr., Werner Ekau, Marion Glaser, Heike K. Lotze, Catharina J.M. Philippart, Penjai Sompongchaiyakul, Edy Yuwono, Jerry M. Melillo, Michel Meybeck, Nancy N. Rabalais, and Jing Zhang

Introduction

The global ocean faces increasing pressures as human populations grow and demand for marine resources expands. In semi-enclosed marine systems (SEMSs), stressors from human activities are concentrated and numerous pressures overlap in space and time, with complex, interacting effects on coastal and marine ecosystems, and feedbacks into the social realm. Chapter 2 provides greater detail of the stresses that impact SEMSs, and in this chapter we focus primarily on how these stresses affect the ability of humans to govern SEMSs effectively.

Within the last decade there has been significant recognition of the role that ecosystem services play in providing key benefits to human society (Daily 1997; Snelgrove et al. 1997; Millennium Ecosystem Assessment 2005; Duffy and Stachowicz 2006). The importance of ecosystem services of SEMSs, their regional differences, and long-term changes have been reviewed by Lotze and Glaser (Chapter 12, this volume). Moreover, recent experimental work has established the linkages between biodiversity and ecosystem services (Lohrer et al. 2004; Waldbusser et al. 2004). Considered in tandem with global biodiversity losses (Worm et al. 2006), the challenges faced in maintaining ecosystem services in SEMSs are considerable.

The goals of this chapter are threefold. First, we identify important threats to and vulnerabilities of ecosystem services in SEMSs. Second, we describe how SEMSs have been managed to address losses of ecosystem services, and strategies that have been invoked to reverse the losses. Finally, we provide case studies that show how different governance and management strategies have succeeded and failed, and we discuss how these efforts can be used to guide future efforts to maintain ecosystem services in SEMSs around the world.

Threats to Ecosystem Services in Semi-Enclosed Marine Systems

Ecosystem services are the benefits that people obtain from ecosystems (Millenium Ecosystem Assessment 2005; Chapter 12, this volume). For SEMSs, the most important provisioning services include food resources from fisheries, hunting, and aquaculture, as well as genetic resources (the genes and genetic material contained in individuals and within populations that offer potential future benefits to humans, such as novel genes and gene products), biochemical resources, and aesthetic and cultural resources (for more detail see Chapter 12). Important "regulating services" of natural systems include climate regulation, water purification, waste treatment, erosion control, and natural recycling of elements. Key "cultural services" include cultural identification, spiritual enrichment, and aesthetic values. "Supporting services" are those necessary for the production of all other ecosystem services and include photosynthesis and primary production, nutrient and water cycling, and habitat provisioning. While we acknowledge the importance of this diversity of services, we focus here primarily on those services of most direct benefit to humans, including fisheries, habitat, and water quality.

Ecosystem services are subject to many threats or drivers of ecosystem change (Figure 4-1) that operate on different spatial and temporal scales and are often not independent of one another. In order to identify the most urgent threats to ecosystem services in SEMSs, we canvased regional experts to provide a relative comparison of the importance and spatial dimension of different threats within their systems (Figure 4-1). In a general sense, human activities that alter species composition and abundance have the capacity to compromise any of these ecosystem services.

We discuss the major threats to ecosystem services and how they manifest themselves at different spatial scales. The summaries are brief and draw on diverse examples, in order to place some of the more detailed case studies that follow within the context of multiple drivers of change, which often act synergistically rather than additively. Most threats operate at multiple spatial scales, and we therefore focus on the most relevant scale for any specific driver. From a governance perspective, this organizational scheme reflects which threats may be dealt with by local users and which require complex solutions that include international cooperation and management strategies at larger scales.

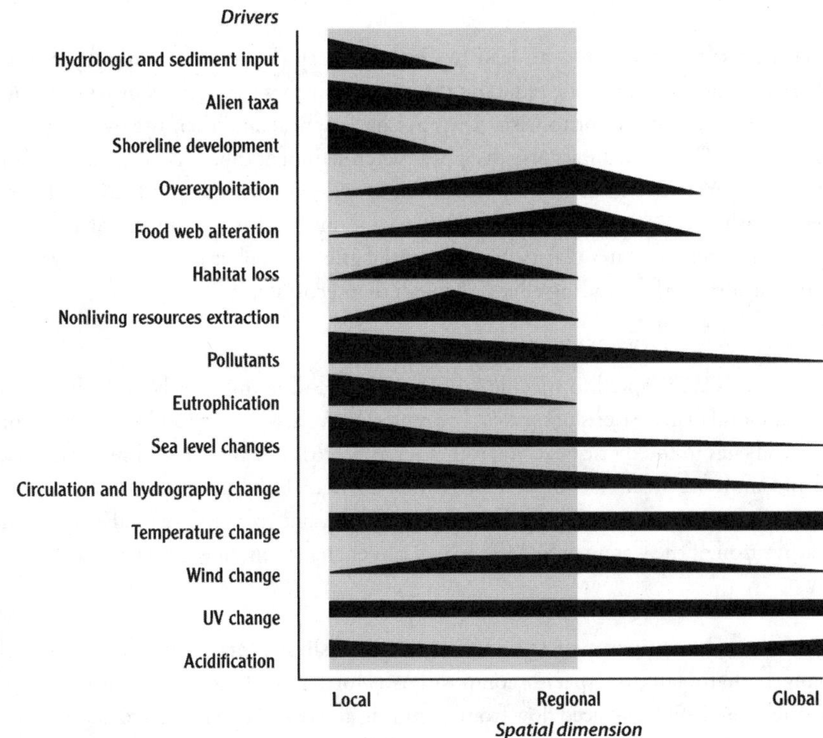

Figure 4-1. Summary of perceived threats to ecosystem services and the spatial scales at which they operate. Wide areas indicate scales at which threat is predominant. The gray box illustrates the spatial dimension of SEMSs, where *local* is defined as a single bay, *regional* refers to the scale of one SEMS, and *global* is defined as ocean scale. Bar height indicates perceived magnitude of threat at the different scales.

Local (Bay and Coastal Ocean) Drivers

Drivers of changes in ecological services that occur on local scales include changes in inputs of freshwater, nutrients, and sediments; biological invasions and stock enhancements; and development of shoreline and nearshore areas.

HYDROLOGIC CHANGE AND SEDIMENT INPUT

Decreases in freshwater and changes in hydroperiod from the diversion and damming of freshwater sources can affect the amount and timing of freshwater and sediment inputs to SEMSs. SEMSs (including their landward components) can act as filters of material entering the ocean, which is described in detail in Chapter 7.

CHANGES IN SEDIMENT INPUTS

Decreased sediment inputs can lead to erosion of delta systems (Barmawidjaja et al. 1995), marshes, and beaches, reducing the protection these structures provide against storms, their usefulness for tourism, their aesthetic appeal, and their use by indigenous populations. Coastal habitats also disappear when sediment inputs decrease, decreasing associated services such as water purification or waste treatment, production of commercial fish species, and maintenance of marine biodiversity. Increases of sediment inputs can decrease water clarity, with potential effects on phytoplankton, sea grasses, kelp, and coral with cascading effects through their ecosystems.

CHANGES IN FRESHWATER INPUTS

Most species are adapted to specific salinity ranges, and salinity levels limit the spatial and temporal distributions of species. The reproductive activity, egg and larval transport (e.g., cod eggs in the Baltic Sea; MacKenzie et al. 1996), and growth and survival of larval and adult organisms are affected by salinity changes. Fish kills can result from freshwater release from dams. Changes in input of freshwater to SEMSs can also affect the stratification of the waters in SEMSs, which can change their biological productivity.

CHANGES IN NUTRIENT INPUTS

Dams can reduce silica supply to SEMSs by trapping silica-containing sediments, which changes phytoplankton species composition (Dortch et al. 2001) and productivity (Turner et al. 1998). Reduced flow from damming and diversion can also reduce nutrient supply to estuaries and thus the primary and secondary production capacity, including commercial fisheries (Nixon 2003, 2004). In contrast, increases in nutrients can influence food web structure and exacerbate toxic blooms that can limit fish and shellfish harvesting (Rabalais 2004). Changes in freshwater and associated nutrients reaching SEMSs also can eliminate wetlands (Levin et al. 2001) if conditions are too saline or fresh or if nutrient supply decreases or increases.

INVASIVE SPECIES AND STOCK ENHANCEMENT

Human activities (e.g., shipping, aquaculture, deliberate introductions) have rapidly promoted the dispersal of nonindigenous marine organisms (Carlton 1996; Ruiz et al. 1997), especially in SEMSs. Invading species can change the structure and function of ecosystems in SEMSs. Invasive predators can suppress populations of native species more than native predators (Salo et al. 2007) and have other negative effects on natural and aquaculture systems as trophic competitors, predators, or disease vectors (Decottignies et al. 2007). Typically, invasive species expand their ranges until they reach environmental and/or ecological conditions that limit their growth or reproduction, and they become nearly impossible to eradicate. Deliberate introductions to increase productivity have led to invasions of the seaweed *Caulerpa taxifolia* in the Mediterranean Sea and the Pacific oyster (*Crassostrea gigas*) in the North Sea (Van der Weijden et al. 2007). Some native species have been cultured and reintroduced to the wild to help in the recovery of endangered species and boost natural production. For example, off the Zhejiang coast, jellyfish that are a traditional dish are seeded from captive populations each year (Qu et al. 2005; Zhang et al. 2006). By introducing large numbers of individuals with a limited number of parents, such activity can change the genetic structure of wild populations.

Rapid warming (and presumably ice loss in the Arctic) can accelerate invasion and establishment of nonindigenous species, many of which have fared better under recent warmer conditions (Stachowicz et al. 2002). Recent observations indicate that the Atlantic Ocean has already been invaded by a Pacific plankton species through the Arctic Ocean (Reid et al. 2007).

SHORELINE AND NEARSHORE DEVELOPMENT

Shore development, including industrialization, urbanization, dike building, and a range of other activities, is often driven by economic forces. It can reduce a wide array of provisioning services, including fish harvests or shrimp aquaculture, when habitat destruction removes areas used for fish and shrimp spawning and/or larval development. Oil and gas extraction from coastal seabeds cause disturbance through drill cuttings and produced water, associated transport issues (tanker traffic and pipelines), and spills. All have potentially negative effects on fisheries. Shore development can also disrupt regulating services that would result from natural habitats, such as climate regulation, water regulation, water purification and waste treatment, and erosion regulation. Cultural services such as heritage, artistic, and aesthetic values in recreation and ecotourism are particularly vulnerable from shoreline and nearshore development, because it is so visible to humans. Such development can also change sediment, water, and nutrient inputs, with the negative impacts described earlier.

Medium-Scale (Regional) or Semi-Enclosed System Effects

Changes that are manifested at regional scales (i.e., beyond individual bays) include living resource overexploitation, food web alterations, habitat loss, and nutrient loading.

OVEREXPLOITATION

Overexploitation of fish populations by recreational, subsistence, and commercial fisheries has become a major problem at global spatial scales. Many fisheries are in major decline from historic baseline levels or have collapsed (Myers and Worm 2003), and many areas have experienced "fishing down the food web" (sequential reduction of the largest species in an ecosystem) (Pauly et al. 1998). Marine living resources are harvested usually for human consumption (including fish oils) but also for animal feeds (e.g., fish meal), aquariums, and clothing (e.g., furs, shark and eel leathers). Bycatch (incidental catch of unwanted species) can reduce populations, as well as genetic and species diversity.

Suspension-feeding organisms, such as oysters, purify water in estuarine and mangrove environments by removing particulate matter, thus improving water clarity, biogeochemical regulation (e.g., cycling of organic carbon and particle-associated elements and compounds), and sedimentation. Reduction of suspension feeders can increase turbidity, decrease submerged aquatic vegetation, change zooplankton populations, and increase ctenophores (Newell 1988).

Spiritually symbolic species (e.g., sharks, whales, and dolphins) embody cultural services that are lost to indigenous cultures (e.g., Inuit in Hudson Bay) when stocks collapse and when conflicts develop among subsistence, recreational, and commercial fisheries. Overexploitation and fisheries closures can lead to loss of cultural traditions, aesthetic values, recreational fishing and (eco)tourism (e.g., bird and whale watching, diving).

Overexploitation of living resources can affect the supporting service of primary production. For example, overexploitation of menhaden and other algal grazers may increase the amount of particulate carbon in the pelagic system, whereas removal of predators may allow grazers to increase, and thus reduce phytoplankton.

FOOD WEB ALTERATION

Overexploitation can alter food webs, with cascading effects through the food web that can have dramatic consequences on production services (e.g., Frank et al. 2007). From a human perspective, effects may be positive or negative. For example, Myers and colleagues (2007) describe cascading declines in shellfish through removal of top predators, whereas Worm and Myers (2003) document a widespread positive effect of cod decline on shrimp. Fishing down the food web (Pauly et al. 1998) reduces populations of large fish but can increase populations of smaller (usually less valuable) fish. Diversity changes may reduce the resilience of estuarine ecosystems (Cohen and Carlton 1998; Stachowicz et al. 1999), especially when entire functional groups (multiple species that perform a specific ecological function, such as nitrogen fixation) are lost. Blooms of gelatinous zooplankton or other undesirable species can result from trophic cascades, yielding negative and unexpected consequences on recreation and ecotourism.

HABITAT LOSS

Habitat loss can influence production services by reducing critical habitat for potential food resources. Loss of species that create physical structure (e.g., coral and oyster reefs, mangroves, kelp forests) can cascade to species that depend on those habitats as adults, juveniles, or larvae. For example, loss of cold-water corals (e.g., Costello et al. 2005) and sea grasses (Gonzalez-Correa et al. 2005) as a result of bottom-trawl fisheries has cascading effects on other species. Loss of biodiversity is often linked to specific habitat loss, and declines in specific habitat such as wetlands may affect climate-regulating services, as discussed earlier.

NUTRIENT LOADING

Increasing human populations and associated agricultural needs have significantly altered global nutrient cycles over the last fifty years and increased nitrogen and phosphorus flux to the coastal ocean (Vitousek et al. 1997; Bennett et al. 2001). Widespread coastal eutrophication (Rosenberg 1985; Nixon 1995; Cloern 2001; Schindler 2006) has resulted in poor water quality, noxious algal blooms, oxygen depletion, and in some cases, loss of sea grasses and fisheries production (Rabalais 2002, 2004; see Chapter 11); much effort has therefore been devoted to reducing nitrogen and phosphorus discharges (National Research Council 2000; Boesch 2002).

The reduced flushing and smaller size of SEMSs often exacerbate effects of nutrient loading, so impacts may prolong effects (see Chapter 3, this volume). For example, following the collapse (circa 1990) of agriculture in the former Soviet republics of Estonia, Latvia, and Lithuania, although fertilizer application fell to 1950s levels, downstream concentrations of inorganic phosphorus and nitrogen were similar in 1994 and 1987 (Löfgren et al. 1999), presumably because nutrients stored in sediments continued to leach out.

Eutrophication can influence the atmosphere by stimulating growth of algae such as *Phaeocystis* species that emit dimethyl sulfide (DMS), which can enhance cloud production. Similarly, methane and other greenhouse gases are produced in larger amounts in anoxic conditions like those that develop during eutrophication. Nutrient loading can also enhance disease organisms such as *Vibrio cholerae* (National Research Council 1999). Aesthetic values are compromised by algal blooms and oxygen depletion that cause foul odors, fish kills, foam accumulation, algal debris on beaches, and "dead zones" (see Chapter 11). All of these outcomes compromise recreation and tourism opportunities.

Large-Scale (Ocean Basin and Global) Effects

Sea level change and ocean acidification are stressors on drivers of ecosystem services that operate at global scales, and thus they represent major challenges for management.

SEA LEVEL CHANGE

Production services in low-lying coastal areas are particularly vulnerable to sea level rises, in that storm surges, salinity increases in estuaries, and rising water tables can cause salt poisoning of terrestrial plants and agricultural land. Loss of wetlands as sea level rises will also impact fisheries for species that depend on wetlands for critical habitat. Rising sea levels alter the extent of wetlands in SEMSs and associated net greenhouse gas sequestration/production, a climate-regulating service. Loss of wetlands with sea level rise could also compromise water filtration and protection from natural hazards such as storm surges and tsunamis. Inevitably, sea level rise will alter natural shorelines and the many recreational and tourism economies they support. The loss of wetlands will also compromise the multiple supporting services that they contribute.

ACIDIFICATION

Ocean acidification has broad-scale ramifications for production services, especially for organisms with calcium carbonate structures that are particularly sensitive to pH changes. Photosynthetic coccolithophores, larval (and even adult) bivalves (Gazeau et al.

2007), and corals are all vulnerable to acidification, with possible direct effects on bivalve and coral reef fisheries and indirect effects on food webs. Genetic, biochemical, and ornamental resources are potentially vulnerable to pH change, which will decrease biodiversity and population levels. Acidification of seawater and seafloor sediments can also alter the chemical reactions that control trace metal cycling, with ramifications for primary production and toxicity to humans and other organisms. Decreasing pH can shift phytoplankton composition, with effects on oceanic CO_2 and subsequent sequestration into seafloor sediments. Shoreline protection by corals and carbonate sands would also be compromised by pH decreases, with negative ramifications for recreation, ecotourism, and habitat provisioning.

Case Studies

Below we present a series of case studies on how human activities have resulted in losses of ecosystem services, on governance strategies to address these dynamics, and on lessons that can be derived from these examples. We have organized these case studies around scales of effect and response, in order to illustrate how different scales of change may require very different types of response in terms of actions and governance bodies. We recognize, however, that the scales of drivers represent a continuum and that threats that are initiated at one scale may manifest themselves at smaller or larger scales, depending on management response.

Small-Scale (Individual Bays and Beyond) Examples

HYDROLOGIC CHANGE AND SEDIMENTATION

The East China Sea provides an excellent example of the complex problems that arise with changes in hydrologic input and sedimentation. The Changjiang (Yangtze) River accounts for 90%–95% (1.0×10^{12} m^3 y^{-1}) of freshwater input to the East China Sea, representing the world's largest river discharging directly into a SEMS. The Changjiang watershed has been affected by accelerating human activities over the last century, particularly through damming that has reduced sediment loading and freshwater flow. The amount of sediment discharged from the Changjiang into the East China Sea in the 1990s was 30%–40% lower than that discharged in the 1960s, and the building of the Three Gorges Dam has reduced that loading by an additional factor of 2 (Yang et al. 2006).

These discharge changes have starved the Changjiang delta region of sediment over the past fifty years, especially since 2000, when changes in the seaward extension of the delta reversed, to retreat and erosion, with profound consequences for the ecosystem and the socioeconomics of the region (S.L. Yang, unpublished data). Examples of deteriorating ecosystem services include major loss of critical habitat, such as salt marshes, that once represented an important spawning ground for commercial species and a key habitat for migratory birds and associated tourism activities (Zhang et al. 2006), as well as

Lessons Learned

Upstream effects can have dramatic consequences for downstream environments and human populations, and managers must consider potential impacts of decisions that extend beyond their immediate geographic locales.

Restoration efforts are expensive and complex because they can add problems such as movement of contaminants, spreading of diseases and invasive species, and loss of genetic diversity.

---

increased seawater intrusion into freshwater supplies in urban areas that may be exacerbated in drought years associated with climate change (Yang et al. 2006).

Management efforts to protect the Changjiang delta, its human population, and natural resources include relocation of sediments dredged from navigation channels to tidal flats in order to offset sediment loss, establishing protected areas in the delta region (e.g., near Shanghai) to limit further habitat loss to urbanization, and transplanting of marsh plants to shallow areas in order to reduce sediment loss from the delta and to restore habitat. Other projects have focused on enhancement of endangered stocks (e.g., the Chinese sturgeon *Acipenser sinensis)*, by releasing cultured larvae or juveniles into the wild.

Since the mid-1990s these efforts have improved substantially the number of species and populations of migratory birds that utilize the coastal wetlands, but new problems have appeared, such as the movement of contaminants in dredged sediments and loss of genetic diversity in stocking programs that utilize offspring of captive populations to seed wild populations.

INVASIVE SPECIES

In the early 1980s, an unknown ship dumped tons of ballast water into the Black Sea. That water, picked up in a distant ocean, contained the comb jelly (ctenophore) *Mnemiopsis leidyi*, which reproduced rapidly, feeding on fish eggs and larvae as well as crustaceans and other food previously eaten by finfish. By 1990, the total biomass of *Mnemiopsis* in the Black Sea was estimated at 2 billion tons, peaking at an average of about 4.5 kg m^{-2} in 1989/1990 and August 1994 (Shiganova et al. 2001). Another gelatinous carnivore, *Beroe ovata*, which preys mainly on *Mnemiopsis* and was introduced into the Black Sea with ballast waters in 1998, contributed to recovery of the ecosystem at the end of the 1990s. Its introduction was immediately followed by a two- to threefold increase in mesozooplankton biomass, ichthyoplankton biomass, and fish stocks (Kideys 2002; Shiganova et al. 2003). *M. leidyi* has now reached high biomass levels in the most important commercial areas of the Caspian Sea and is jeopardizing fisheries by altering the entire food chain, particularly pelagic fish (Ivanov et al. 2000). Very recently, this species has also been observed in the Baltic Sea (Hansson 2006) and the southern North Sea (Faasse and Bayha 2006).

Lessons Learned

Ballast water should be treated as already advocated by the International Convention for the Control and Management of Ships' Ballast Water & Sediments, adopted by consensus by the IMO in 2004.

Deliberate introductions should be considered with extreme caution, even where they appear to be safe and beneficial, because adaptation and changing environments may produce unexpected negative results.

Some marine nonindigeneous species have been introduced deliberately in an attempt to improve ecosystem services. For example, Dutch oyster farmers imported spat of the Pacific oyster (*Crassostrea gigas*) from British Columbia to the coastal North Sea for aquaculture in 1964 after the indigenous oyster (*Ostrea edulis*) was wiped out by diseases and overfishing. At that time, *C. gigas* was not considered to be potentially invasive because water temperatures were too cold for its reproduction. However, the combination of warming waters and local adaptation has resulted in the spreading of Pacific oysters and a subsequent decline in indigenous bivalves (mussels and cockles). The oyster also provided a conduit for other invasive species (Wolff and Reise 2002; Nehring 2006). Recently, several studies have been funded to advise state regulators in Maryland and Virginia (USA) on the costs and benefits of introducing *Crassostrea ariakensis* to the Chesapeake Bay to reestablish oyster fisheries decimated by disease, habitat destruction, and overfishing (e.g., National Research Council 2004). Considerable controversy remains regarding the desirability of this introduction, in view of previous unintended consequences that have resulted from other introductions.

On February 13, 2004, the International Maritime Organization (IMO) adopted the International Convention for the Control and Management of Ships' Ballast Water & Sediments by consensus at a diplomatic conference in London. The convention's goal is to prevent, minimize, and ultimately eliminate the transfer of harmful aquatic organisms and pathogens through the control and management of ships' ballast water and sediments (International Maritime Organization 2008). There are discussions on whether to introduce the predatory comb jelly *B. ovata* to the Caspian Sea, in order to reduce the invasive *M. leidyi* (Volovik and Korpakova 2004), as occurred fortuitously in the Black Sea.

Medium-Scale (Regional, Semi-Enclosed System) Effects

OVEREXPLOITATION

Overexploitation (including bycatch) of many fishery resources in SEMSs has occurred over the last four decades as a result of growth in commercial fishing and destructive fishing practices. In the Bay of Bengal, the operation of about 4,000 mechanized boats and 26,000 traditional craft has contributed to overfishing (Vijayan et al. 2000), which has been compounded by destructive fishing methods that include bottom trawling, blast fishing, and fishing with poisons. Collectively, these activities have contributed to significant declines in fish and shrimp populations. Gill nets with finer mesh to collect smaller fish also capture juveniles of some taxa, leading to depletion of fish stocks such as the frigate tuna (*Auxis thazard*) (Jude et al. 2002). Overfishing results in significant impacts on marine biodiversity, reducing human food resources. At least 20% of coastal communities in Bangladesh depend on living resources from the Bay of Bengal for their livelihood (Roy 2001) and as a primary source of animal protein. Open-access management schemes in the countries that surround the Bay of Bengal have contributed to overfishing. Developing mariculture will probably worsen rather than solve the problem in the near future, given that production of 3 metric tonnes of finfish require 5 metric tonnes of fish meal (Tacon and Barg 1998), though there have been some developments in feeds that have the potential to reduce waste and utilize nonfish protein.

Until 2003, Bangladesh, India, and Sri Lanka did not apply principles of participatory management and sustainability in their fisheries policies, and funding was not available to implement sustainable fisheries management. In response to declining fisheries, the Food and Agricultural Organization (FAO) developed the Bay of Bengal Programme (BOBP), which employs an innovative participatory approach to resolving fisheries management issues. The BOBP is an intergovernmental program that includes eight countries around the Bay of Bengal and focuses on the development of sustainable fisheries and capacity building (Bay of Bengal Programme 2008). The program evaluates the needs of countries that depend on Bay of Bengal resources and involves member countries in management. Through the BOBP, the countries that surround the Bay of Bengal will consider policies to ensure sustainable use of the coastal zone and marine resources without compromising the integrity of the natural environment.

In 2005, the government of Bangladesh developed a Coastal Zone Policy, in which it declares its intention to develop integrated coastal zone management and to contribute to the sustainable utilization of fishing resources in the Bay of Bengal (Ministry of Water Resources 2005).

HABITAT LOSS

Regional seas provide a variety of habitats to species or groups of species, and the loss of area or function of these habitats represents a major threat to ecosystem services in SEMSs. We summarize contrasting examples from the Wadden Sea in the North Sea, the *bodden* areas in the Baltic Sea, and the mangrove forests along the Bay of Bengal to illustrate different solutions to habitat loss at different scales.

The Wadden Sea, which comprises the southern and eastern coast of the North Sea, with adjacent estuaries, is an important area for recruitment of commercial fishes. Other important ecosystem services include an extensive filtering function for nutrients and other material exported from land and acting as a store or sink for metabolic products.

Lessons Learned

Because of the novelty of the BOBP, it is difficult to predict whether it will be successful. Nonetheless, the failure of past practices suggests that independent management of shared ecosystems is problematic and that the needs and goals of user countries must be considered if an effective and sustainable management scheme is to be developed. To date, the BOBP has not succeeded in establishing any form of transnational governance or management of the Bay of Bengal. Beyond the clearly pressing need to achieve an agreement on maritime boundaries among India, Bangladesh, and Myanmar, the needs and priorities of the different user groups and other stakeholders in and adjoining the bay must be considered if effective and sustainable management of this SEMS is to be developed.

In developing countries, in particular, the scarcity of financial resources creates an additional challenge for the development of sustainable fisheries. Independent bodies that receive outside funding provide a potential tool for cooperative and participatory governance.

The system is dominated by tidal currents and high turnover rates that create a very productive and efficient system for water purification and food production. Increasing activities—such as shipping, dredging for sand mining and navigation, dumping of dredged material, and emplacement of man-made devices for shoreline protection—influence nearshore currents and bottom characteristics that can lead to habitat loss for flatfish and brown shrimp.

The *bodden* areas of the Baltic Sea are similar in function to the Wadden Sea; however, tidal influences are minimal in the former. Shallow, sheltered areas provide habitat for young fish and a productive benthic community, and sea grass and reed meadows also serve as filter systems that improve water quality. Like the Wadden, the *bodden* area supports tourism activities such as boating, sailing, fishing, and swimming, resulting in potential conflict with other uses and maintenance of ecological function such as spawning grounds, nursery areas, or shelters from predation.

Mangrove forests such as those in coastal areas of the Bay of Bengal represent tropical counterparts to the Wadden and *bodden* seas. These tidal forests and associated tidal mudflats provide fisheries, recruitment areas for food fishes, filtering function for water purification, and a sink for land-derived sediments and organic and inorganic matter. Although of no direct importance for tourism, mangroves compete for space with tourism (hotel and resort developments) and other land-based activities such as aquaculture. The genetic resources of mangrove forests are potentially very valuable (e.g., for pharmaceuticals) but are poorly assessed and understood. Future loss of many services in these systems is expected if they are destroyed.

Lessons Learned

Governance of similar resources may sometimes occur most effectively at very different spatial scales, and in instances where relatively few countries share a resource, it may be possible to develop effective comanagement schemes in which resource users are involved in the management.

Although certainly fraught with difficulties, the inclusion of upstream countries that share an SEMS's watershed, but not its coastline, in its governance and management is needed (e.g., Hungary for drainage to the Black Sea or Laos for drainage to the Gulf of Thailand/South China Sea), to ensure comprehensive management of factors such as sediment load, harmful substances, and freshwater inputs.

Restoration efforts can help to offset habitat loss, but if loss exceeds replacement, then decline in ecosystem services can be expected.

Management of these ecosystems works on different scales. In the North Sea, regional conventions and agreements (e.g., the North Sea Conference, Wattenmeerforum, Aktionskonferenz Nordsee, OSPAR) attempt to manage the Wadden Sea as a whole and include representation from the countries that surround it (Germany, The Netherlands, Denmark, the United Kingdom). Problems in the *bodden* areas occur on a smaller scale and can therefore be managed at a local level. Bilateral projects between Germany and Poland, such as Research for an Integrated Coastal Zone Management in the German Oder Estuary Region (IKZM-Oder 2008) have been established to improve the management of areas in the fragile border zone. Research institutes and nongovernmental organizations (NGOs) are active in both regions and attempting to establish integrated management tools and programs.

Mangrove problems in the Bay of Bengal are mainly dealt with on national levels, and management typically operates on relatively small scales. Reforestation programs are underway in Bangladesh to counteract degradation and loss of mangrove areas. Driven by governmental programs and supported by international funds (e.g., the Maturing Mangrove Plantations of the Coastal Afforestation Project and the FAO/UNDP Project; Drigo et al. 1987), the Forest Research Institute in Chittagong runs a reforestation project that has replanted approximately 170,000 ha of mangrove since 1966. Unfortunately, this effort is sometimes offset by clearing of mangroves over the same time period at other places, and the overall area of mangroves in Bangladesh has changed little during the last decades (average = 6,000 km^2; Wilkie and Fortuna 2003). Intergovernmental organizations such as the BOBP have had some success in promoting regional initiatives on sustainable resource use. By contrast, an initiative in Thailand begun in 2003 to frame the management of the Thai waters in the Gulf of Thailand by formulating a national marine policy has been stalled indefinitely by internal political changes.

NUTRIENT LOADING

Nutrient enrichment is becoming increasingly widespread in the scales of impact and the regions affected. We summarize below the effect of nutrient loading on the North Sea, coastal Florida, and the Baltic Sea and the progress in efforts to reduce those loadings. Although it is possible to reduce nutrient loading to pre-eutrophied levels, changes in coastal nutrient concentrations have seldom resulted in the intended management goals (i.e., improvement of particular ecosystem services).

The coastal waters of the North Sea have been subjected to many decades of nutrient enrichment followed by subsequent efforts to reduce nutrient loading. Changes in nitrogen and phosphorus concentrations in coastal waters during these periods were reflected in phytoplankton biomass, production, and community structure. Enrichment and subsequent reduction of phosphorus was followed by corresponding shifts in other coastal ecosystem components (i.e., macrozoobenthos and birds). Although phosphate concentrations are now substantially lower than they have been in the recent past, the ecosystem has not reverted to its previous state. This shift may be a result of nitrogen and silicon concentrations that have remained relatively high or a result of compound effects of overfishing and climate change. These changes may have shifted threshold values for restoration, thereby altering relationships between nutrient loading and ecosystem services (Philippart et al. 2007).

During the 1980s, nitrogen loading in Tampa Bay and Sarasota Bay (Gulf of Mexico), primarily by domestic wastewater, was reduced by 57% and 46%, respectively, as a result of decisions made primarily at state and municipal levels. In 2002, both bays had lower phytoplankton concentrations, greater water clarity, and more extensive sea grass coverage than in the early 1980s (Tomasko et al. 2005). Given that storm water runoff is currently the primary source of nutrient input, year-to-year variation in nitrogen loads will be strongly related to annual rainfall. Phytoplankton abundance, rainfall, and nitrogen loads in southwest Florida's estuaries will all influence water clarity and therefore sea grass growth (Tomasko et al. 2005). Atmospheric nitrogen now represents a major source of nutrients into these coastal waters and also needs to be monitored.

Between 1970 and 1985 there was a threefold increase in surface nitrate and phosphate concentrations in the Baltic Sea (Nehring and Matthaus 1991), resulting in increased toxic or noxious algal blooms. One major problem is cyanobacterial blooms in the open sea, particularly of the toxic, nitrogen-fixing genus *Nodularia*. In addition, multiple fish kills by the phytoplankton *Prymnesium parvum* have been reported from the Baltic coastal zone. In order to mitigate the problem of nutrient loading in the Baltic Sea, the countries with significant riparian loading agreed to reduce river nutrient loads by 50% (Neumann et al. 2002). Modeling studies indicated that most countries would gain net economic benefits from the 50% nitrogen and phosphorus reduction policy (Turner et al. 1999).

Lessons Learned

Reduction of riverine inputs of nutrients in overloaded systems will not likely result in "pristine" conditions, because SEMSs have compounding factors such as overfishing, atmospheric deposition, climate change, habitat destruction, and nutrients stored in bottom sediments that slow down or even make impossible restoration of pre-impact ecosystem services.

Nutrient reduction is often easier (more affordable) for any one nutrient than for all; this results in changes in nutrient ratios that affect phytoplankton species composition and subsequently ecosystem services. Closer attention should be paid to the importance of balanced nutrient composition, as well as nutrient supply dynamics, for the development of eutrophication versus efficient trophic transfer and fish production in nutrient-enriched systems.

When nutrient management measures such as reduction of loads are taken, all ecosystem services that are likely to be influenced by intervention (e.g., food, oxygen production through photosynthesis, primary production, nutrient cycling, and habitat provisioning) should be considered.

POLLUTION WITH INDUSTRIAL WASTES

The Gulf of Thailand (GOT) is part of the Sunda Shelf, with water depths that vary between 45 m and 85 m and a shallow sill that limits water exchanges with the South China Sea. Twenty-three rivers drain large amounts of freshwater and pollutants into the gulf. Major land uses of the catchment are agriculture and related agro-industries, and there has been increasing oil and gas production in the last decade, along with new deep-water ports to accommodate these activities. Untreated municipal and industrial organic wastewater from large cities is causing harmful algal blooms and oxygen depletions. The major pollutants of the estuaries and GOT include nitrate, phosphate, and silicate, but heavy metals are also a problem. Pollution includes, for example, mercury loading, small oil tanker spills, and industrial waste from coastal development to accommodate tourism demands.

These threats have reduced the quality and availability of seafood in Thailand. Monitoring by the Pollution Control Department (PCD) of Thailand during 1995–1998 indicated risk imposed on humans from seafood consumption, and there have been incidents of food poisoning and illness associated with seafood consumption. Though mercury is elevated, a public health threat from seafood contamination does not yet appear to be significant. Pollution has also resulted in deterioration of coastal water quality (e.g., from fecal coliform bacteria) and beach appearance.

Thailand is working to alleviate the pollution problem and to restore ecosystem services in the GOT. Water treatment facilities are being built in more and more communities, funded initially by the Asian Development Bank and later by the Thailand Environ-

Lessons Learned

Carefully designed monitoring programs are essential for effective management, but enforcement is challenging. International support to kick-start new monitoring programs and provide proof of concept can lead to expansion through domestic programs.

---

mental Fund. The Thai PCD also monitors water quality twice a year along the entire coastline of Thailand, including monitoring of mercury levels in fish.

The Thai government has also established regulations such as the Swine Effluent Standard, Pier Effluence Standard, and Shrimp Farm Effluence Standard. Operators of coastal aquaculture, fishery activities, pig farms, and factories are now required to treat wastewater to these standards before releasing it. More measures have to be established and enforced effectively to solve the industrial water pollution problem. New standards for mercury emissions have been established, and a collaborative working group has been set up between government agencies and university scientists to ensure that an appropriate strategy is developed to resolve the mercury problem and to carry out capacity and risk assessment analyses. There has also been development of Association of Southeast Asian Nations (ASEAN) marine water quality criteria (for oil, grease, and metals such as zinc) as part of the ASEAN–Canada Cooperative Programme on Marine Science. The PCD has organized workshops to accelerate implementation of an action plan on water quality and established an Environmental Quality Index for Tourist Beaches and Islands to evaluate suitability for swimming.

These actions have yielded positive results. Coastal water quality in the GOT and Andaman Sea has generally improved in recent years. Surveys of solid waste found on beaches, land use, conditions of sand dunes, erosion, and coral reef health show that environmental quality is "good," with slight improvement in some tourist areas. Because of its importance, the coastal water quality monitoring program will continue in the future. Finally, recent surveys (2001–2003) showed declines in total mercury concentrations in coastal water, and all areas were in compliance with the national standard.

Large-Scale (Ocean Basin to Global) Effects

ACIDIFICATION

The uptake of atmospheric CO_2 by the ocean helps to reduce greenhouse effects but this "service" of the ocean in global temperature regulation also lowers the pH and thus acidifies the surface ocean. This is an example of a global environmental issue that will be expressed in different ways in different regions. The biological effects of ocean acidification are only beginning to be studied (see Orr et al. 2005b; Royal Society of London 2005), although it is predicted that decreasing ocean pH will damage (and potentially eliminate) warm-water corals in SEMSs and other tropical coastal areas. Ocean acidifica-

Lessons Learned

Because of the large scales involved, efforts to mitigate problems such as ocean acidification will require substantial proactive measures through international initiatives such as the Kyoto Protocol. The large scale of the problem also creates a strong momentum in which reversal will be difficult and slow to achieve.

In the absence of clear scientific consensus that a given activity will not cause harm, the burden of proof falls on those advocating it (Raffensperger and Tickner 1999). Precautionary approaches are recommended in areas such as management of marine fisheries and coral reefs because specific responses of ocean ecosystems to acidification are difficult to predict.

Careful and integrated monitoring of particularly vulnerable systems, such as coral reefs, is recommended.

---

tion will add to several other human impacts on corals, the most important of which is increased ocean temperatures. Ocean acidification also will affect the production of calcareous plankton, including coccolithophores and pteropods (Riebesell et al. 2000; Feely et al. 2004), which form the base of some ocean food webs, resulting in changes in the quantity and quality of food for commercial fish and marine mammals. Because the solubility of CO_2 in seawater increases in colder water, ocean acidification effects are expected to be particularly serious and to occur sooner in high-latitude areas (Orr et al. 2005a) including, potentially, Hudson Bay, the Laptev Sea, and the Kara Sea. Ocean acidification may seriously affect the development of larval marine organisms because many larvae in the ocean surface layer form calcium carbonate skeletons.

Management options to cope with ocean acidification are limited. Large-scale preventive measures, such as global treaties to limit human inputs of CO_2 (most notably the Kyoto Protocol) and CO_2 emission–trading systems, are most likely to be effective because of the global nature of the problem. Mitigative management is untried but might focus on reducing other human pressures on coral systems and fisheries stressed by ocean acidification. Research on biological effects of ocean acidification and on management approaches is urgently needed (Cicerone et al. 2004; Kleypas et al. 2006). Specifically, national and international fisheries management in coral reef and high-latitude areas likely to be affected by ocean acidification need to incorporate the effects of all stressors. Presently (at least in the best single-species management systems), total allowable catch (TAC) levels are set each year, based on the fishing level that managers believe will sustain fish populations at desirable levels. TAC levels regulate future fishing efforts and are mostly derived from past population and harvest levels. Under increased climate change, fisheries management could be more effective if it builds in a precautionary cushion to account for the fact that stocks will be increasingly stressed by warmer water temperatures, changing pH, and, potentially, changes in food supply and recruitment failure.

Governance and Management of SEMSs

Multiple human threats in SEMSs require an integrated management approach over different spatial scales (e.g., watershed, coastal zone, and adjacent ocean), temporal scales (e.g., historical changes, current and future threats), and ecosystem aspects (e.g., the major physical, chemical, and biological variables). Multiple threats also put new demands on the larger governance framework, that is, the institutional structures through which diverse social actors, including public authorities, influence and enact policies and decisions in public life (Risse 2007).

Past attempts to manage coastal and marine ecosystems have often been fragmented and ineffective. More recently, broad, integrated management plans have developed from the 1960s and strengthened over the past two decades (Sorensen 2002; Millennium Ecosystem Assessment 2005). A recent survey counted 698 such initiatives in 145 countries, including 76 initiatives at the international level (Sorensen 2002). Still, many countries with long-established and well-designed coastal management plans cannot halt or even reverse overexploitation, habitat loss, and pollution (Millennium Ecosystem Assessment 2005), in part because degradation processes started centuries ago and underlay current ecosystem states (Jackson et al. 2001; Lotze et al. 2006). Moreover, degradation often occurs faster than management and governance can respond, and efforts are undermined by conflicts among multiple stakeholders (Millennium Ecosystem Assessment 2005).

Ideally, integrated management would address human threats at different spatial scales and place management in a comprehensive governance framework that defines the fundamental objectives, policies, laws, and institutions. For example, nutrient loading through watersheds and the coastal zone results from land runoff, groundwater discharges, sewage outflows, and municipal and industrial discharges. Thus, a range of actors and institutions would be involved in integrated management, each with different interests, perspectives, and knowledge. Nutrient loading via atmospheric deposition ultimately requires a larger, supraregional, or even global-scale approach.

Scale-related management issues also occur with overexploitation. Management must address problems ranging from diadromous fish populations to major commercial species (fish, invertebrates, and plants) that inhabit coastal waters and continental shelves. These areas often straddle or fall within national exclusive economic zones or areas subject to regional multilateral agreements, while the high seas mostly extend beyond national jurisdictions and require international or even global coordination efforts.

On the supranational level, a set of important framework instruments, including the following, has developed over recent decades:

- the Ramsar Convention on Wetlands (1971)
- UNEP's Regional Seas Programme (1974) and its Action Plans
- the UN Convention on the Law of the Sea (UNCLOS 1982)

- the UN Conference on Environment and Development's Agenda 21, particularly Chapter 17 (1992)
- UNEP's Global Programme of Action for the Protection of the Marine Environment from Land-Based Activities (GPA 1995)
- the Convention on Biological Diversity (CBD 1992) with its Jakarta Mandate on the Conservation and Sustainable Use of Marine and Coastal Biological Diversity (1995)
- the Plan of Implementation of the World Summit on Sustainable Development (2002)

These global treaties and nonbinding multilateral agreements set the stage for national and regional management efforts, and they are often complemented by more specific protocols, annexes, or action plans, such as the UN Environment Programme Global Programme of Action for the Protection of the Marine Environment from Land-Based Activities (see UNEP/GPA 2008).

The development of these international framework agreements confirms two recent trends in coastal and marine governance (Figure 4-2). First, there has been a long-term shift from top-down, centralized approaches, in which national authorities (advised by scientists) develop and enact rules, to decentralized and more participatory forms of management ("community-centered management"). This trend has problems, such as an often limited knowledge base, an embedding in local power structures, and a lack of reliable financing. In addition, bottom-up approaches are often ill equipped to handle major infractions and macro developments whose social and economic drivers are far beyond their reach. Thus, the pendulum has begun to swing back to hybrid forms of comanagement that involve national and regional authorities, local participants and structures, and often international civil society players.

The second trend is an emerging shift from single-issue or single-species, yield-oriented approaches to ecosystem-based management that aims to maintain the continued functioning of whole ecosystems. Earlier versions of ecosystem management, such as the U.S. Federal Ecosystem Management Initiative, focused largely on the natural components and their scientific management. More recently, integrated ecosystem approaches, as promoted under the CBD—in the context of Integrated Marine and Coastal Area Management (IMCAM)—are characterized by their inclusion of stakeholders, their iterative procedural setup, and their commitment to adaptive management.

Available evidence suggests that resource management strategies that consider the consequences of resource removal on ecosystems and human well-being are more effective than sectoral or single-species management (Kay and Alder 2004). Fisheries management agencies and NGOs increasingly promote ecosystem-based fisheries management that addresses multispecies interactions as well as habitat and environmental (e.g., water quality) requirements for survival and reproduction (Pikitch et al. 2004). These approaches call not only for more-effective regulations of exploitation, but also for pol-

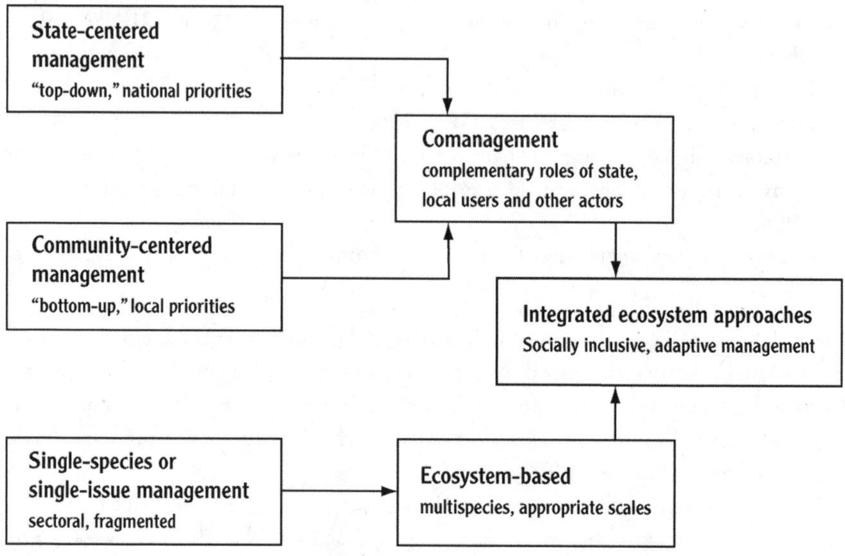

Figure 4-2. Flowchart of the development of management approaches.

lution controls and the protection of coastal and marine habitats. Examples of successful pollution control include wastewater treatment to reduce nutrients and pollutants from point sources such as municipal and industrial outflows, changes in land-use practices that include buffer strips to prevent non-point land runoff, and restoration or construction of new wetlands to enhance filter and storage capacity. However, management interventions to control pollution have often failed, and no country has succeeded in comprehensively limiting pollution of nearshore environments (Millennium Ecosystem Assessment 2005). Costs for habitat restoration can be extremely high and thus unrealistic for developing countries (Millennium Ecosystem Assessment 2005), and not all habitats can be effectively restored.

In the framework of ecosystem-based management, there is increasing interest in marine protected areas as a tool for halting the overexploitation of resources (U.S. Commission on Ocean Policy 2004). Worldwide, there were an estimated 4,116 coastal and marine protected areas in 2003 (Spalding et al. 2003), but despite the large number of individual sites, their coverage accounted for < 1% of the global ocean (Millennium Ecosystem Assessment 2005). Current marine protected areas range from many small fisheries reserves to a few larger networks of marine reserves such as the Great Barrier Reef Marine Park in Australia (Murray et al. 1999; Day 2002; Pauly et al. 2002). A recent analysis of the effects of forty-eight marine reserves worldwide showed that compared with unprotected areas, average diversity, productivity, and resilience are enhanced and variability of fish biomass is reduced in protected areas (Worm et al. 2006). However, the effectiveness of many protected areas is limited because cooperation and enforcement are lacking at local and regional scales, and most protected areas are only partially protected. Thus, marine protected areas as a tool have not been used to their full potential so far (Agardy et al. 2003).

Recommendations

SEMSs are especially vulnerable to human disturbances because of their limited exchange with the open ocean, and many also support dense human populations. As in many other natural systems, the interacting effects of multiple drivers further complicate management strategies. As the examples above illustrate, management of drivers that operate at smaller spatial and temporal scales is relatively easier to achieve than management of drivers that operate at large scales. As many problems in SEMSs are associated with intermediate spatial and governance scales, their management is particularly dependent on successful regional and sectoral integration but may also hold particular chances for creative, regionally specific approaches.

Local intervention at the level of communities and local stakeholders can be effective in reducing some drivers of ecosystem service loss. With the exception of large-scale trawling disturbance, many types of habitat destruction can be managed through sets of rules at the local level or habitat restoration initiatives. Restoration, however, can rarely bring back all the lost services. Moreover, restoration has to outpace rates of habitat loss, which does not always happen. Restoration can be expensive, and it runs the risk of introducing invasive species, spreading disease, and reducing genetic diversity. Clearly, reducing habitat destruction is a more effective and promising strategy for preserving ecosystem services.

Prevention and coordinated national-level intervention are the best strategies for dealing with invasive species, even though invasive species typically cause local problems initially. Given the modes of transit for invasive species and the mobility of the ships involved, local initiatives are unlikely to be effective. Because invasive species are extremely difficult to manage once they have become established, efforts to prevent transport must be prioritized. National laws that regulate ballast water disposal represent one of the best tools to reduce invasions, but because SEMSs are typically bordered by multiple countries, cooperation and parallel regulations are necessary for real effectiveness. Because the full impacts of invasive species are very difficult to predict in advance, deliberate introductions should be avoided until all other potential solutions have been examined.

Scale is also important to consider in managing pollution. Point source pollution can be resolved by local governance where individual polluters (e.g., industries) are induced to reduce emissions by legal bans or economic incentives. Problems such as nutrient enrichment, however, often involve diverse stakeholders who may live long distances from the coastal zone where impacts are most severe. In this instance, nutrient sources may come from another city, county, or even country, and stakeholders from those sources must be made aware of the consequences of their actions and assisted in finding alternative, less destructive approaches. These types of problems require regional cooperation that may include multiple nations, demonstrating that independent management of shared ecosystems is problematic, and the needs and goals of user countries must be considered if an effective and sustainable management scheme is to be developed.

The problem of shared resources is common in marine fisheries because resources often straddle and move across regional and international boundaries. Management of these types of fisheries without consultation among stakeholder groups is doomed to fail, and establishment of an independent body that comanages the resource, taking the interests of all stakeholders into account, is the most promising solution.

Problems manifested at global scales are particularly challenging as they require cooperation and initiative by many countries and stakeholders around the world. These needs are pressing because large-scale disturbances have a strong momentum that is very difficult to reverse, or even slow down. Efforts to mitigate problems such as ocean acidification will thus require substantial proactive measures through international initiatives such as the Kyoto Protocol, but lack of cooperation from a few key nations will reduce the likelihood that disturbing trends can be reversed.

In summary, effective governance and management of SEMSs require an integrated approach that considers all ecosystem services and their interaction. The development of ecosystem-based management in fisheries that is gaining support in many areas of the world is promising in that it aims at ecosystems as a whole. This is also one of the central tenets of the ecosystem approach as it is being developed under the Convention on Biological Diversity. In a similar vein, there is growing recognition of the necessity to couple coastal zone management with watershed management, as under the European Water Framework Directive, aided by recent research initiatives (e.g., Land–Ocean Interactions in the Coastal Zone, LOICZ).

Carefully designed monitoring programs (e.g., for pollution, overfishing, ballast water) are also essential for effective management, despite the challenges of cost and enforcement. International programs that can help to fund establishment of comanagement bodies and related capacity-building measures are particularly important for developing countries that may need funding to kick-start new monitoring programs and establish domestic management. Given the many losses of ecosystem services that have occurred over the last century, a precautionary approach is recommended. SEMSs provide vital resources to many people, and their continued functioning must be a top priority for the many stakeholders who benefit from them.

References

Agardy, T., P. Bridgewater, M.P. Crosby, J. Day, P.K. Dayton, R. Kenchington, D. Laffoley, P. McConney, P.A. Murray, J.E. Parks, and L. Peau. 2003. Dangerous targets? Unresolved issues and ideological clashes around marine protected areas. *Aquatic Conservation—Marine and Freshwater Ecosystems* 13:353–367.

Barmawidjaja, D.M., G.J. van der Zwaan, F.J. Jorissen, and S. Puskaric. 1995. 150 years of eutrophication in the Northern Adriatic Sea: Evidence from a benthic foraminiferal record. *Marine Geology* 122:367–384.

Bay of Bengal Programme. 2008. http://www.bobpigo.org.

Bennett, E.M., S.R. Carpenter, and N.F. Caraco. 2001. Human impact on erodable phosphorus and eutrophication: A global perspective. *BioScience* 51:227–234.

Boesch, D.F. 2002. Challenges and opportunities for science in reducing nutrient overenrichment of coastal ecosystems. *Estuaries* 25:744–758.

Carlton, J.T. 1996. Pattern, process, and prediction in marine invasion ecology. *Biological Conservation* 78:97–106.

Cicerone, R., J. Orr, P. Brewer, P. Haugan, L. Merlivat, T. Ohsumi, S. Pantoja, H.-O. Poertner, M. Hood, and E. Urban. 2004. The ocean in a high-CO_2 world. *Oceanography* 17:72–78.

Cloern, J.E. 2001. Our evolving conceptual model of the coastal eutrophication problem. *Marine Ecology Progress Series* 210:223–253.

Cohen, A.H., and J.T. Carlton. 1998. Accelerating invasion rate in a highly invaded estuary. *Science* 279:555–558.

Costello, M.J., M. McCre, A. Freiwald, T. Lundalv, L. Jonsson, B.J. Bett, T. van Weering, H. de Haas, J.M. Roberts, and D. Allen. 2005. Role of deep-sea cold-water *Lophelia* coral reefs as fish habitat in the north-eastern Atlantic. In *Cold-Water Corals & Ecosystems*, edited by A. Freiwald and J.M. Roberts, 771–805. Berlin: Springer.

Daily, G.C. 1997. *Nature's Services: Societal Dependence on Natural Ecosystems*. Washington, DC: Island Press.

Day, J.C. 2002. Zoning—lessons from the Great Barrier Reef Marine Park. *Ocean & Coastal Management* 45:139–156.

Decottignies, P., P.G. Beninger, Y. Rincé, and P. Riera. 2007. Trophic interactions between two introduced suspension-feeders, *Crepidula fornicata* and *Crassostrea gigas*, are influenced by seasonal effects and qualitative selection capacity. *Journal of Experimental Marine Biology and Ecology* 342:231–241.

Dortch, Q., N.N. Rabalais, R.E. Turner, and N.A. Qureshi. 2001. Impacts of changing Si/N ratios and phytoplankton species composition. In *Coastal Hypoxia: Consequences for Living Resources and Ecosystems*, edited by N.N. Rabalais and R.E. Turner, 37–48. Coastal and Estuarine Studies Vol. 58. Washington, DC: American Geophysical Union.

Drigo, R., M.A. Latif, J.A. Chowdhury, and M. Shaheduzzaman. 1987. *The Maturing Mangrove Plantations of the Coastal Afforestation Project*. Field Document No. 2. Chittagong City: Bangladesh Forest Research Institute (BFRI).

Duffy, J.E., and J.J. Stachowicz. 2006. Why biodiversity is important to oceanography: Potential roles of genetic, species, and trophic diversity in pelagic ecosystem processes. *Marine Ecology Progress Series* 311:179–189.

Faasse, M.A., and K.M. Bayha. 2006. The ctenophore *Mnemiopsis leidyi* A. Agassiz 1865 in coastal waters of The Netherlands: An unrecognized invasion? *Aquatic Invasions* 1:270–277.

Feely, R.A., C.L. Sabine, K. Lee, W. Berelson, J. Kleypas, V.J. Fabry, and F.J. Millero. 2004. Impact of anthropogenic CO_2 on the $CaCO_3$ system in the oceans. *Science* 305:362–366.

Frank, K.T., B. Petrie, and N.L. Shackell. 2007. The ups and downs of trophic control in continental shelf systems. *Trends in Ecology and Evolution* 22:236–242.

Gazeau, F., C. Quiblier, J.M. Jansen, J.-P. Gattuso, J.J. Middelburg, and C.H.R. Heip.

2007. Impact of elevated CO_2 on shellfish calcification. *Geophysical Research Letters* 34:L07603, doi: 10.1029/2006GL028554, 2007.

Gonzalez-Correa, J.M., J.T. Bayle, J.L. Sanchez-Lizaso, C. Valle, P. Sanchez-Jerez, and J.M. Pablo. 2005. Recovery of deep *Posidonia oceanica* meadows degraded by trawling. *Journal of Experimental Marine Biology and Ecology* 320:65–76.

Hansson, H.G. 2006. Ctenophores of the Baltic and adjacent seas: The invader *Mnemiopsis* is here! *Aquatic Invasions* 1:295–298.

IKZM-Oder. 2008. Integrated Coastal Zone Management in the German Oder Estuary Region. http://www.ikzm-oder.de/en/projekt-ikzm-oder.html.

International Maritime Organization. 2008. http://globallast.imo.org/index.asp?page =mepc.htm&menu=true.

Ivanov, V.P., A.M. Kamakin, V.B. Ushivtzev, T. Shiganova, O. Zhukova, N. Aladin, S.I. Wilson, G.R. Harbison, and H.J. Dumont. 2000. Invasion of the Caspian Sea by the comb jellyfish *Mnemiopsis leidyi* (Ctenophora). *Biological Invasions* 2:255–258.

Jackson, J.B.C., M.X. Kirby, W.H. Berger, K.A. Bjorndal, L.W. Botsford, B.J. Bourque, R.H. Bradbury, R. Cooke, J. Erlandson, J.A. Estes, T.P. Hughes, S. Kidwell, C.B. Lange, H.S. Lenihan, J.M. Pandolfi, C.H. Peterson, R.S. Steneck, M.J. Tegner, and R.R. Warner. 2001. Historical overfishing and the recent collapse of coastal ecosystems. *Science* 293:629–638.

Jude, D., N. Neethiselvan, P. Gopalakrishnan, and G. Sugumar. 2002. Gill net selectivity studies for fishing frigate tuna, *Auxis thazard* Lacepede (Perciformes/Scombridae) in Thoothukkudi (Tuticorin) waters, southeast coast of India. *Indian Journal of Marine Sciences* 31 (4):329–333.

Kay, R., and J. Alder. 2004. *Coastal Planning and Management.* 2nd ed. London: EF&N Spoon.

Kideys, A.E. 2002. Fall *and* rise of the Black Sea ecosystem. *Science* 297:1,482–1,484.

Kleypas, J.A., R.A. Feely, V.J. Fabry, C. Langdon, C.L. Sabine, and L.L. Robbins. 2006. *Impacts of Ocean Acidification on Coral Reefs and Other Marine Calcifiers: A Guide for Future Research.* Report of a workshop held 18–20 April 2005, St. Petersburg, FL, sponsored by the National Science Foundation (NSF), National Oceanic and Atmospheric Administration (NOAA), and U.S. Geological Survey (USGS). Boulder, CO: University Corporation for Atmospheric Research.

Levin, L.A., D. Boesch, A. Covich, C. Dahm, C. Erseus, K. Ewel, R. Kneib, M. Palmer, and P. Snelgrove. 2001. The role of biodiversity in the function of coastal transition zones. *Ecosystems* 4:430–451.

Löfgren, S., A. Gustafson, S. Steineck, and P. Stählnacke. 1999. Agricultural development and nutrient flows in the Baltic states and Sweden after 1988. *Ambio* 28:320–327.

Lohrer, A.M., S.F. Thrush, and M.M. Gibbs. 2004. Bioturbators enhance ecosystem function through complex biogeochemical interactions. *Nature* 431:1,092–1,095.

Lotze, H.K., H.S. Lenihan, B.J. Bourque, R.H. Bradbury, R.G. Cooke, M.C. Kay, S.M. Kidwell, M.X. Kirby, C.H. Peterson, and J.B.C. Jackson. 2006. Depletion, degradation, and recovery potential of estuaries and coastal seas. *Science* 312:1,806–1,809.

MacKenzie, B.R., M.A. St. John, and K. Wieland. 1996. Eastern Baltic cod: Perspectives from existing data on processes affecting growth and survival of eggs and larvae. *Marine Ecology Progress Series* 134:265–281.

Millennium Ecosystem Assessment. 2005. Coastal systems. In *Ecosystems and Human Well-Being: Current State and Trends*, 513–549. Washington, DC: Island Press.

Ministry of Water Resources, Government of the People's Republic of Bangladesh. 2005. *Coastal Zone Policy 2005*. http://www.lcgbangladesh.org/WaterManagement/reports/Coastal%20Zone%20Policy%202005_English.pdf.

Murray, S.N., R.F. Ambrose, J.A. Bohnsack, L.W. Botsford, M.H. Carr, G.E. Davis, P.K. Dayton, D. Gotshall, D.R. Gunderson, M.A. Hixon, J. Lubchenco, M. Mangel, A. MacCall, D.A. McArdle, J.C. Ogden, J. Roughgarden, R.M. Starr, M.J. Tegner, and M.M. Yoklavitch. 1999. No-take reserve networks: Sustaining fishery populations and marine ecosystems. *Fisheries* 24:11–25.

Myers, R.A., J.K. Baum, T.D. Shepherd, S.P. Powers, and C.H. Peterson. 2007. Cascading effects of the loss of apex predatory sharks from a coastal ocean. *Science* 315:1,846–1,850.

Myers, R.A., and B. Worm. 2003. Rapid worldwide depletion of predatory fish communities. *Nature* 423:280–283.

National Research Council. 2000. *Clean Coastal Waters: Understanding and Reducing the Effects of Nutrient Pollution*. Washington, DC: National Academies Press.

National Research Council. 2004. *Nonnative Oysters in the Chesapeake Bay*. Washington, DC: National Academies Press.

Nehring, D., and W. Matthaus. 1991. Current trends in hydrographic and chemical parameters and eutrophication in the Baltic Sea. *Internationale Revue der Gesamten Hydrobiologie IGHYAZ* 76:297–316.

Nehring, S. 2006. *NOBANIS—Invasive Alien Species Fact Sheet*: Crassostrea gigas. From online database of the North European and Baltic Network on Invasive Alien Species (NOBANIS). http://www.nobanis.org/files/factsheets/Crassostrea_gigas.pdf.

Neumann, T., W. Fennel, and C. Kremp. 2002. Experimental simulations with an ecosystem model of the Baltic Sea: A nutrient load reduction experiment. *Global Biogeochemical Cycles* 16 (3):1,003, doi: 10.1029/2001GB001450, 2002.

Newell, R. 1988. Ecological changes in Chesapeake Bay: Are they a result of overharvesting the American oyster, *Crassostrea virginica*? In *Understanding the Estuary: Advances in Chesapeake Bay Research*, 536–546. Chesapeake Research Consortium Publication 129. Annapolis, MD: Chesapeake Bay Program.

Nixon, S.W. 1995. Coastal marine eutrophication: A definition, social causes, and future concerns. *Ophelia* 41:199–219.

Nixon, S.W. 2003. Replacing the Nile: Are anthropogenic nutrients providing the fertility once brought to the Mediterranean by a great river? *Ambio* 32:33–39.

Nixon, S.W. 2004. The artificial Nile. *American Scientist* 92:158–165.

Orr, J.C., V.J. Fabry, O. Aumont, L. Bopp, S.C. Doney, R.A. Feely, A. Gnanadesikan, N. Gruber, A. Ishida, F. Joos, R.M. Key, K. Lindsay, E. Maier-Reimer, R. Matear, P. Monfray, A. Mouchet, R.G. Najjar, G.-K. Plattner, K.B. Rodgers, C.L. Sabine, J.L. Sarmiento, R. Schlitzer, R.D. Slater, I.J. Totterdell, M.-F. Weirig, Y. Yamanaka, and A. Yool. 2005a. Anthropogenic ocean acidification over the twenty-first century and its impact on calcifying organisms. *Nature* 437:681–686.

Orr, J.C., S. Pantoja, and H.-O. Pörtner. 2005b. Introduction to special section: The ocean in a high-CO_2 world. *Journal of Geophysical Research* 110:C09S01, doi: 10.1029/2005JC003086.

Pauly, D., V. Christensen, J. Dalsgaard, R. Froese, and F. Torres, Jr. 1998. Fishing down marine food webs. *Science* 279:860–863.

Pauly, D., V. Christensen, S. Guenette, T.J. Pitcher, U.R. Sumaila, C.J. Walters, R. Watson, and D. Zeller. 2002. Towards sustainability in world fisheries. *Nature* 418:689–695.

Philippart, C.J.M., J.J. Beukema, G.C. Cadee, R. Dekker, P.W. Goedhart, J.M. van Iperen, M.F. Leopold, and P.M.J. Herman. 2007. Impact of nutrient reduction on coastal communities. *Ecosystems* 10:96–119.

Pikitch, E.K., C. Santora, E.A. Babcock, A. Bakun, R. Bonfil, D.O. Conover, P. Dayton, P. Doukakis, D. Fluharty, B. Heneman, E.D. Houde, J. Link, P.A. Livingston, M. Mangel, M.K. McAllister, J. Pope, and K.J. Sainsbury. 2004. Ecosystem-based fishery management. *Science* 305:346–347.

Qu, J.G., Z.L. Xu, Q. Long, L. Wang, X.M. Shen, J. Zhang, and Y.L. Cai. 2005. *East China Sea*. GIWA (Global International Waters Assessment) Regional Assessment 36. Kalmar, Sweden: United Nations Environment Programme (UNEP).

Rabalais, N.N. 2002. Nitrogen in aquatic ecosystems. *Ambio* 31 (2):102–112.

Rabalais, N.N. 2004. Eutrophication. In Vol. 13 of *The Sea*, edited by A.R. Robinson, J. McCarthy, and B.J. Rothschild, 819–865. Cambridge, MA: Harvard University Press.

Raffensperger, C., and J. Tickner (eds.). 1999. *Protecting Public Health and the Environment: Implementing the Precautionary Principle.* Washington, DC: Island Press.

Reid, P.C., D.G. Johns, M. Edwards, M. Starr, M. Poulin, and P. Snoeijs. 2007. A biological consequence of reducing Arctic ice cover: Arrival of the Pacific diatom *Neodenticula seminae* in the North Atlantic for the first time in 800,000 years. *Global Change Biology* 13:1,910–1,921.

Riebesell, U., I. Zondervan, B. Rost, P.D. Tortell, R.E. Zeebe, and F.M.M. Morel. 2000. Reduced calcification of marine plankton in response to increased atmospheric CO_2. *Nature* 407:364–367.

Risse, T. 2007. Regieren in Räumen begrenzter Staatlichkeit: Zur Reisefähigkeit des Governance-Konzeptes, SFB-Governance Working Paper Series, Nr. 5, DFG Sonderforschungsbereich 700, Berlin, April 2007.

Rosenberg, R. 1985. Eutrophication: The future marine coastal nuisance? *Marine Pollution Bulletin* 16:227–231.

Roy, P.K. 2001. Coastal resource degradation and user-right abuse in Bangladesh: An overview of the challenges in user-based community management. In *Forging Unity: Coastal Communities and the Indian Ocean's Future*, 160–171. Conference proceedings of the International Collective in Support of Fishworkers (ICSF) organized at IIT Madras, Chennai, India, 9–13 October 2001. http://www.bdix.net/sdnbd_org/world_env_day/2004/bangladesh/document/roy.pdf.

Royal Society of London. 2005. *Impacts of Surface Ocean Acidification from Rising Atmospheric Carbon Dioxide.* London: Royal Society.

Ruiz, G.M., J.T. Carlton, E.D. Grosholz, and A.H. Hines. 1997. Global invasions of marine and estuarine habitats by non-indigenous species: Mechanisms, extent, and consequences. *American Zoologist* 37:621–632.

Salo, P., E. Korpimäki, P.B. Banks, M. Nordström, and C.R. Dickman. 2007. Alien predators are more dangerous than native predators to prey populations. *Proceedings of the Royal Society B* 274:1,237–1,243.

Schindler, D.W. 2006. Recent advances in the understanding and management of eutrophication. *Limnology and Oceanography* 51:356–63.

Shiganova, T.A., Z.A. Mirzoyan, E.A. Studenikina, S.P. Volovik, I. Siokou-Frangou, S. Zervoudaki, E.D. Christou, A.Y. Skirta, and H.J. Dumont. 2001. Population development of the invader ctenophore *Mnemiopsis leidyi* in the Black Sea and in other seas of the Mediterranean basin. *Marine Biology* 139:431–445.

Shiganova, T.A., E.I. Musaeva, Yu. V. Bulgakova, Z.A. Mirzoyan, and M.L. Martynyuk. 2003. Invaders ctenophores *Mnemiopsis leidyi* (A. Agassiz) and *Beroe ovata* Mayer 1912, and their influence on the pelagic ecosystem of northeastern Black Sea. *Oceanology* 30:180–190.

Snelgrove, P.V.R., T.H. Blackburn, P. Hutchings, D. Alongi, J.F. Grassle, H. Hummel, G. King, I. Koike, P.J.D. Lambshead, N.B. Ramsing, V. Solis-Weiss, and D.W. Freckman. 1997. The importance of marine sediment biodiversity in ecosystem processes. *Ambio* 26:578–583.

Sorensen, J. 2002. *Baseline 2000 Background Report: The Status of Integrated Coastal Management as an International Practice (Second Iteration).* www.uhi.umb.edu/b2k/baseline2000.pdf.

Spalding, M., S. Chape, and M. Jenkins. 2003. *State of the World's Protected Areas.* http://valhalla.unepwcmc.org/wdbpa/sowpr/Introduction.pdf.

Stachowicz, J.J., J.R. Terwin, R.B. Whitlatch, and R.W. Osman. 2002. Linking climate change and biological invasions: Ocean warming facilitates non-indigenous species invasion. *Proceedings of the National Academy of Science USA* 99:15,497–15,500.

Stachowicz, J.J., R.B. Whitlatch, and R.W. Osman. 1999. Species diversity and invasion resistance in a marine ecosystem. *Science* 286:1,577–1,579.

Tacon, A.G.J., and U.B. Barg. 1998. Major challenges to feed development for marine and diadromous finfish and crustacean species. In *Tropical Mariculture,* edited by S.S. De Silva. Oxford: Academic Press/Elsevier.

Tomasko, D.A., C.A. Corbett, H.S. Greening, and G.E. Raulerson. 2005. Spatial and temporal variations in seagrass coverage in southwest Florida: Assessing the relative effects of anthropogenic nutrient load reductions and rainfall in four contiguous estuaries. *Marine Pollution Bulletin* 50:797–805.

Turner, R.E., N. Qureshi, N.N. Rabalais, Q. Dortch, D. Justic, R.F. Shaw, and J. Cope. 1998. Fluctuating silicate:nitrate ratios and coastal plankton food webs. *Proceedings of the National Academy of Science USA* 95:13,048–13,051.

Turner, R.K., S. Georgiou, I.-M. Gren, F. Wulff, S. Barrett, T. Soderqvist, I.J. Bateman, C. Folke, S. Langaas, T. Zylicz, K.-G. Maler, and A. Markowska. 1999. Managing nutrient fluxes and pollution in the Baltic: An interdisciplinary simulation study. *Ecological Economics* 30:333–352.

UNEP/GPA. 2008. UN Environment Programme, Global Programme of Action for the Protection of the Marine Environment from Land-Based Activities. http://www.gpa.unep.org/.

U.S. Commission on Ocean Policy. 2004. *An Ocean Blueprint for the 21st Century.* Washington, DC: U.S. Commission on Ocean Policy.

Van der Weijden, W., R. Leeuwis, and P. Bol. 2007. *Biological globalisation: Bioinvasions and their impact on nature, the economy and public health.* Zeist, The Netherlands: KNNV.

Vijayan, V., L. Edwin, and K. Ravindran. 2000. Conservation and management of marine fishery resources of Kerala State, India. *Naga, The ICLARM Quarterly* 23 (3):6–9.

Vitousek, P.M., J.D. Aber, R.W. Howarth, G.E. Likens, P.A. Matson, D.W. Schindler, W.H. Schlesinger, and D.G. Tilman. 1997. Human alterations of the global nitrogen cycle: Sources and consequences. *Ecological Appliations* 7:737–750.

Volovik, S.P., and I.G. Korpakova. 2004. Introduction of *Beroe* cf *ovata* to the Caspian Sea needed to control *Mnemiopsis leidyi*. In *Aquatic Invasions in the Black, Caspian, and*

Mediterranean Seas, edited by H. Dumont, T.A. Shiganova, and U. Niermann, 177–192. Vol. 35 of NATO Science Series 4: Earth and Environmental Sciences. Berlin: Springer.

Waldbusser, G.G., R.L. Marinelli, R.B. Whitlatch, and P.T. Visscher. 2004. The effects of infaunal biodiversity on biogeochemistry of coastal marine sediments. *Limnology and Oceanography* 49:1,482–1,492.

Wilkie, M.L., and S. Fortuna. 2003. *Status and Trends in Mangrove Area Extent Worldwide.* Forest Resources Assessment Working Paper No. 63. Rome: Food and Agriculture Organization (FAO). http://www.fao.org/docrep/007/j1533e/j1533e00.htm.

Wolff, W.J., and K. Reise. 2002. Oyster imports as a vector for the introduction of alien species into northern and western Europe coastal waters. In *Invasive Aquatic Species of Europe: Distribution, Impacts and Management*, edited by E. Leppäkoski, S. Gollasch, and S. Olenin, 193–205. Berlin: Springer.

World Summit on Sustainable Development (WSSD). 2002. *Plan of Implementation of the World Summit on Sustainable Development*, paragraph 29. United Nations. 2004. http://www.un.org/esa/sustdev/documents/WSSD_POI_PD/English/WSSD_PlanImpl.pdf.

Worm, B., E.B. Barbier, N. Beaumont, J.E. Duffy, C. Folke, B.S. Halpern, J.B.C. Jackson, H.K. Lotze, F. Micheli, S.R. Palumbi, E. Sala, K.A. Selkoe, J.J. Stachowicz, and R. Watson. 2006. Impacts of biodiversity loss on ocean ecosystem services. *Science* 314:787–790.

Worm, B., and R.A. Myers. 2003. Meta-analysis of cod–shrimp interactions reveals top-down control in oceanic food webs. *Ecology* 84:162–173.

Yang, S.L., M. Li, S.B. Dai, Z. Liu, J. Zhang, and P.X. Ding. 2006. Drastic decrease in sediment supply from the Yangtze River and its challenge to coastal wetland management. *Geophysical Research Letters* 33:L06408, doi: 10.1029/2005GL025507.

Zhang, J., S.L. Yang, Z.L. Xu, and Y. Wu. 2006. Impact of human activities on the health of ecosystems in the Changjiang delta region. In *The Environment in Asia Pacific Harbours*, edited by E. Wolanski, 93–111. Berlin: Springer.

5

Integrating Tools to Assess Changes in Semi-Enclosed Marine Systems

Carolien Kroeze, Jack Middelburg, Rik Leemans, Elva Escobar-Briones, Wolfgang Fennel, Marion Glaser, Akira Harashima, Kon-Kee Liu, and Michel Meybeck

Semi-enclosed marine systems (SEMSs) are affected by natural processes and their variability, by climate change and sea level rise, and by a multitude of human activities. This chapter discusses and reviews the available multidisciplinary integrating tools that are available to advance our understanding of the complex dynamics of and interactions within SEMSs. We identify major components of a conceptual framework for an integrative approach, which takes into account interactions between, for example, an aquatic system and its adjacent land ecosystems, and between local people and ecosystems, and local and international markets. We argue that the development of a comprehensive conceptual framework is a major step but only an initial one. Interactions among the various components are relatively well known in qualitative terms. The real scientific challenge occurs when the concepts have to be implemented in quantitative models. The results of integrative modeling can serve the scientific community, as well as decision makers and other stakeholders. The integrated tools described here should help to shape and develop policies designed to solve the problems of SEMSs.

Introduction

The SEMSs included in this book are characterized by a relatively high input of freshwater from river runoff and groundwater discharge and by restricted exchange with the adjacent open ocean. These systems include relatively landlocked seas (such as the Baltic Sea and the Black Sea) as well as seas with wide openings to the open ocean (such as the North Sea, the Bay of Bengal, and the East China Sea) (see Chapter 1, this volume). In the latter systems, the inner boundaries are always the estuaries and the coastlines. The outer boundaries for open seas are fronts separating the less saline nearshore water from the more saline offshore water. These frontal boundaries may vary seasonally, under different runoff or wind conditions. Throughout this chapter, SEMSs will be used as a generic term.

Freshwater residence time plays a critical role in the biogeochemistry and ecology of SEMSs. These systems are thus strongly influenced by processes on land and in the adjacent open ocean, as well as by climate. The characteristics and dynamics of coastal seas can therefore not be understood in isolation.

Many changes in SEMSs are caused by natural processes and their variability, by climate change and sea level rise, and by a multitude of human activities (Chapters 2 and 4, this volume). These changes can lead to problems such as marine pollution, as with persistent organic pollutants (POPs), and to eutrophication, harmful algal blooms, anoxia, biodiversity loss, and fish stock depletion, as well as to economic opportunities, such as aquaculture and wind parks. Some of the human activities are directly linked to the seas, such as the extraction of fish, shipping, and drilling for oil and gas, but many are only remotely linked (e.g., agriculture or mining) to the seas or occur along the coasts (e.g., coastal urbanization). Nutrient and sediment loads delivered to the seas are controlled by the catchments (as a provider) and by the estuarine zone, where river waters mix with ocean water and the estuary acts as a filter for nutrients, carbon, pollutants, and sediments (see Chapter 7, this volume).

SEMSs are influenced by all these impacts and their interactions. The factors and processes involved in individual influences and the consequences of individual changes are relatively well understood, but the resulting effects on SEMSs are extremely difficult to predict because of systemic interactions and feedbacks with other processes and changes and also time lags in response to management actions. To enhance understanding of the dynamic behavior of the seas, a highly integrated approach is needed. This approach should not only allow description of the ongoing changes and their causes and effects, but also allow managers to project future behavior under assumed plausible changes in human activities or management. The latter capability helps policy makers to appraise effective strategies for mitigating problems and a secured future sustainable use and to assess the trade-offs among different options.

This chapter aims to discuss and review available multidisciplinary integrating tools that are instrumental in advancing our understanding of the complex dynamics of and interactions within SEMSs. These tools should be scientifically robust. The required observation and monitoring data for testing and validation will also be described.

The Conceptual Framework

Different Approaches to Integration

There are many different approaches to integrating scientific knowledge and understanding from different disciplines (Committee on Facilitating Interdisciplinary Research 2004). The most common approach is ad hoc mutual exchange of information between disciplines or studies focused at specific pieces. Although some insights can emerge, this approach rarely leads to improved understanding of the main systemic properties and behavior of the entire system. Many multidisciplinary studies therefore aim to integrate more strongly and add an integrative activity toward the end of the project. Although this frequently used approach aims at integration, the results are limited because such integration will rarely identify and quantify the proper linkages and interactions. To accomplish such integration, it should form the starting point so that concepts, components, major linkages, and possible interactions are identified at the beginning and can be further elaborated during the different studies. This means that the aim to integrate helps to structure the studies of the different disciplines involved. This is an effective approach to integrate different disciplines (Leemans 2006). An additional advantage is that the conceptual model can be communicated to users and other stakeholders, such as local and national decision makers. Their desires can then be better incorporated, which enhances the final applicability.

A Comprehensive Conceptual Framework

To enable understanding of the complex relationships between the different factors that influence SEMSs, the different components must be identified. The full domain is already described in Chapter 1. A SEMS includes the coasts with beaches, wetlands and estuaries, the continental shelf, and the deep water, but its dynamic behavior can be understood only if the interacting components are considered as well. These are the different catchments surrounding the sea that deliver nutrients, sediments, and freshwater; the connected ocean; and the atmosphere and its climate patterns. All these components drive major aspects of the physical, chemical, and ecological processes. The interactions and exchanges among these components determine the natural controls of these marine systems (I–VI in Figure 5-1).

Additionally, people and their activities (i.e., the anthroposphere) have to be considered. They form the major direct human controls (A–F in Figure 5-1). This human dimension adds strongly to the complexity. People carry out many diverse activities within catchments, at and along the coasts, and at sea and in the ocean. The activities have two major effects. First, they satisfy major human needs—they provide food through catching fish and farming shrimp, they protect coastlines, and they provide

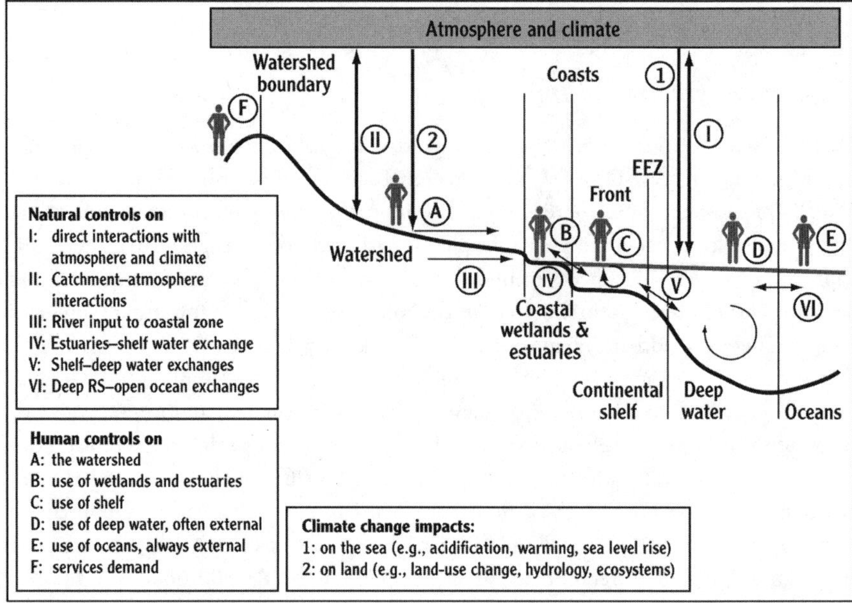

Figure 5-1. Major drivers and controls of the functioning of SEMSs.

opportunities for transport. All these benefits to people are provided through services from SEMSs (Chapters 4 and 12, this volume). Some of these services are not delivered directly to people within the system boundaries but are delivered to those far beyond. The socioeconomic drivers are thus anchored in international markets but often regulated by national laws. This holds especially for some of the fish and shrimp farming and the fisheries inside exclusive economic zones (EEZs). Local fisheries along the coast, however, provide services primarily to local communities. These connections crossing the boundaries of regional sea systems are explicitly addressed in the conceptual framework (Figure 5-1). The delivery and sustainability of all these services are possible because of all the physical, chemical, and ecological processes and diversity within the regional sea systems, but these can impair the sustainability if thresholds are passed (Chapter 3, this volume). Second, these human activities indirectly, and often unintentionally, change the amount and composition of riverine fluxes from the catchments (see Chapter 7, this volume; Seitzinger et al. 2005), deplete resources from the seas and alter their major biogeochemical processes, and change atmospheric composition and climate.

Finally, regional sea systems are influenced by the atmospheric composition, weather, climate, and changes therein (1 and 2 in Figure 5-1). Such changes are driven by greenhouse gas emissions worldwide and cause changes in atmospheric temperature, precipitation, and winds and consequently change sea surface water temperature, sea level, and

Box 5-1. Examples of Characteristics of SEMSs that Complicate the Development of Integrated Models

- River export of organic carbon, suspended matter, and nutrients change, with major consequences for regional seas, including changes in the metabolic balance, oxygen status, carbon dioxide air–sea exchange, light climate, and eutrophication.
- There are synergistic effects, for example, between changes in the physics of a regional sea and changes in riverine nutrient inputs. Changes in water column stratification, as well as the ratio of dissolved Si to dissolved N, may have consequences for phytoplankton biomass, composition, and production. Also, the accumulation of anthropogenic carbon dioxide in the atmosphere results in acidification of seawater, modifying coastal biodiversity and damaging sensitive ecosystems, such as coral-based ones.
- Alien species may outcompete indigenous species, with major implications for biodiversity and ecosystem functioning.
- Land-use change along the shoreline may have consequences beyond its direct domain.
- The pelagic and benthic components of the marine ecosystem are linked.
- Natural variability and direct and indirect human drivers must be considered.

water column stratification. Additionally, the increase in atmospheric CO_2 concentrations and the consequent uptake by marine waters decrease their pH. This acidification will lead to considerable changes in the chemistry and biology of SEMSs and the ocean (Orr et al. 2005), with consequences for economic activities such as tourism (coral destruction) and shellfishery (Gazeau et al. 2007; Chapters 2 and 4, this volume).

This all-encompassing conceptual framework illustrates the complexity and highlights the interactions between, for example, the seas and their adjacent land ecosystems, and between local people and the ecosystems and global fish markets they depend on that operate outside the EEZs. All these different components, with their specific dimensions and typical process scales and interactions (feedbacks, synergies, and trade-offs), are comprehensively included in this framework. The challenge is not to further detail the framework, but to develop models that describe and quantify the connections and interactions and integrate all the apparent dimensions.

The various compartments within a SEMS are intimately connected (Box 5-1). These interactions can be one-way, that is, with one compartment driving another one, or two-way, in which the two compartments mutually affect each other. Interactions can be direct or indirect; for instance, changes in subsystem A have a consequence for subsystem B only via their effect on subsystem C. Compartments may respond immediately or with a delay. The response time depends on a number of factors, including the size of the perturbation, the size of the stock in the reservoir, and characteristics of the subsystem (e.g., water residence time). The interaction may be linear or nonlinear. Nonlinear interactions combined with two-way interactions may result in hysteresis, the differential response upon positive and negative forcings and bifurcations, and eventually in alternative stable states (Chapter 3, this volume).

Developing and Applying an Integrated Model

Integrating Existing Models

State-of-the-art models of SEMSs can now be integrated into systems that link watershed components and river runoff to oceanographic models forced by atmospheric processes and acting back to the atmosphere.

Ocean and atmosphere models are based on first principles by using the equations of fluid dynamics and thermodynamics. Models of ocean physics are virtually independent of the chemical and biological processes in the water, apart from a few examples of weak interactions (e.g., control of sea surface temperature by phytoplankton). These models can therefore be run without considering these interactions. Moreover, advective and diffusive (turbulent diffusion, often mediated by vigorous small-scale mixing) transport derived from physics-only models can then be used to transport biological and chemical components, either off-line or fully coupled.

Food webs are complex systems that transfer nutrients to phytoplankton and bacteria via zooplankton to fish. Phytoplankton (and cyanobacteria) represent a spectrum of microscopic single-cell primary producers, which build up organic compounds directly from carbon dioxide and the nutrients dissolved in the water. The captured matter is consumed by secondary producers, the zooplankton, encompassing various size classes. The larger zooplankton, for example, copepods, are eaten by fish, which are caught by humans. There is a further pathway for nutrients through recycling, that is, through "sloppy feeding" and excretion as well as via microbial mineralization of detrital material.

The range of institutions and stakeholders, also with their different rationales and motivations, is similarly complex, and social models reflect such complexity inadequately. Nonetheless, the connection between social and ecological models (here the term *social* includes the economic) offers innovative scope for assessing social–ecological dynamics.

Since it is obviously impossible to include the full food web and the full range of controlling physical processes within one generic model, the complexity of nature is strongly reduced in models. The degree of idealization achieved by isolating parts of the food chain, truncating upper and/or lower trophic levels by effective parameterization, and simplifying controlling physics depends on the problem to be considered. Hence, there is a variety of model types; they range from those with aggregated state variables that interact as in chemical reactions, to those with variables referring to organisms' life cycles and size class stages, to those tracking individuals propagating through the model system. A comprehensive overview of the building and application of coupled physical–biological models was recently given in Fennel and Neumann (2004).

A great challenge is the development of a new generation of models, bridging the gap between fish production models and biogeochemical models. Current biogeochemical models include the upper food web only implicitly, in terms of zooplankton mortality rates (model closure terms), while fish production models include the lower food web through prescribed food concentrations (implying unlimited primary production). There are attempts to include fish aspects by merging nutrient-phytoplankton-zooplankton-detritus types of lower–food web models with single-species fish models (see Megrey et al. 2007a, 2007b). A quantitative, mass-conserving coupled model reaching from nutrients to predator fish, an end-to-end food web model, was recently presented by Fennel (2008). The quantitative role of fish and fisheries on the nutrient balance in the Baltic Sea was documented by Hjerne and Hanson (2002). These attempts linking nutrient flows to end-to-end food web dynamics are long overdue, but their generality, applicability, behavior, and performance need study. Moreover, our knowledge of marine food web functioning is still very limited, with novel organisms and key functions (archael nitrification; Wuchter et al. 2006) being discovered on a regular basis; it is clear that existing model formulations might need revision. Regional sea biogeochemical or ecological models often do not resolve the interactions of organisms living in the sediments (Soetaert et al. 2006). This is unfortunate because many ecosystem functions and services (e.g., nutrient recycling, denitrification, harvestable fish and shellfish resources) are unique to the sediment compartment. When seabed processes are explicitly included, the physical processes in the bottom boundary layer must be parameterized as well.

Although much effort has been devoted to accurate modeling of sediment deposition, resuspension, and transport processes in regional seas and along shorelines, these models still have many caveats, such as inadequate representation of grain size spectra, biologically induced stabilization, and erosion, and dealing with various time scales (from less than tidal for deposition/erosion to more than decadal for sediment burial). Moreover, such coastal sediment transport models are normally not coupled to biogeochemical models, because the latter incorporate empirical, often static relationships for light-attenuation patterns. Assessment of changes in primary production due to changes in sediment dynamics is therefore not straightforward with these models, and one has to revert to empirical approaches.

Watershed models depend to a larger degree on data, or data-derived relationships, than the ocean models. For the formulation of model dynamics, needed relationships can be derived through statistical analysis and combined with spatially resolved patterns of controlling factors, such as land use and fertilization. This information is, however, not always available for all world regions.

These watershed models should be coupled to those for SEMSs; this implies that estuarine processes should be resolved. Estuaries function as an efficient filter for watershed-derived materials and provide many ecosystem functions and services, including habitats for resources (see Chapter 7, this volume). Given the diversity of estuarine systems (e.g., fjords, tidal-dominated estuaries, salt-wedge estuaries), there is a clear need to develop generic estuarine system models for use in integrated assessment studies. Integrated models do not need all the complexity of the underlying models of subsystems. Depending on the aims of such integrated models, one may, for instance, need only a simplified version of watershed models, providing inputs of nutrients and sediments into the estuaries.

There is still a relative lack of models dealing with the estuarine filters in which the land-based fluxes can be markedly decreased and modified. Due to the heterogeneity of estuarine types (e.g., fjord, delta, lagoon, mangrove), each river input should be linked to a specific estuarine model. Outputs from estuarine models are then used as inputs to continental shelf models. However, such coupled, comprehensive river–estuary–coastal–sea models are so far limited to specific regions because of their high data requirement and the need to have detailed subdomain models available. As a consequence, global models do not take into account the overall filter functions of estuaries, although the estuarine filter function is an emergent property of the system that may develop in time (Soetaert et al. 2006). Models can be down- and upscaled in the physical space, that is, grid size and time step, but also in the phase space of state variables. Models can also be applied separately or with dynamical interaction included. The choice of the model type also depends on the questions being addressed. Not every problem requires a full three-dimensional interactive model system. Scientific problems posed at multiannual time scales or dealing with long-living organisms (e.g., mangroves, whales) do not require models that resolve temporal variability at tidal cycles. For instance, models dealing with coastal sediment dynamics and their consequences for fish and bird habitats do not have to be based on first-principle physical laws resolving movement of individual particles on a time scale of minutes.

Integration of disciplinary models implies that they should together and individually obey certain basic laws, such as mass conservation. Moreover, the various components of integrated models should operate on the same temporal and spatial scales; otherwise, specific modules, such as model couplers, should explicitly account for this. For instance, models for atmospheric processes operate at much shorter time scales and larger spatial scales than models based on dynamic geographic information systems for mangrove ecosystem or sea grass ecosystem dynamics. Integration of various models also implies that identical material or information is exchanged among the models. Exchange can occur only if the same units or currencies are used or if the above-mentioned model couplers take care of this.

Data Collection, Handling, and Use

Integrated modeling (Figure 5-2) requires information on physical, chemical, and biological characteristics, as well as on the socioeconomic drivers of the system. Lack of data and inconsistency between data sets are problems typical of integrating modeling.

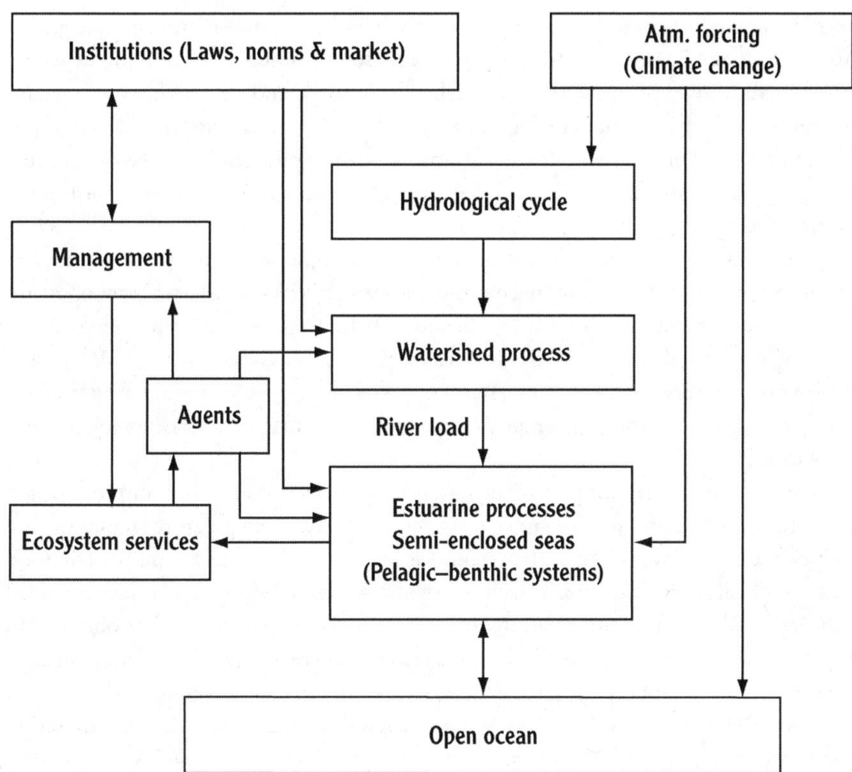

Figure 5-2. A schematic diagram of a possible integrated model. Each of the boxes or any combination of the linked boxes can be a module of the integrated model. Each module needs various types of data.

To serve the data needs, we may rely on several approaches—acquiring data from ongoing monitoring programs and field studies, deriving data from remotely sensed observations, generating data by model simulation, and retrieving data from data archives. We may also combine the above approaches, as discussed below.

For atmospheric forcing, data are available from various weather forecast services, such as the National Centers for Environmental Prediction (NCEP) in the United States and the European Centre for Medium-Range Weather Forecasts (ECMWF). Precipitation data, which are the most important to feed into watershed models for prediction of river runoff and chemical loads, are also available from weather services. Alternatively, global riverine discharge data are available from hydrological models, such as the STN30 (Vörösmarty et al. 2000a, b). The chemical loads of river discharges, such as nutrient loads, can be derived from the global Nutrient Export from Watersheds (NEWS) models (Seitzinger et al. 2005). For climate change scenarios, atmospheric forcing must come from climate model simulations.

Marine environmental data are often provided by coastal monitoring programs. Monitoring of SEMSs has traditionally been based on shipboard sampling or shore-based small-scale campaigns. This approach will continue, and it has to be continued for many variables (chemical and biological stock and flux measurements) that cannot yet be measured in situ or remotely. The disadvantage of this approach is the level of staffing required, the lack of continuity due to logistic and financial restrictions, and the limited spatial coverage.

Regional sea and coastal monitoring is increasingly based on an integration of remote sensing tools combined with online monitoring using ferries and other ships of opportunity and autonomous moorings at key locations. While online monitoring observations have high temporal resolution and increasing coverage of chemically and biologically relevant parameters, they are limited in their spatial coverage. Once established, a monitoring network may provide a comprehensive suite of useful data for the integrated modeling tools.

Remote sensing techniques offer synoptic spatial coverage with unprecedented opportunities for assessment of larger areas; remotely sensed temperature and sea surface elevation are key to validate circulation models for SEMSs. Remotely sensed distribution patterns of chlorophyll, colored dissolved organic matter (CDOM), and total suspended matter provide essential information on spatial distribution that cannot be obtained in any other way. Moreover, some key ecosystems can be remotely sensed for changes in coverage and functioning, in particular, systems such as mangroves and other coastal wetlands. Remote sensing does, however, have a few drawbacks, including the still rather coarse resolution available to resolve key coastal features, extensive cloud coverage at key boundaries (land–sea and shelf break), and the optical characteristics of coastal waters rich in suspended matter and CDOM.

A Coastal Ocean Observations Module of the Global Ocean Observing System (GOOS) is being developed. A network for coastal observation is being built with synergy of moored automatic instruments, shipboard observations, and remote sensing techniques. If successfully implemented, the GOOS coastal module could be a very useful data source for the integrated modeling tools.

Open-ocean conditions are usually less variable than those of the coastal ocean. Therefore, the data for model initiation can often rely on climatology data, which represent the long-term average condition. In fact, the monthly climatology data are often used when observations are not available. Such data are often available from large data centers around the world, such as the National Oceanographic Data Center (NODC) of the United States or the World Data Center for Marine Environmental Sciences (WDC-MARE) located in Germany.

Information on socioeconomic drivers is typically available at the national level, and not on the level of river basins of coastal fringe or SEMSs. The information is typically collected for administrative reasons, and not necessarily for management or scientific reasons. Nevertheless, databases from, for instance, the Food and Agricultural Organization (FAO) or national planning bureaus can be useful for integrated analyses of aquatic systems. Several of these national-scale databases have been downscaled to the grid level (e.g., EDGAR; Van Aardenne et al. 2001) and can be used in integrated modeling exercises. In recent years, large multidisciplinary data archives, which cover both natural and socioeconomic information, are being established in data centers, such as the Center for International Earth Science Information Network (CIESIN) of Columbia University. They could prove useful to the integrated modeling tools.

While there are many good data sources available, we still lack enough data to understand response curves between management decisions (e.g., different types of land-use change or different fishery regulatory systems) and ecosystem processes.

Spatial Complexity of Semi-Enclosed Marine Systems

A typical problem of integrated modeling of SEMSs is associated with spatial issues. First, there is a mismatch of natural and political boundaries, which may cause management problems. The environmental impacts usually result from processes occurring in zones defined by natural boundaries, such as a catchment basin on land or an estuary with its adjacent plume in the sea, whereas the management units are confined to zones defined by political boundaries, such as a national border or the EEZ limits. International or jurisdictional agreements are needed for transboundary management actions.

Second, there are transboundary and remote drivers. Drivers of environmental changes in a regional sea are often located outside of its geographic limits, such as climate change or pollutants from dry and wet precipitation.

Third, national versus international access may be an issue. Some SEMSs (e.g., the Gulf of St. Lawrence) are completely contained within a single country's EEZ, but others may have considerable areas of international water for open access (e.g., the Gulf of Mexico) or are split among the EEZs of several countries (e.g., the Baltic Sea).

Solving these spatial issues allows for the development of models that address major scientific questions dealing with all the different ecological boundaries and the assessment of management options, taking into account the inherent ecological, physical, and institutional complexities (for an elaborated example, see Young et al. 2007).

Testing and Validation: Uncertainty Analysis

It is clear that grid cell–oriented spatial models cannot provide a one-to-one correspondence with observations made in a location within the corresponding grid cell. Models should reproduce response patterns to forcing, evolution of structures, and time scales.

Sources of model error can include weaknesses in model formulations, inadequate spatial resolution, and aggregation of state variables, but they can also be errors in driving forces and fluxes specified in the models. For example, the quality of coastal circulation modeling depends strongly on the type of meteorological data used to drive the ocean model. Coarse wind fields are generally too smooth and miss strong short-term wind events, which are important for vertical mixing or, in shallow waters, for resuspension of sedimentary material at the water–seafloor interface.

Uncertainties can be assessed with the help of model runs by injecting defined errors and following how the corresponding signal propagates through the model in order to quantify the consequences.

Scenarios

In this section we address two questions: How we can use the conceptual framework as a basis for integrated analyses and exploration of future trends? And what descriptors are required and which indicators are worthwhile to analyze for what types of scenarios?

In scenario analysis, we explore possible futures or analyze how to realize desirable futures. In these analyses, we may distinguish among three types of human influences on SEMSs: (1) human activities on the land, changing the inputs of nutrients and sediments into the SEMSs; (2) human activities along the shore and in estuaries, shelves, and the open ocean, influencing biological functions of SEMSs; and (3) human-induced atmospheric changes and climate change, affecting chemical, physical, and biological processes in SEMSs, both on land and at sea (Table 5-1). Moreover, SEMSs affect human well-being through ecosystem services (Chapters 4 and 12, this volume).

Exploring future trends in natural and social dynamics surrounding SEMSs requires, first of all, "story lines" or narratives of the future scenarios to be explored. These narratives need to at least describe future trends in relevant human activities in the river basins draining into the SEMSs (agriculture, industry, population density, sewage treatment, reservoir and dam building), as well as relevant human activities in the SEMSs (in particular fishing, related to the demand for fish). These narratives are relatively easily developed. To assess the impact of climate change on SEMSs requires that the narratives be accompanied with information on the associated changes in precipitation, temperature, and wind patterns. Of these, changes in wind patterns may be the most difficult to obtain at the level of individual SEMSs, yet these are expected to have major influences on marine primary and secondary production.

Identification of the most appropriate indicators for scenario analyses of human–nature dynamics in SEMSs depends on the purpose of the analyses. Studies aiming to increase our understanding of changes in SEMSs typically require a full description of the biological, chemical, and physical processes involved. For studies aiming to provide a basis for decision making or management, it could be sufficient to assess the environmental pressure (e.g., fishing intensity or consumptive water use) or state (e.g., nutrient loading) of the system or to use simple indicators for human impacts (e.g., based on critical loads; Sverdrup and Barkman, 1994). Because of the complexity of the system, indicators for pressure and state may be easier to model than indicators for impact.

Table 5-1. Examples of Different Types of Human Influences on SEMSs

Human Activity	*Environmental Pressure*	*Threats*
Type 1: Human activities on the land, including human responses to global change	Changes in inputs of nutrients and sediments to SEMSs	Pollution, eutrophication, disconnectivity between rivers and seas, sediment imbalance
Type 2: Human activities in SEMSs, including human responses to global change	Overexploitation, shore developments, shipping, coastal defense	Reduced carrying capacity for fishing, habitat loss, alien species
Type 3: Human-induced climate change impact on seas	Changes in temperature, precipitation, wind	Changes in turbidity, having ecological consequences

Specific types of decision-making scenario analyses may explore possible restoration of disturbed systems. For instance, policy measures may restore the environmental conditions in terms of physical and chemical characteristics, but whether this will also restore the ecosystem functions or biodiversity is not easily determined. Some damages may be irreversible, or systems may change to another (potentially less desirable) state (see Chapter 3, this volume). It should be noted that these types of analyses require region-specific approaches, since most marine systems have their specific characteristics (Meybeck et al. 2006), and it is not easy to draw general conclusions for future trends of all SEMSs.

The Definition and Role of Indicators

The term *indicator* is used here for variables that indicate a system's position in relation to predefined desirable and undesirable conditions. Indicators may provide information about important changes to guide future decisions and actions. Indicators are more valuable when spatial and temporal coverage is sufficient to allow meaningful comparisons. Criteria for good indicators are scientific validity, reliability of statistical measurement, acceptability for decision makers and other users, responsiveness to change, and data availability at an affordable cost.

Choosing an indicator implies that it provides the most pertinent information on the criterion to be assessed. If indicator systems are to encompass local, national, and global sustainability needs, not only user participation is essential, but also science-based, system-focused indicator development. In relation to human–nature relations and their effect on

Table 5-2. Indicators/Descriptors of a System's Abilities to Provide Services (based on Bossel 1999)

Question	*Indicator/ Descriptor Type*
Is system service compatible with its environment?	Existence
Does the system effectively and efficiently provide services?	Effectiveness
Can the system adapt to new challenges?	Adaptability
Is the system compatible with interacting subsystems?	Coexistence
Is the system able to reproduce at sufficient rates?	Reproduction
Is the system compatible with psychological needs and culture?	Psychological needs

the state of environmental systems, four types of approaches can be differentiated: (1) ecocentric, (2) anthropocentric, (3) interdisciplinary, and (4) complex system-based.

1. The first social indicators in environmental management addressed human uses of ecosystem services. Such ecocentric indicators, for instance, on how fast fish reproduce or corals grow, are combined with social indicators of human population size and growth and of resource use rates in order to assess the joint outcomes of natural resource reproduction and human consumption rates.
2. Anthropocentric indicators assess ecosystem contributions to human well-being. They take the improvement of the human condition as their overall goal and concentrate on the outcomes of ecosystem use for the state of humanity.
3. Interdisciplinary indicator systems combine ecocentric and anthropocentric indicators in relation to the dynamics surrounding specific natural resources (rather than whole ecosystems) to assess biological, economic, and social change. Liu and colleagues (2004) suggest different sets of indicators for the ecological, socioeconomic, community, and institutional sustainability of fisheries. These indicators address the internal dynamics of social (including institutional), ecological, and economic subsystems.
4. System-based indicator systems address a clearly defined human–nature complex (such as a SEMS) as an integrated dynamic system. A major challenge here is to select the minimum number of indicators to assess system viability and performance. Bossel (1999) proposed generic questions to enable an overall assessment of integrated human–nature systems (Table 5-2).

System-based indicator development defines the conditions under which integrated human–nature systems are viable and perform in desired ways. It offers a way to achieve comprehensiveness while avoiding data overload and inefficiency. An indicator-based system monitoring framework offers a path toward the integrated assessment of human–nature dynamics for SEMSs.

System-based approaches have been criticized for the lack of involvement of directly affected stakeholders in the determination of management priorities (Rotmans and van Asselt 2002). This lack of involvement can result in negative outcomes that include non-compliance, social and economic impoverishment, polarization of interest groups, and normative insecurity and conflict (Ghimire and Pimbert 1997; Govan et al. 1998; Glaser et al. 2003; Manuel-Navarrete et al. 2004; Schöler 2005). While system-based approaches to indicator development provide the essential analytic core to integrated systems, they lack the participatory dimension essential to prevent counterproductive social reactions. Methods for the participatory derivation of indicators for sustainable system management are under development (Fontalvo et al. 2007).

Table 5-3 lists possible descriptors/indicators to assess human-induced changes in SEMSs (see Chapter 4, this volume, for details on possible threats to SEMSs).

The Way Forward: Opportunities and Challenges

Applications of Integrating Tools

Integrating tools such as those described in this chapter can be used for different purposes. SEMSs are complex because of the different components, dimensions and interactions, feedbacks, and synergies at various temporal and spatial scales. To fully understand the complexity, we need comprehensive integrating tools. These tools need to account for the drivers inside and outside SEMSs.

An important scientific application is the quantification of the role and strengths of the different interactions and feedbacks. Simulation experiments can be set up in which specific processes are switched off or added. This provides insights into the relevance of these processes or their parameterization. Examples of these approaches are already available (e.g., Xiao et al. 1998; Schaeffer et al. 2006). Because such an integrated tool includes the hydrosphere, the biosphere, the atmosphere, and the anthroposphere, new insights can be obtained that are impossible with disciplinary models or only partially integrated models. However, integrating tools by their very nature do not represent the details within the individual components, and specific questions within a domain (e.g., the ocean or human society) can better be addressed by disciplinary models.

A major societal application of integrating tools would be to determine the relative importance of the various different natural and anthropogenic drivers that alter SEMSs. Such an approach facilitates the identification of major threats that lead to degradation or collapse of (parts of) individual marine systems. Applying such tools allows an early identification of the consequences of these threats and hopefully will facilitate the development of the appropriate responses. The integrating tools should thus provide the capacity to assess changes and improve possibilities for managing such systems. Developing and implementing scenarios using such tools help to advance such assessment capability. Given the wide variety in characteristics of different seas, these assessments need

Table 5-3. Possible Descriptors/Indicators to Assess Human-Induced Changes in SEMSs

Drivers of Disruption of Ecosystem Services	*Possible Descriptors/ Indicators for Change*	*Interesting Trends to Analyze*
	Related to Human Activities in the River Basins Draining into the SEMSs	
Hydrologic sediment input	Sediment input to the sea	Trend in sediment input
Habitat loss	Coastal area use for human activities	Trend in % of coastal area used over time
Eutrophication	Nutrient export by rivers	Trend in nutrient inputs over time; exceedance of critical levels/loads
Shoreline development	Length of shoreline developed	Trend in % of shoreline developed
Pollutants	Intensity of chemical industrial activities along the coast	Trend in concentrations relative to allowable levels
	Related to Human Activities in SEMSs	
Alien taxa	Intensity of long-distance shipping	Trend in number of ships from other continents
Overexploitation	Fishing intensity	Trend in catch of sensitive species
Food web alteration	Functional biodiversity	Trend in (functional) biodiversity loss
Nonliving resource extraction	Area use for resource extraction	Trend in % of area used
	Related to External Factors	
Sea level rise	Sea level	Trend in sea level rise
Circulation and hydrography changes	Current patterns and speeds, precipitation	Trend in current speed and directions, and rainfall
Temperature changes	Regional temperature	Trend in regional temperature
Wind changes	Regional wind patterns	Trend in regional wind patterns
UV changes	UV radiation	Trend in UV radiation
Acidification	Atmospheric CO_2 concentrations	Global trend in pH

to be done at the level of the individual marine system. The management relevance of such integrated models is a unique strength of these models (Alcamo et al. 1998a).

Communicating Results to Broader Audiences

The results of integrated modeling can serve the scientific community as well as decision makers and other stakeholders. Participation of the envisaged users, explaining the underlying conceptual framework, and presenting the model results will increase the relevance of the model results for these users. Such dialogue will also help to identify the relevant and appropriate indicators; this will no doubt accelerate the acceptance of the results of integrated models (Van Daalen et al. 1998).

Possible messages to communicate are associated with managing SEMSs in relation to the activities in the catchments. The model will clarify the downstream implications of changes in upstream management. In addition, integrating tools can be used to assess policies to reduce specific threats as well as the interactions between policies designed to address different threats.

Research Needs and Challenges

Developing a comprehensive conceptual framework is a major step but only the initial one. This chapter has shown that the different interactions among the various components are relatively well known in qualitative terms. The real scientific challenge occurs when the concepts have to be implemented in quantitative models. This means that the integration of all the different units and dimensions has to be resolved. Few of the important components have been linked before, but the real challenge is in integrating the physical, chemical, and ecological models, with their specific spatial and temporal dimensions, with the socioeconomic models that focus on the choices of individuals and communities. Earlier examples of successful approaches are available (e.g., Alcamo et al. 1998b; Prinn et al. 1999; Fischer et al. 2001), but these are still strongly based on and biased toward natural processes. Recently, the emergence of agent-based models (part of the family of lattice automaton models) and spatial socioeconomic databases provide new opportunities (e.g., Janssen and Ostrom 2006). A strong focus on improving these natural science and social science linkages will definitely enhance the rigor of integrated models. It should be noted, however, that lattice automaton–type models are not directly applicable to hydrologic and open-ocean systems. They could work for ecosystems along the coast, and probably for benthic/sediment ecosystems, but not for processes and organisms in seas where Langrangian approaches are often necessary (i.e., where ecosystems move with water masses).

There are many threats and problems in and around SEMSs. These cannot be solved in isolation. The governance of all these systems remains complex, with the involvement of local municipal authorities, national governments, international treaties, and open access. The Millennium Ecosystem Assessment (2005) showed that successful responses to deal with all these problems involved proactive participatory policies with the appropriate monitoring and assessment tools. The integrated tools described here should help to shape and develop such policies. The necessity for applying such models for policy support is thus huge and urgent.

References

Alcamo, J., G.J.J. Kreileman, and R. Leemans. 1998a. Global models meet global policy: How can global and regional modellers connect with environmental policy makers? What has hindered them? What has helped? In *Global Change Scenarios of the 21st Century: Results from the IMAGE 2.1 Model*, edited by J. Alcamo, R. Leemans, and G.J.J. Kreileman, 261–265. Paris: Pergamon/Elsevier.

Alcamo, J., R. Leemans, and G.J.J. Kreileman (eds.). 1998b. *Global Change Scenarios of the 21st Century: Results from the IMAGE 2.1 Model.* Paris: Pergamon/Elsevier.

Bossel, H. 1999. *Indicators for Sustainable Development: Theory, Method and Applications.* Winnipeg, Canada: International Institute for Sustainable Development.

Committee on Facilitating Interdisciplinary Research. 2004. *Facilitating Interdisciplinary Research.* Washington, DC: National Academies Press.

Fennel, W. 2008. Towards bridging biogeochemical and fish production models. *Journal of Marine Systems* 71:171–195, doi: 10.1016/j.jmarsys.2007.06.008.

Fennel, W., and T. Neumann. 2004. *Introduction to the Modelling of Marine Ecosystems.* 1st Ed., in Elsevier Oceanography Series Vol. 72. Paris: Elsevier.

Fischer, G., M.M. Shah, H. van Velthuizen, and F.O. Nachtergaele. 2001. Global agro-ecological assessment for agriculture in the 21st century. Laxenburg, Austria: International Institute for Applied Systems Analysis (IIASA).

Fontalvo, M., M. Glaser, and A. Ribeiro. 2007. A method for the participatory design of an indicator system as a tool for local coastal management. *Ocean and Coastal Management* 50:779–795.

Gazeau, F., C. Quiblier, J.M. Jansen, J.-P. Gattuso, J.J. Middelburg, and C.H.R. Heip. 2007. Impact of elevated CO_2 on shellfish calcification. *Geophysical Research Letters* 34:L07603, doi: 10.1029/2006GL028554 2007.

Ghimire, K.B., and M.P. Pimbert. 1997. *Social Change and Conservation.* London: Earthscan.

Glaser, M., U. Berger, and R. Macedo. 2003. Local vulnerability as an advantage: Mangrove forest management in Pará state, North Brazil under conditions of illegality. *Regional Environmental Change* 3:162–172.

Govan, H., A. Inglis, J. Pretty, M. Harrison, and A. Wightman. 1998. Best practice in community participation in national parks. *Scottish Natural Heritage Review* 107:1–75.

Hjerne, O., and S. Hansson. 2002. The role of fish and fisheries in Baltic Sea nutrient dynamics. *Limnology and Oceanography* 47:1,023–1,032.

Janssen, M.A., and E. Ostrom. 2006. Empirically based, agent-based models. *Ecology and Society* 11:37. http://www.ecologyandsociety.org/vol11/iss2/art37/.

Leemans, R. 2006. Scientific challenges for anthropogenic research in the 21st century: Problems of scale. In *Earth System Science in the Anthropocene: Emerging Issues and Problems*, edited by E. Ehlers and T. Krafft, 249–262. Berlin: Springer.

Liu, W.H., C.H. Ou, and K.H. Ting. 2004. Sustainable coastal fishery indicator system: A case of Gungliua, Taiwan. *Marine Policy* 29:199–210.

Manuel-Navarrete, D., J.J. Kay, and D. Dolderman. 2004. Ecological integrity discourses. *Human Ecology Review* 11:215–229.

Megrey, B.A., K.A. Rose, S.-I. Ito, D.E. Hay, F.E. Werner, Y. Yamanaka, and M.N. Aita. 2007a. North Pacific basin-scale differences in lower and higher trophic level marine ecosystem responses to climate impacts using a nutrient–phytoplankton–zooplankton model coupled to a fish bioenergetics model. *Ecological Modelling* 202:190–210.

Megrey, B.A., K.A. Rose, R.A. Klumb, D.E. Hay, F.E. Werner, D.L. Eslinger, and S.L. Smith. 2007b. A bioenergetics-based population model of Pacific herring (*Clupea harengus pallasi*) coupled to a lower trophic level nutrient–phytoplankton–zooplankton model: Description, calibration, and sensitivity analysis. *Ecological Modelling* 202:144–164.

Meybeck, M., H.H. Dürr, and C.J. Vörösmarty. 2006. Global coastal segmentation and its river catchment contribution: A new look at land–ocean linkage. *Global Biogeochemical Cycles* 20:GBIS90, doi: 10.1029/2005GB002540.

Millennium Ecosystem Assessment. 2005. *Ecosystems and Human Well-Being: Synthesis.* Washington, DC: Island Press.

Orr, J.C., V.J. Fabry, O. Aumont, L. Bopp, S.C. Doney, R.A. Feely, A. Gnanadesikan, N. Gruber, A. Ishida, F. Joos, R.M. Key, K. Lindsay, E. Maier-Reimer, R. Matear, P. Monfray, A. Mouchet, R.G. Najjar, G.-K. Plattner, K.B. Rodgers, C.L. Sabine, J.L. Sarmiento, R. Schlitzer, R.D. Slater, I.J. Totterdell, M.-F. Weirig, Y. Yamanaka, and A. Yool. 2005. Anthropogenic ocean acidification over the twenty-first century and its impact on calcifying organisms. *Nature* 437:681–686.

Prinn, R.G., H. Jacoby, A.P. Sokolov, C. Wang, X. Xiao, Z. Yang, R.S. Eckhaus, P. Stone, A.D. Ellerman, J.M. Melillo, J. Fitzmaurice, D.W. Kicklighter, G. Holian, and Y. Liu. 1999. Integrated global system model for climate policy assessment: Feedbacks and sensitivity studies. *Climatic Change* 41:469–546.

Rotmans, J., and M. van Asselt. 2002. Integrated assessment: Current practices and challenges for the future. In *Implementing Sustainable Development: Integrated Assessment and Participatory Decision-Making Processes*, edited by H. Abaza and A. Baranzini, 78–116. Cheltenham, UK: Edward Elgar.

Schaeffer, M., B. Eickhout, M. Hoogwijk, B. Strengers, D. van Vuuren, R. Leemans, and T. Opsteegh. 2006. CO_2 and albedo climate impacts of extratropical carbon and biomass plantations. *Global Biogeochemical Cycles* 20, doi: 10.1029/2005GB002581.

Schöler, Y. 2005. Allmendetragik in der Ostsee? Mecklenburg-Vorpommerns Kutter- und Küstenfischer in den Zwängen der EU-Fischereipolitik. In *Küste, Ökologie und Mensch. Integriertes Küstenmanagement als Instrument nachhaltiger Entwicklung*, edited by B. Glaeser, 157–172. Munich: Oekom.

Seitzinger, S.P., J.A. Harrison, E. Dumont, A.H.W. Beusen, and A.F. Bouwman. 2005. Sources and delivery of carbon, nitrogen, and phosphorus to the coastal zone: An overview of global Nutrient Export from Watersheds (NEWS) models and their application. *Global Biogeochemical Cycles* 19:GB4S01, doi: 10.1029/2005GB002606.

Soetaert, K., J.J. Middelburg, C. Heip, P. Meire, S. Van Damme, and T. Maris. 2006. Long-term change in dissolved inorganic nutrients in the heterotrophic Scheldt estuary (Belgium/The Netherlands). *Limnology and Oceanography* 51:409–423.

Sverdrup, H., and A. Barkman. 1994. Critical loads of nitrogen for marine ecosystems:

Suggesting and applying a simple method to the Bothnian Sea. In *Mapping and Modeling of Critical Loads: A Workshop Report*, edited by M. Hornung, M.A. Sutton, and R.B. Wilson, 113–123. Proceedings of the Grange-over-Sands Workshop, 24–26 October 1994, Institute of Terrestrial Ecology, UK.

Van Aardenne, J.A., F.J. Dentener, J.G.J. Olivier, C. Goldewijk, and J. Lelieveld. 2001. A 1 degrees × 1 degrees resolution data set of historical anthropogenic trace gas emissions for the period 1890–1990. *Global Biogeochemical Cycles* 15:909–928.

Van Daalen, C.E., W.A.H. Thissen, and M.M. Berk. 1998. The Delft process: Experiences with a dialogue between policy makers and global modellers. In *Global Change Scenarios of the 21st Century: Results from the IMAGE 2.1 Model*, edited by J. Alcamo, R. Leemans, and G.J.J. Kreileman, 267–285. Paris: Pergamon/Elsevier.

Vörösmarty, C.J., B.M. Fekete, M. Meybeck, and R. Lammers. 2000a. Geomorphometric attributes of the global system of rivers at 30-minute spatial resolution. *Journal of Hydrology* 237:17–39.

Vörösmarty, C.J., B.M. Fekete, M. Meybeck, and R. Lammers. 2000b. Global system of rivers: Its role in organizing continental land mass and defining land-to-ocean linkages. *Global Biogeochemical Cycles* 14 (2):599–622, doi: 10.1029/1999GB900092.

Wuchter, C., B. Abbas, M.J.L. Coolen, L. Herfort, J. Van Bleijswijk, P. Timmers, M. Strous, E. Teira, G.J. Herndl, J.J. Middelburg, S. Schouten, and J.S. Sinninghe Damsté. 2006. Archael nitrification in the ocean. *Proceedings of the National Academy of Science USA* 103:12,317–12,322.

Xiao, X., J.M. Melillo, D.W. Kicklighter, A.D. McGuire, R.G. Prinn, C. Wang, P.H. Stone, and A.P. Sokolov. 1998. Transient climate change and net ecosystem production of the terrestrial biosphere. *Global Biogeochemical Cycles* 12:345–360.

Young, M.N., R. Leemans, R.M.J. Boumans, R. Costanza, B.J.M. de Vries, J. Finnigan, U. Svedin, and M.D. Young. 2007. Future scenarios of human environment systems. In *Sustainability or Collapse? An Integrated History and Future of People on Earth*, edited by R. Costanza, L.J. Graumlich, and W. Steffen, 446–470. Cambridge, MA: MIT Press.

6

Physical Processes in Semi-Enclosed Marine Systems

Wolfgang Fennel, Denis Gilbert, and Jilan Su

Physical Processes and Changes

The most important characteristics of regional, marginal, and semi-enclosed seas are the ways they mediate the transformation of inputs from land and interact with the ocean. The land aspect of this interaction is mainly related to river discharges into the sea, while the interactions with the ocean or neighboring seas involve various water exchange processes between the seas and with the adjacent ocean. In semi-enclosed marine systems (SEMSs) the transformations of river runoff and ocean waters are controlled by circulation and mixing processes, driven externally by atmospheric forces and interaction with the ocean, including tides. The role of river plumes for the transformation of water masses also depends on the intensity and geographical distribution of discharges, for example, from a single large river versus many smaller rivers distributed around the coast of the system.

An overall measure of the intensity of water mass transformations of a regional sea is the typical residence time, which is defined by the ratio of the total water volume to the sum of the river inputs and water exchange with neighboring seas and the ocean. The residence time depends strongly on the widths and depths of the openings to the ocean, ranging from open shelves to narrow connections through straits and sill areas. Also important is the interaction of neighboring seas within regional systems (e.g., Bohai Sea and Yellow Sea, Yellow Sea and East China Sea, Gulf of Mexico and Caribbean Sea) or, within a regional sea, between different subbasins (e.g., in the Baltic Sea) or channels (e.g., in the Gulf of St. Lawrence). However, the residence time of a water parcel does not affect all other constituents, for example, dissolved nutrients. Apart from physical transport and mixing processes, the characteristic time scale of nutrients depends also on the way organic matter cycles through the food web. Hence, attempts to characterize SEMSs should involve the consideration of physical–biological interactions.

One of the characteristic features of water mass transformations is the formation of persistent horizontal and vertical density gradients due to temperature and/or salinity, which are important habitat properties that affect biodiversity along the gradients because of the response of metabolic processes in organisms to temperature and salinity changes. Stratification often reduces the vertical transfer of oxygen to deeper water, leading to hypoxia or even anoxia (see Chapter 11, this volume). On seasonal and shorter time scales, the formation and destruction of the seasonal thermocline* is controlled by solar heating and mixing events through waves, tides, and winter convection. Formation of the seasonal thermocline is important for the onset of spring phytoplankton blooms. In systems with marked flood seasons, enhanced runoffs leave their signature on stratification offshore of river mouths. Mesoscale current patterns, filaments, and eddies modify stratification.

Relevant time and space scales of physical processes are set by driving forces, geometry, and the internal properties of the system, such as the Rossby radius of deformation and phase speeds. The wave phase and wave group speeds are relevant for the propagation of inertial waves, Kelvin waves, and other coastally trapped waves. The response of a coastal sea to the sudden onset of alongshore winds produces inertial waves that accompany the processes of geostrophic adjustment, where a cross-shore pressure gradient drives an alongshore coastal jet. Alongshore-propagating coastally trapped waves set up alongshore pressure gradients and undercurrents. The travel times of these waves set further time scales of the system response, depending on bathymetry, stratification, and extent of the coast along the wind direction.

Interactions of SEMSs with the ocean depend on the nature of their connections. For open shelves, shoreward near-bottom transports during upwelling events are important. In systems with wide and deep connecting channels, such as in the Gulf of St. Lawrence, persistent intrusions of ocean slope waters can be maintained by estuarine circulation. Frontal instabilities of strong slope currents can shed intrusions of surface and subsurface waters as, for example, in the East China Sea or in the Gulf of Mexico. For systems with narrow connections to the ocean, water exchange is weak and often intermittent as, for example, in the major saltwater inflows into the Baltic Sea.

The physical factors shown in Tables 6-1 and 6-2 are found in different combinations in each SEMS. While Table 6-1 gives a rough overview for all SEMSs considered in this book, Table 6-2 summarizes more detailed features of the examples illustrated in the following section.

Freshwater inputs are structured in buoyant plumes and modified by winds, tidal mixing, and tide-induced fronts. The Baltic Sea is an example where many rivers are distributed around the coast, while the East China Sea is mainly affected by the Changjiang (Yangtze) River, which is the third largest in the world. In the Gulf of St. Lawrence, the freshwater discharge of the St. Lawrence River enters the system through a big river mouth and generates the mainly buoyancy-driven Gaspe Current.

Atmospheric forcing (e.g., wind fields) and solar radiation have seasonally varying

*A thermocline is a layer of water where the temperature changes rapidly with depth.

Table 6-1. Comparison of the Relative Strengths of Important Physical Processes of Several SEMSs

SEMS	*Ice*	*Freshwater Inflow*	*Ocean Water Inflow*	*Stratification*
Black Sea	0	+	+	+++
Baltic Sea	++	++	+	+++
Hudson Bay	+++	++	++	++
Gulf of St. Lawrence	++	+++	++	++
Northern Adriatic Sea	0	++	++	++
Gulf of Thailand	0	+++	++	++
North Sea	+	++	++	+
Sea of Okhotsk	++	+	+++	+
East China Sea	0	++	+++	++
Gulf of Mexico	0	++	+++	+
Laptev Sea	+++	+	+++	+
Kara Sea	+++	+	+++	+
Bay of Bengal	0	++	+++	++

Relative strength: 0 = nonexistent, + = weak, ++ = moderate, +++ = strong

strength and also vary considerably across different seas. The Baltic Sea extends from 54°N to 66°N, stretching from subarctic to temperate climatic conditions, and is exposed to highly variable winds and many storm events. The northern parts of the Baltic Sea and most of the Gulf of St. Lawrence are regularly covered by ice during the winter. For the Bohai Sea, Yellow Sea, and East China Sea, atmospheric forcing switches from summer to winter monsoons, with strong northeasterly winds in winter and mild southwesterly winds in summer. In the shallow regions, strong wind mixing often homogenizes the entire water column.

There are virtually no tides in the Baltic Sea, weak tides in the Gulf of Mexico, medium tides in the Gulf of St. Lawrence, and strong tides in the Bohai Sea, Yellow Sea, and East China Sea, but there are very strong tides in the upper St. Lawrence Estuary and parts of both the Yellow Sea and East China Sea (Table 6-2). With regard to tidal dissipation, the Bohai Sea, Yellow Sea, and East China Sea are among the most important shallow seas in the world, where significant residual currents are found.

Example Regions

Baltic Sea

The Baltic Sea is a tideless semi-enclosed sea consisting of several basins, which are connected by several channels, some with sills. The bathymetry of the Baltic is shown in

Table 6-2. Physical Descriptors of Regional Seas Described in Chapter 6

Descriptor	*Gulf of St. Lawrence*	*Gulf of Mexico*	*Baltic Sea*	*Bohai Sea*	*Yellow Sea*	*East China Sea*
Size/Morphology						
Volume, V (1,000 km^3)	34	2,580	21.6	1.39	16.7	398
Sea area, As (Mkm^2)	0.23	1.6	0.412	0.077	0.38	0.77
Catchment area, Ac (Mkm^2)	1.6	3.2	1.745	1.336	0.334	2.044
Catchment/sea area ratio, Ac/As	7	2	4.23	20.8	0.76	2.5
Mean depth (m)	152	1,615	52	18	44	370
Maximum depth (m)	540	3,800	459	83	140	2,719
% shelf area (depth < 200 m)	67	32	0.3	100	100	~67
Sill depth (m)	450	2,000	18			
Tidal range (m)	0.5–6	0.1–0.5	0.0	0.8–3.2	0.5–8.1	1.3–6.1
Water Balance						
Total annual runoff (km^3)	600	950	540	75	56	1,200
Water balance (+ vs. –)	+		+	–		–
River water residence time (y)	0.5–1		15			
Precipitation minus evaporation (km^3)	120		60			

Temperature/Salinity/O_2 Regime						
Sea surface temperature, T (°C)	-2–22	12–28	0–23	-1–27	1–27	9–29
Sea surface seasonal ΔT (°C)	6–22		20–23	21–27	16–24	5–20
% ice coverage during winter	30–100	0	20–60	11–65	0–1	0
Latitudinal sea surface ΔT (°C)	15		3–10	3–4	7–12	3–18
Sea surface salinity	25–32	10–36.5	6–10	26–31	30–34	12–34.75
Stratification types	Mainly salinity stratified	Mainly temperature stratified	Mainly salinity stratified	Mainly temperature stratified	Mainly temperature stratified	Mainly temperature stratified
Bottom water residence time (y)	3–6		30	1–2	4–5	1–2 (shelf)
Oxygenation saturation (%)	15–120	0–120	0–120			1–120

Figures for the Bohai Sea, Yellow Sea, and East China Sea are from Su and Yuan 2005 and Wang et al. 2007.

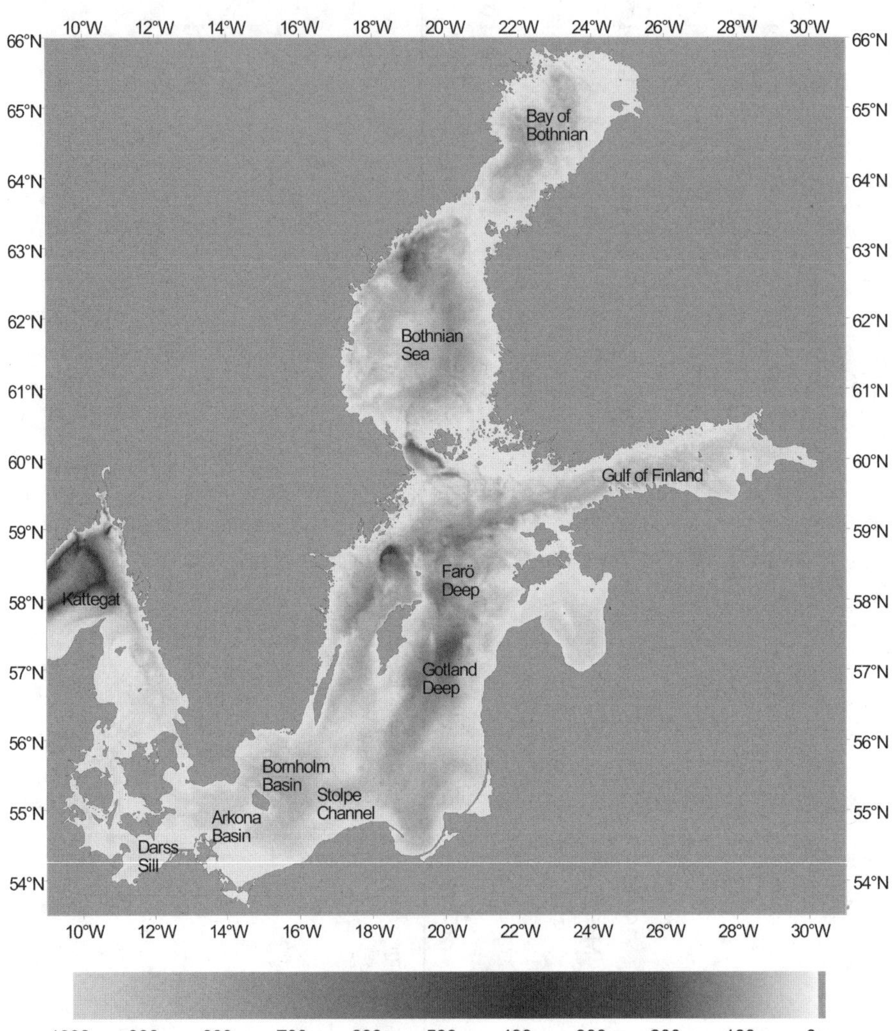

Figure 6-1. Topographic map of the Baltic Sea.

Figure 6-1. The Baltic has only narrow connections to the ocean through the Danish straits, and water exchange with the North Sea is hindered by the shallow Darss sill (maximum depth 18 m). General overviews of the oceanography of the Baltic Sea are given, for example, in Omstedt and colleagues (2004) and Kullenberg (1981).

As a result of the substantial freshwater surplus and the restricted water exchange, the Baltic is the largest brackish sea in the world, with a water volume of about 21,000 km^3. The annual total contribution of all rivers is about 540 km^3, while the excess of precipitation over evaporation is 60 km^3 per year. In the eastern and northern parts of the Baltic

Sea, the surface salinity is very low, but it increases toward the Danish straits. A strong halocline* separates the brackish surface water from the saltier bottom water.

The halocline is associated with strong vertical stability and thus a weak vertical exchange through the halocline. As a consequence, hypoxia or even anoxia occurs frequently in the deep waters below the halocline. The general structures of the salinity and oxygen distributions are shown in Plate 1. From the conservation of mass, it can be estimated that the annual outflow of brackish water is about 1,400 km^3, while the near-bottom inflow of saline water is about 830 km^3 per year. Based on the estimates of total water volume, inputs, and outputs given above, the resulting time scales, which are described as flushing time or residence time, vary between 15 y and 35 y.

The salt balance in the Baltic Sea is maintained by a dynamic equilibrium between near-bottom inflow of saline water, outflow of brackish surface water, and a weak vertical salt diffusion through the halocline. The deeper water can only be ventilated through horizontal propagation of dense saline bottom water, which is formed in the transition areas between the North Sea and the Baltic and cascades over the sills until it arrives in the central basin. Major inflows of dense saline deep water near the bottom occur when a substantial water volume passes the Darss sill and advances through the Arkona Basin and into the Bornholm Basin, where the dense water must first fill the basin to the depth level of the Stolpe Channel before it can propagate further into the central Baltic to renew the bottom water. Since major inflows are initiated by strong winds, which imply strong vertical mixing, a substantial part of the saline water will be mixed with brackish surface water in the Arkona Basin and leave the Baltic (Matthäus and Lass 1995). The time scale of the propagation of inflowing dense bottom water toward The Gotland Basin ranges from a couple of weeks to several months.

The vertical temperature distribution shows a strong seasonal cycle, with the spring and summer formation of a warm upper layer 10 to 20 m thick and a strong thermocline below. The temperature distribution along the deepest part of the Baltic in summer is also shown in Plate 1. During winter the temperatures are uniform in the top 60 m, near 4°C. Irregular weather patterns and frequent storms, which are typical for the latitudes from 50°N to 60°N, generate a variety of mesoscale currents, such as eddies, coastal jets, and associated up- and downwelling patterns.

The permanent halocline and seasonal thermocline are relevant for shaping the ecosystem of the Baltic Sea. The horizontal and vertical salinity gradients set the long-term conditions for the marine ecosystem. Since salinity gradients affect osmoregulation of living cells, the number of species in the Baltic is relatively small. Changes in temperature on seasonal to interannual time scales affect fish metabolism, growth, and fecundity.

Bohai Sea, Yellow Sea, and East China Sea

The Bohai Sea, the Yellow Sea, and the shelf of the East China Sea (ECS) represent a broad epicontinental sea; the bathymetry is shown in Figure 6-2A. The Bohai and Yellow

*A halocline is a layer of water where the salinity changes rapidly with depth.

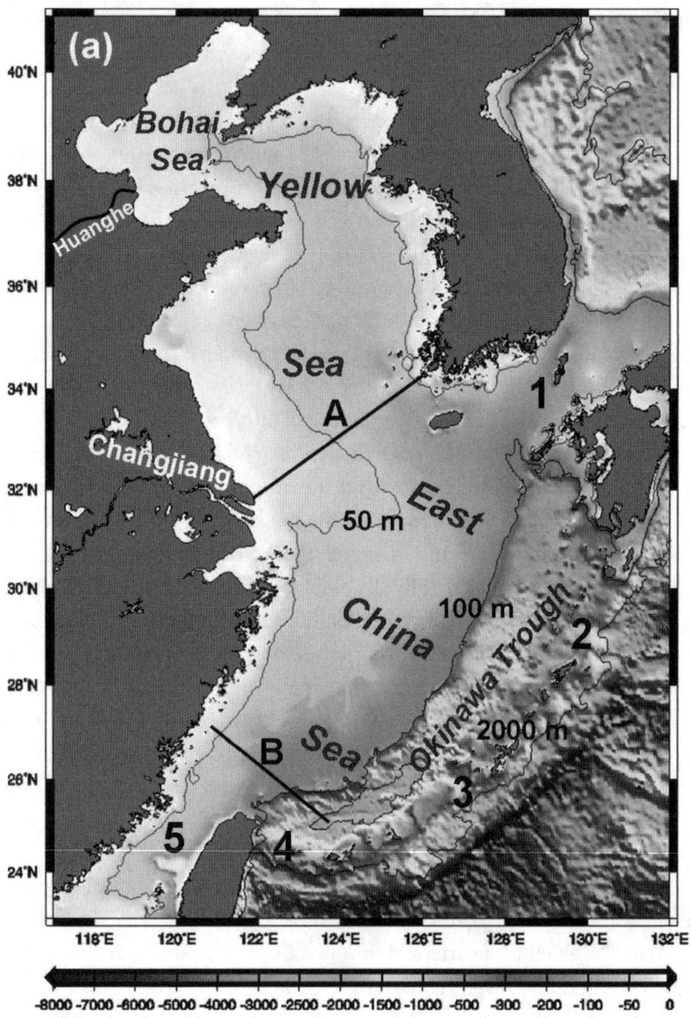

Figure 6-2. *Panel A:* Bathymetry of the Bohai and the Yellow and East China seas. 1—Korea/Tsushima Strait; 2—Tokara Strait; 3—Okinawa Valley; 4—East Taiwan Channel; 5—Taiwan Strait. The transects A and B are indicated for later reference. Base map courtesy of the Continental Margins Task Team, http://cmtt.pangaea.de/2_03_East_China_Sea.pdf.

seas (BYS) form a shallow gulf opening to the ECS. A trough runs through the BYS and leads down to the northern end of the Okinawa Trough in the ECS. The Okinawa Trough, with a maximum depth of about 2,300 m, borders the Pacific Ocean through the Ryukyu Archipelago. There are three deep depressions along the Ryukyu Ridge, namely, the Tokara Strait (with a maximum depth less than 1,000 m), the Okinawa Val-

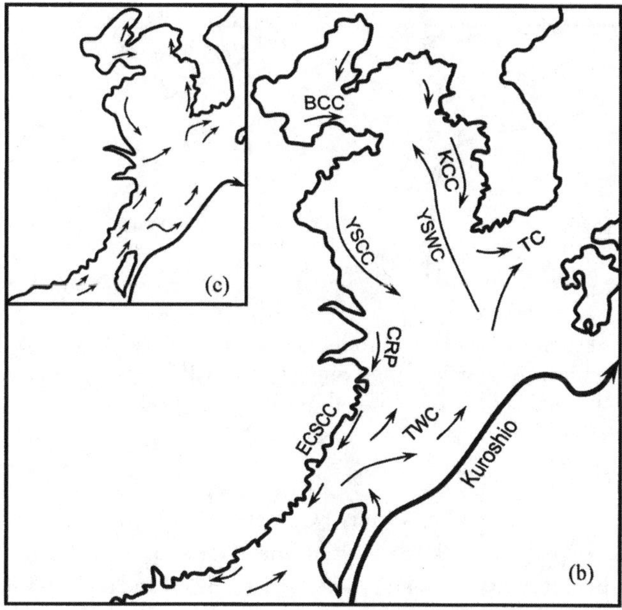

Figure 6-2. *Panels B and C:* Sketches of the circulation in winter (B) and summer (C) (after Su 1998). BCC—Bohai Coastal Current; YSCC—Yellow Sea Coastal Current; KCC—Korean Coastal Current; YSWC—Yellow Sea Warm Current; TC— Tsushima Current; CRP—Changjiang River plume; ECSCC—East China Sea Coastal Current; TWC—Taiwan Warm Current. Reproduced with permission from Harvard University Press.

ley (over 1,000 m deep), and the East Taiwan Channel (about 700 m deep). The ECS is also connected to the Japan/East Sea to the northeast through the Korea/Tsushima Strait and to the South China Sea to the south through the Taiwan Strait.

Circulation in the BYS is mainly driven by monsoonal winds, whereas for the circulation in the ECS, the monsoonal winds and the Kuroshio (a strong western boundary current in the North Pacific) are equally important (Figure 6-2). In addition, river discharges (in particular the Changjiang River) and tides also affect the circulation in these seas. The flushing of the BYS is mainly driven by winds, while the water renewal on the ECS shelf is also controlled by Kuroshio intrusions. For the renewal of the bottom layer water in the Okinawa Trough in the ECS, intrusion of the Ryukyu Current water through the Okinawa Valley is also important (Yu et al. 1994). The residence times for the different basins range from one to five years (see Table 6-2).

The northward current through the Taiwan Strait brings high-salinity water from the South China Sea into the ECS (Figure 6-2). The net transport through the Taiwan Strait reaches 3 sverdrups (Sv; 1 Sv = 10^6 m^3 s^{-1}) in summer but is small in winter because of the strong northerly winds (Su and Wang 1987; Fang and Zhao 1988; Su 1998). The Tsu-

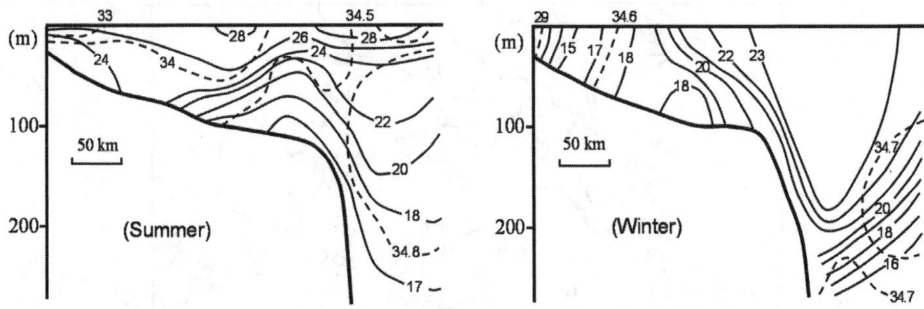

Figure 6-3. Shelf intrusion of the Kuroshio across transect B (see Figure 6-2) northeast of Taiwan in summer (August 1984) and winter (January 1985), as indicated by isotherms (solid) and isohalines (dashed). After Su and Pan 1987; reproduced with permission from *Acta Oceanologica Sinica.*

shima Current (TC) exports low-salinity ECS water to the Japan/East Sea. The average volume transport of the TC is 1.3 Sv, with an annual variation of 2.0 Sv and higher levels in summer. These currents are important for water renewal and replenishment of nutrients in the ECS.

Intrusions of the Kuroshio in the ECS occur mainly over the shelf northeast of Taiwan, when the Kuroshio enters the ECS from the eastern side of the Taiwan Island (Su and Pan 1987; Liang and Su 1994; Su 1998). The intrusions have seasonal characteristics; during winter, both the surface and subsurface Kuroshio waters are involved, while summer intrusions consist of only subsurface Kuroshio water (Figure 6-3). Intrusions of Kuroshio subsurface water can be traced to the areas offshore of the Changjiang River mouth. Another mode of Kuroshio intrusions is mediated by frontal eddies, which are often detected along the shelf breaks in the ECS.

The northward current through the Taiwan Strait, direct intrusion of the Kuroshio northeast of Taiwan, and the frontal eddies along the Kuroshio front all contribute to the transformation of the shelf water in the ECS. As for the Okinawa Trough in the ECS, renewal of its upper-layer water is carried out directly by the Kuroshio, whereas inflow of Pacific water through the Okinawa Valley renews the lower-layer water of the Okinawa Trough (Yu et al. 1994).

In winter, the strong northerly monsoon results in well-mixed or weakly stratified water in areas with water depth shallower than about 70 m in the BYS and the northern ECS. The winds drive southward-flowing coastal jets along both sides of the BYS. There is a weak compensating mean northward subsurface current, the Yellow Sea Warm Current (YSWC) (e.g., see Su 1998) (Figure 6-4). This winter circulation is important for water exchange in the BYS.

The Changjiang River discharges directly into the middle of the ECS. In winter, the freshwater plume flows alongshore in a southward direction and merges with discharges

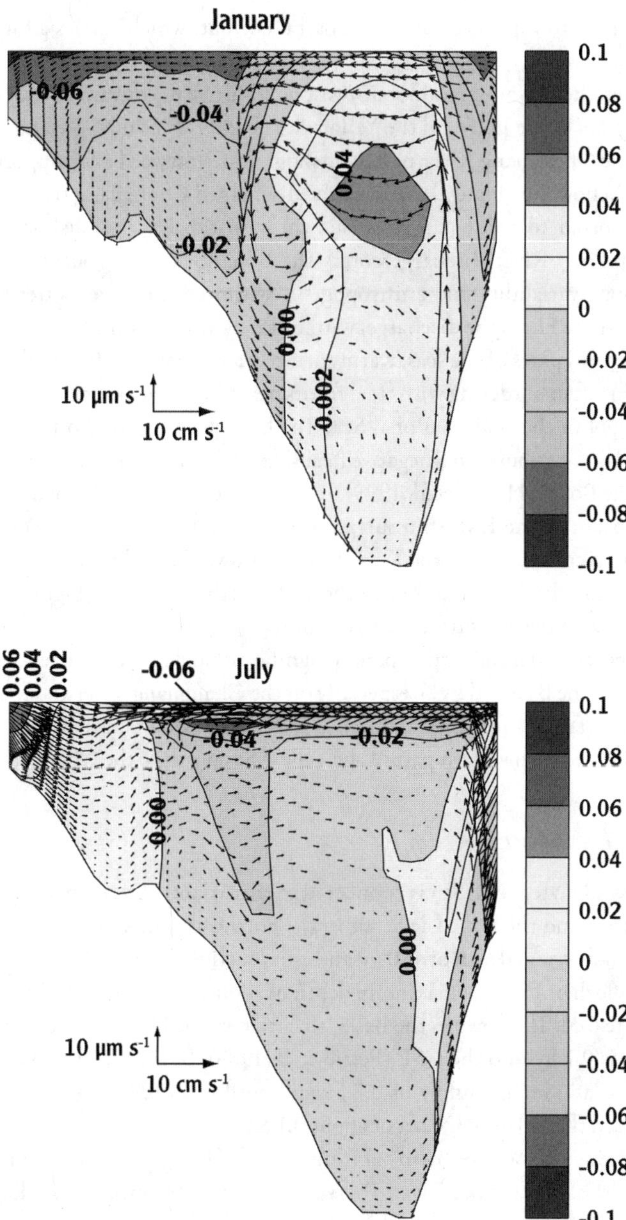

Figure 6-4. Exchange between the Yellow Sea and East China Sea. Plotted are contours (vectors) of velocity that are normal (tangent) to transect A (see Figure 6-2). Positive normal velocity is northward, that is, into the page. Reprinted from Naimie et al. 2001, copyright 2001, with permission from Elsevier.

from other rivers to form the Zhe-Min coastal current, which carries a high sediment load (Plate 2).

In summer, the prevailing mild southwest winds drive northward coastal currents on the western sides of the ECS and the Yellow Sea (Figures 6-2 and 6-3). The wind-driven flows push the Changjiang River plume offshore northeastward and support its spreading into the Yellow Sea (Plate 2). Most of the discharged Changjiang freshwater is eventually transported to the Japan/East Sea through the Korea/Tsushima Strait. The Huanghe (Yellow) River discharge enters the Bohai Sea at its southern shore. Both decadal climate variability and anthropogenic withdrawal of river water have significantly reduced the Huanghe discharge, with unknown effects on the coastal currents.

Over extended parts of the seas, warm surface waters overlay winter cold-water pools at the bottom, often bordering with tidal fronts next to the coasts. There is also a summer cold-water pool in the shallow Bohai Sea. The large pool of cold bottom water in the Yellow Sea also contributes to a negative heat flux to the Bohai Sea through tidal dispersion across the Bohai (Huang et al. 1999). In summer the warm ECS shelf water inhibits onshelf intrusion of the Kuroshio surface water, except where frontal eddies develop. However, interaction between the Kuroshio and Taiwan Island results in the upwelling of the Kuroshio subsurface water onto the shelf north of Taiwan (Figure 6-3), as indicated by patches of low sea surface temperature.

With respect to tidal energy dissipation, significant tidal residual currents are found in shallow areas of the BYS and ECS, especially off the Changjiang River and the Hangzhou Bay south of it (famous for its tidal bore), as well as in the bays near Korea and over the tidal ridges in the southwestern part of the Yellow Sea north of the Changjiang River.

Gulf of St. Lawrence

The Gulf of St. Lawrence (GSL) is connected to the Atlantic Ocean by the Cabot Strait in the southeast and the Belle Isle Strait in the northeast. The Cabot Strait has a width of 104 km, a maximum depth of 480 m, and cross-sectional area of 35 km^2; the Belle Isle Strait has a width of 15 km, a maximum depth of about 60 m, and cross-sectional area of 1 km^2 (Figure 6-5). The flows in the Belle Isle Strait can be bidirectional; the net annual average flow is 0.1 Sv into the GSL (Petrie et al. 1988). In the wider and deeper Cabot Strait, there is an average inflow of 0.7 Sv and outflow of 0.8 Sv, giving a net average outflow toward the Atlantic Ocean of about 0.1 Sv.

The GSL receives about 600 km^3 of freshwater discharge per year, with roughly 70% coming from the St. Lawrence River. This single source of freshwater is so dominant that its plume can be traced on the Magdalen Shallows all the way to Cabot Strait, where it exits the GSL. The annual freshwater input by precipitation minus evaporation is about 120 km^3 over the GSL (Table 6-2; Koutitonsky and Bugden 1991).

Although located farther south (from 45.5°N to 51.5°N) than the Baltic Sea, the GSL is characterized by cold winters, which cause ice formation over most of its surface. The

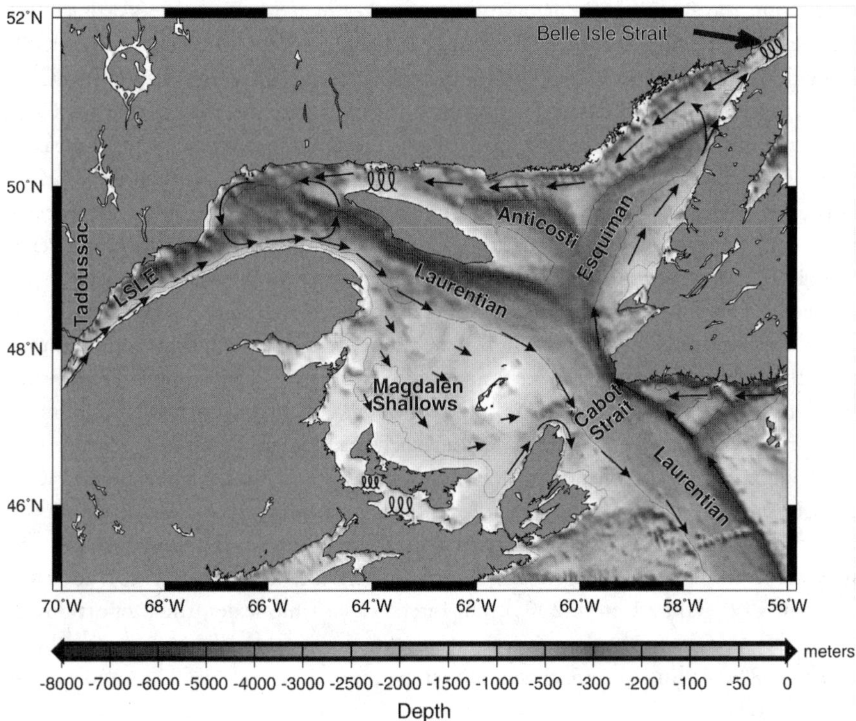

Figure 6-5. Schematics of the surface circulation in the Gulf of St. Lawrence. The curly symbols indicate areas with strong tidal mixing. The names of the deep channels (Laurentian, Anticosti, Esquiman) are indicated, together with key geographical names; LSLE—Lower St. Lawrence Estuary. Base map courtesy of the Continental Margins Task Team, http://cmtt.pangaea.de/4_09_Gulf_of_St_Lawrence.pdf.

ice extent has a peak in March (Saucier et al. 2003). The presence of ice during winter dampens surface waves and the associated vertical mixing. This factor, combined with the fact that brine rejection makes a small direct contribution to the winter erosion of the halocline (Saucier et al. 2003), leads to a relatively shallow (generally less than 100 m thick) surface mixed layer during winter (Galbraith 2006). Intense stratification occurs in late summer, followed by significant mixing in the fall caused by episodic storms.

The annual cycle of surface temperature has its largest amplitude over the Magdalen Shallows (-2°C in winter; 20°C in summer). The other extreme is found in the Lower St. Lawrence Estuary (LSLE), near Tadoussac, where very intense vertical mixing and upwelling caused by strong tidal currents over an abrupt change in topography reduce the annual cycle of surface temperatures to a range between -1°C in winter to about 6°C in summer. In the northeast gulf, where the influence of inflowing salty ($S > 32.35$)

water from the Belle Isle Strait is strongest, winter mixed-layer thickness sometimes exceeds 100 m (Galbraith 2006). The volume of Labrador Shelf waters entering the GSL through the Belle Isle Strait during winter months varies greatly from year to year. This factor, together with interannual variations in winter surface heat fluxes over the GSL, drives the interannual variability of the volume, thickness, and temperature of the cold intermediate layer (CIL). The CIL results from the formation of a seasonal thermocline in the top 30 m of the water column during the spring, which effectively caps the remnants of the near-freezing winter mixed layer until the following fall (Gilbert and Pettigrew 1997). Observed summer temperature profiles are shown in Plate 3, Panel A.

Between 125 m and 250 m depth, temperature increases with depth (Plate 3), which would lead to unstable stratification if it were not for the stable salinity gradient that more than compensates and generates a stable stratification. The maximum temperature around 250 m depth is characteristic of the continental slope water mass that is found to the south of the Newfoundland and Scotian shelves and north of the Gulf Stream. This slope water mass enters the mouth of the Laurentian Channel and is a mixture of Labrador Current Water (LCW) and North Atlantic Central Water (NACW). As part of the estuarine circulation of the GSL, waters in the 200 to 300 m depth range take three to six years to travel landward from the mouth of the Laurentian Channel to Tadoussac (Bugden 1991; Gilbert et al. 2005). On interdecadal time scales, the proportions of LCW and NACW in the deep waters entering the Laurentian Channel can vary by as much as 18%, leading to large changes in temperature (2°C), salinity (0.3), and oxygen (40 μmol L^{-1}).

Gulf of Mexico

The Gulf of Mexico (GOM) bears closest resemblance to the East China Sea (ECS), since both the GOM and ECS are influenced by freshwater input from a large river (Mississippi and Changjiang, respectively) and by the intrusion of a major ocean current carrying warm and salty water (Loop Current and Kuroshio, respectively). The GOM is the largest of the six SEMSs considered in this chapter, in terms of both its surface area and its volume (Table 6-2), and it is characterized by large-volume transports at its two open boundaries (Figure 6-6). In the Yucatan Strait, the average inflow to the GOM is 23–25 Sv (Sheinbaum et al. 2002). The flow through the Yucatan Channel is the source of the Loop Current that extends 500 to 700 km northward into the GOM before making a loop, as its name implies, and then heading southeast toward the Florida Straits (Figure 6-6). The water that flows eastward through the Florida Straits (sill depth 800 m) is regarded as the beginning of the Gulf Stream, whose name denotes its Gulf of Mexico origin.

For reasons that are still not fully understood and hard to predict (Oey 2004), every three to seventeen months, the Loop Current sheds an anticyclonic (clockwise) warm-core ring. Figure 6-6 shows such a ring about to be shed just south of the Mississippi River delta. Loop Current rings have diameters between 200 km and 300 km, vertical

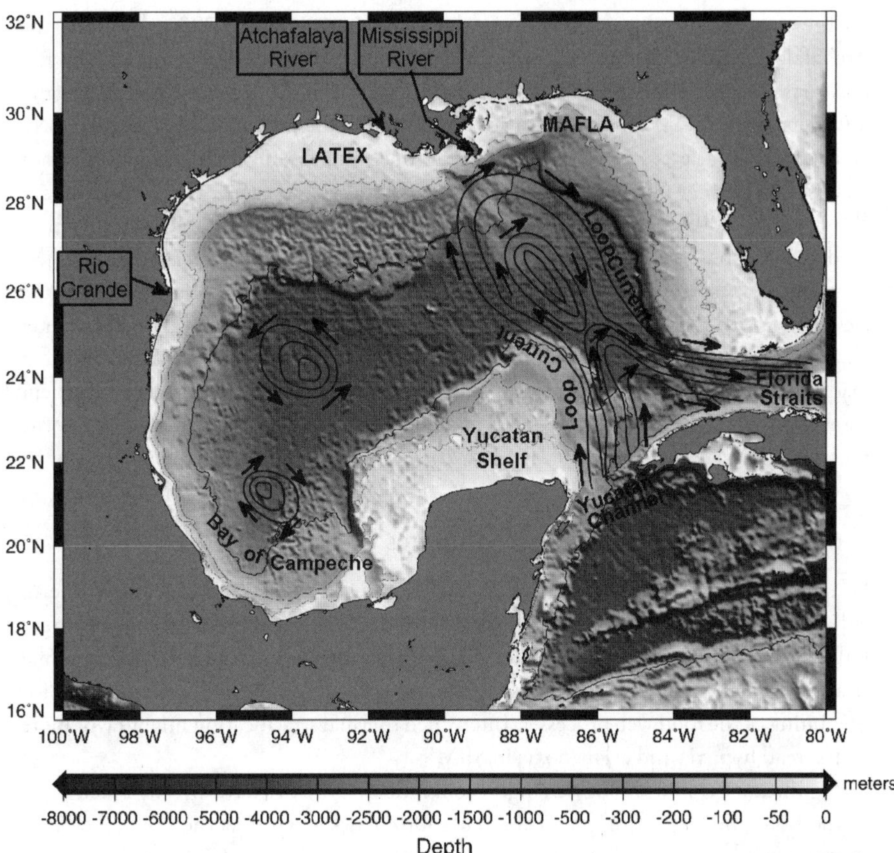

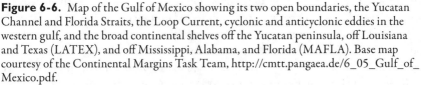

Figure 6-6. Map of the Gulf of Mexico showing its two open boundaries, the Yucatan Channel and Florida Straits, the Loop Current, cyclonic and anticyclonic eddies in the western gulf, and the broad continental shelves off the Yucatan peninsula, off Louisiana and Texas (LATEX), and off Mississippi, Alabama, and Florida (MAFLA). Base map courtesy of the Continental Margins Task Team, http://cmtt.pangaea.de/6_05_Gulf_of_Mexico.pdf.

extent between 800 m and 1,000 m, and volumes between 25,000 km^3 and 70,000 km^3. By all accounts, these rings are enormous, and their quick swirl speeds of 3 to 4 knots (1.5–2 m s^{-1}) represent real hazards to deep oil platform operations. Loop Current rings generally travel westward at about 2–5 km d^{-1} and have lifetimes of several months to approximately a year. The GOM is densely populated with both clockwise and anticlockwise eddies (Figure 6-6). When these rings impinge on the upper continental slope, close to the shelf break, they mediate exchanges of water parcels between the shelf and the deep ocean.

The combined freshwater flow from the Mississippi River (13,700 $m^3 s^{-1}$) and the Atchafalaya River (5,800 $m^3 s^{-1}$) accounts for about two-thirds of the total freshwater flow to the GOM (30,000 $m^3 s^{-1}$) and plays a crucial role in the delivery of sediments and nutrients to the GOM. The spreading of freshwater plumes of the Mississippi and Atchafalaya rivers is affected by winds. During summer months, upwelling-favorable winds push the freshwater plumes offshore while the coastal jets transport the water eastward (Morey et al. 2003). At other times of the year, downwelling-favorable easterly winds predominate, pushing the freshwater plumes onshore while the associated coastal jets drive the plumes westward, toward Texas. On an annual basis, approximately 53% of the freshwater from the Mississippi River is delivered to the Louisiana–Texas Shelf (Etter et al. 2004; Plate 4), where it affects the stratification and delivers substantial nutrient loads, which lead to coastal hypoxia (Rabalais et al. 2002). Freshwater and nutrient transports from the Mississippi and Atchafalaya rivers are important factors for the northern GOM ecosystem (Rabalais et al. 2002; Hetland and DiMarco 2007).

Long-Term Anthropogenic Changes

Long-term changes in response to anticipated increase of ocean temperature due to enhanced greenhouse gas concentrations imply reduction of ice coverage in the northern Baltic Sea and in the Gulf of St. Lawrence. This reduction will cause stronger thermal stratification and earlier onset of phytoplankton spring blooms, with consequences for the timing of the food web processes. Intensified thermal stratification might favor more widespread hypoxia and even anoxia in SEMSs.

Significant geomorphology changes from human activities have already taken place in the Bohai Sea, affecting physical circulation. It is known that the Huanghe River shifts its course on a time scale of the order of ten years within the Huanghe River delta, because of the rather gentle slope of the delta. However, in connection with oil exploration activities, the river course was artificially forced into a fixed position since the early 1970s, resulting in an enormous sand spit protruding across almost one-third of the width of the Laizhou Bay mouth. Such human intervention has significantly altered circulation in the bay. This alteration, in addition to overfishing and damming of other small rivers, is believed to be one of the contributing factors to the failure of the important shrimp (*Penaeus chinensis*) fishery.

Throughout the twentieth century, the U.S. Army Corps of Engineers has increasingly channelized the Mississippi River, built dikes, and stabilized the diversion of a fraction of its outflow to the Atchafalaya River, with consequences for circulation patterns and water residence times on the Louisiana–Texas Shelf. Similarly, the Chinese government is constructing engineering works that will divert about 5% of the Changjiang River discharge to the north when completed around 2050. In the drainage basin of the Gulf of St. Lawrence, major dams for generation of hydroelectric power were built along the Saguenay, Manicouagan, St. Lawrence, and other rivers during the twentieth century. Since the major electricity demand occurs in winter, this has led to increased freshwater inflow during winter months and a reduced spring freshet; the consequences on ecosystem functioning are not well understood. The freshwater and sediment budgets of SEMSs are explored in greater detail in the next chapter.

References

Bugden, G.L. 1991. Changes in the temperature–salinity characteristics of the deeper waters of the Gulf of St. Lawrence over the past several decades. In *The Gulf of St. Lawrence: Small Ocean or Big Estuary?*, edited by J.-C. Therriault, 139–147. *Canadian Special Publication of Fisheries and Aquatic Sciences* 113. Ottawa: National Research Council of Canada.

Etter, P.C., M.K. Howard, and J.D. Cochrane. 2004. Heat and freshwater budgets of the Texas–Louisiana Shelf. *Journal of Geophysical Research* 109:C02024, doi: 10.1029/2003JC001820.

Fang, G.H., and B.R. Zhao. 1988. A note on the main forcing of the northeastward flowing current off the southeast China coast. *Progress in Oceanography* 21:363–372.

Galbraith, P.S. 2006. Winter water masses in the Gulf of St. Lawrence. *Journal of Geophysical Research* 111:C06022.

Gilbert, D., and B. Pettigrew. 1997. Interannual variability (1948–1994) of the CIL core temperature in the Gulf of St. Lawrence. *Canadian Journal of Fisheries and Aquatic Sciences* 54 (Suppl. 1):57–67.

Gilbert, D., B. Sundby, C. Gobeil, A. Mucci, and G.-H. Tremblay. 2005. A seventy-two-year record of diminishing deep-water oxygen in the St. Lawrence Estuary: The northwest Atlantic connection. *Limnology and Oceanography* 50 (5):1,654–1,666.

Hetland, R.D., and S.F. DiMarco. 2007. How does the character of oxygen demand control the structure of hypoxia on the Texas–Louisiana continental shelf? *Journal of Marine Systems* 70:49–61, doi: 10.1016/j.jmarsys.2007.03.002.

Huang, D.J., J.L. Su, and J.O. Backhaus. 1999. Modelling of the seasonal thermal stratification and baroclinic circulation in the Bohai Sea. *Continental Shelf Research* 19:1,485–1,505.

Koutitonsky, V.G., and G.L. Bugden. 1991. The physical oceanography of the Gulf of St. Lawrence: A review with emphasis on the synoptic variability of the motion. In *The Gulf of St. Lawrence: Small Ocean or Big Estuary?*, edited by J.-C. Therriault, 57–90. *Canadian Special Publication of Fisheries and Aquatic Sciences* 113. Ottawa: National Research Council of Canada.

Kullenberg, G. 1981. Physical oceanography. In *The Baltic Sea*, edited by A.Voipio, 135–181. Paris: Elsevier.

Liang, X., and J. Su. 1994. A two-layer model for the summer circulation of the East China Sea. *Acta Oceanologica Sinica* 13:325–344.

Matthäus,W., and H.U. Lass. 1995. The recent inflow into the Baltic Sea. *Journal of Physical Oceanography* 25:280–286.

Morey, S.L., P.J. Martin, J.J. O'Brien, A.A. Wallcraft, and J. Zavala-Hidalgo. 2003. Export pathways for river discharged fresh water in the northern Gulf of Mexico. *Journal of Geophysical Research* 108:C10, 3303, doi: 1029/2002JC001674.

Naimie, C.E., C.A. Blain, and D.R. Lynch. 2001. Seasonal mean circulation in the Yellow Sea: A model-generated climatology. *Continental Shelf Research* 21:667–695.

Oey, L.-Y. 2004. Vorticity flux through the Yucatan Channel and Loop Current variability in the Gulf of Mexico. *Journal of Geophysical Research* 109:C10004, doi: 10.1029/2004JC002400.

Omstedt, A., J. Elken, A. Lehmann, and J. Piechura. 2004. Knowledge of the Baltic Sea physics gained during the BALTEX and related programmes. *Progress in Oceanography* 63:1–28.

Petrie, B., B. Toulany, and C.J.R. Garrett. 1988. The transport of water, heat and salt through the Strait of Belle Isle. *Atmosphere–Ocean* 26:234–251.

Rabalais, N.N., R.E. Turner, Q. Dortch, D. Justic, V.J. Bierman, and W.J. Wiseman. 2002. Nutrient-enhanced productivity in the northern Gulf of Mexico: Past, present and future. *Hydrobiologia* 475/476:39–63.

Saucier, F.J., F. Roy, D. Gilbert, P. Pellerin, and H. Ritchie. 2003. Modeling the formation and circulation processes of water masses and sea ice in the Gulf of St. Lawrence, Canada. *Journal of Geophysical Research* 108 (C8):3,269, doi: 10.1029/2000JC000686, 2003.

Sheinbaum, J., J. Candela, A. Badan, and J. Ochoa. 2002. Flow structure and transport in Yucatan Channel. *Geophysical Research Letters* 29 (3), doi: 10.1029/2001GL013990, 2002.

Su, J. 1998. Circulation dynamics of the China Seas: North of 18°N. In Vol. 11 of *The Sea*, edited by A.R. Robinson and K. Brink, 483–506. New York: John Wiley & Sons.

Su, J., and Y. Pan. 1987. On the shelf circulation north of Taiwan. *Acta Oceanologica Sinica* 6 (Suppl. 1):1–20.

Su, J., and W. Wang. 1987. On the sources of the Taiwan Warm Current from the South China Sea. *Chinese Journal of Oceanology and Limnology* 5:299–308.

Su, J., and Y. Yuan (eds.). 2005. *Hydrography of the China Seas.* In Chinese. Beijing: Ocean Press.

Wang, Y., G. Fu, and Y. Zhang. 2007. River–sea interactive sedimentation and plain morphological evolution. In Chinese. *Quaternary Sciences* 27:674–689.

Yu, H., J. Su, Y. Miao, and B. Li. 1994. Low salinity water center of Kuroshio in the East China Sea and intrusion of western boundary current east of Ryukyu Islands. In *Proceedings of the Symposium of the China–Japan Joint Research Program on the Kuroshio*, 145–164. Beijing: Ocean Press.

7

Cascading Filters of River Material from Headwaters to Regional Seas: The European Example

Michel Meybeck and Hans H. Dürr

Introduction

River inputs to the ocean have been an important concern of coastal oceanographers and global-scale geochemists, physical geographers, and hydrologists for more than thirty years (Garrels and Mackenzie 1971). In the early literature, only global river fluxes of sediments, major ions, nutrients, and organic carbon could be estimated (Fournier 1960; Livingstone 1963; Alekin and Brazhnikova 1968; Meybeck 1979, 1982; Van Bennekom and Salomons 1981). Regional inputs could not be established for individual oceans and/or regional seas, except for Western Europe, the former Soviet Union, and the conterminous United States, for lack of appropriate databases, GIS techniques, and global-scale models.

Global-scale models of river transfers at the 30 × 30 min resolution or finer are now developing quickly (Beusen et al. 2005; Seitzinger et al. 2005). These models benefit from the development of databases (Meybeck and Ragu 1995; Milliman et al. 1995; UN GEMS/Water Programme 2008) and digitized river networks and from data on river runoff, global climate, lithology, relief, and human impacts (from agriculture, urbanization, and damming) (Vörösmarty et al. 2000a, b; Meybeck et al. 2001; Fekete et al. 2002; Vörösmarty et al. 2003; Dürr et al. 2005).

Recent specific studies of river basins—now completed on all continents—also show that material lost from the continents and carried by rivers in dissolved and particulate forms is exposed to a cascading and nested set of continental filters (Meybeck and Vörösmarty 2005). Other types of filters have been defined at the land–ocean interface, in the estuarine zone, where fresh and ocean waters mix, and on the continental shelf (Mantoura et al. 1991; Crossland et al. 2005). Finally, regional seas can be considered as mega-filters of continental material fluxes at the global scale (Meybeck et al. 2007a). The net inputs of nutrients, carbon, and pollutants to the ocean needed for regional and global biogeochemical models are therefore very different from the riverine fluxes determined on land from observations made at the last gauging stations in any specific watershed.

This chapter puts into perspective the multiple filters of land-derived material carried by rivers, from headwater filters to regional seas. A simple typology of estuarine filters is developed based on their capacity to trap river particulates, their capacity to process river nutrients and organic matter, and their location with regard to the coastline and the shelf break. As an example, the distribution of estuarine filters is then established for European semi-enclosed marine systems (SEMSs). Human influences on fluvial systems, both direct and through global change, are considered.

The inputs of sediments from the land to SEMSs, and the processing of material within them, described in these papers as mega-filters, help determine the turbidity, chemical reactions, and habitability of SEMSs for marine organisms.

Cascading Filters of Continental Material within River Catchments and at the Land–Ocean Interface

River-borne material originates from natural sources through mechanical erosion, atmospheric fixation, soil leaching, and chemical weathering. Any fluvial system—river network plus groundwaters, wetlands, surface waters, and estuaries—can be decomposed into a succession of transfers and filters (Figure 7-1). Filters alter and regulate dissolved and particulate matter fluxes. For example, particulates can be trapped in coastal sediments. Carbon- and nitrogen-containing products can be transformed into greenhouse gases such as CO_2, CH_4, and N_2O, which escape to the atmosphere. Carbon and nutrients may be stored in soils and in terrestrial vegetation (Figure 7-1, SV). Slopes (SL), wetlands (WL), floodplains (FLP), and lakes (LK) act as natural sinks and/or biogeochemical reactors. As a result, river material reaching the river mouth (Figure 7-1, Station S3) may be very different quantitatively and qualitatively from the sum of fluxes measured or modeled from headwaters to the coast.

The retention rate within catchments exceeding 1,000 km^2 is commonly between 50% and 90% for river particulates (Walling 1983). River basins also store a large proportion of particulate organic carbon (Stallard 1998), while a significant portion of nitrate is denitrified in most river basins (Howarth et al. 1996). Lakes may store up to 99% of river detrital material and more than 50% of dissolved silica when a lake's residence time is greater than one year (Meybeck and Ragu 1995). There is also growing evidence that part of the dissolved organic carbon leached from headwater soils and peatland is eventually respired in rivers (Abril et al. 2002). This natural state is greatly impacted by human activities that modify natural river fluxes, accelerating both sources and sinks (Vörösmarty and Meybeck 2004; Meybeck and Vörösmarty 2005). This point is developed further below.

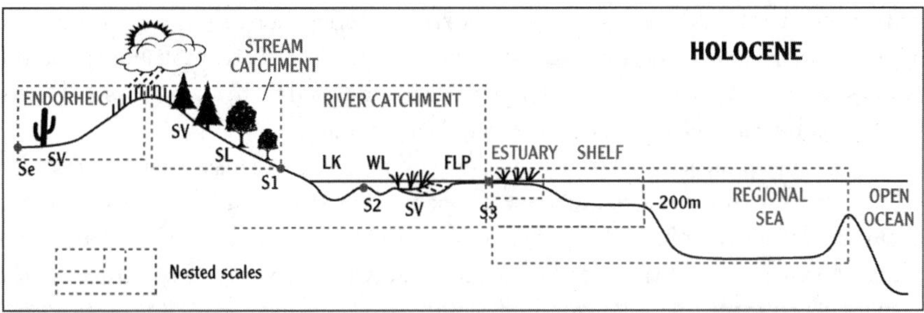

Figure 7-1. Nested filters of river material to oceans in natural Holocene conditions. Headwater filters: processing and storage of waterborne material in soil and vegetation (SV) and on slopes (SL). River catchment filters: lakes (LK), wetlands (WL), and floodplains (FLP). Positions of budget stations: Se, S1, S2, and S3.

Once entering the estuarine filter, river material is again processed, stored, and transferred to the continental shelf through various physical, chemical, and biogeochemical processes, which greatly depend on the type of estuary (Mantoura et al. 1991; Perillo 1995; Perillo et al. 1999; Chen et al. 2003; Syvitski et al. 2005). So far the impact of estuarine filters has not been addressed adequately at regional or global scales, for lack of detailed description and distribution of estuarine filter types for each category of river material (e.g., mud, silt, sand, organic carbon, dissolved nutrients, dissolved and particulate nutrients). Only rough estimates of particulate matter retention have been proposed, and such estimates generally mix the estuarine filter and the shelf filter.

Toward a Typology of Estuarine Filter Functions

Estuarine filters correspond to the interface between land and ocean, where fresh and ocean waters mix. An extensive estuarine literature describes different types of estuaries according to their origins, morphology, dynamics, sediment balance, biogeochemistry, and ecology. The main advantage of a geomorphic classification is that estuarine types can be mapped from the combination of topographic maps at appropriate resolution, tidal amplitude maps, lithological maps (for karst regions), and former glaciation maps (for fjords and fjärds). It will be demonstrated from the European example that rivers can be linked to many estuarine types. These classes are commonly aggregated into less than a dozen major types of estuaries, including fjords, fjärds, rias, macrotidal estuaries (tides > 4 m), coastal lagoons, several types of deltas, and direct inputs of groundwaters. Yet, it is very difficult for modelers to use this information to parameterize filtering rates of river material.

Four main types of estuarine filters are defined here on the basis of the fate of river material: (1) *internal filters*, in which material is trapped/processed within an estuary; (2) *proximal filters*, in which part of the dissolved and fine (clay-size) material escapes the estuary (e.g., through plumes) and is deposited on the shelf, from which it is eventually redistributed by lateral currents (most of the river material not reaching the open ocean); (3) *distal filters*, in which the ultimate deposition of particulates is far offshore in deeper waters, and plumes may extend well beyond the continental shelf; and (4) *limited filters*, which combine miscellaneous types (generally uncommon).

The retention capacity of estuaries with regard to river particulates, and hence which of the four estuary types characterize a specific estuary, depends on several factors, including (1) the morphology of the estuary, particularly in fjords, rias, macrotidal systems, and coastal lagoons, which are mostly located upstream of the coastline; (2) the estuary depth, exceeding 100 m for fjords; and (3) the residence time of fine particles and their settling on tidal mudflats in macrotidal estuaries. The degradation of organic material is enhanced in macrotidal estuaries where the residence time of fine particles is high and depends on the oxygenation of estuaries (in some macrotidal estuaries the maximum turbidity zone is hypoxic). The internal production of organic matter (e.g., phytoplankton, seaweeds, and submerged aquatic vegetation) depends on temperature, light penetration, nutrient circulation, tidal exposure, and water residence time. Most of these control factors depend on the relative positions of (1) the freshwater–seawater mixing zone, which can be defined as the zone where seawater proportion is between 0.1% and 90% of full strength; (2) the coastline; (3) the upper limit of tidal influence; and (4) the location of sand and silt deposition, which greatly depends on the position of waters deeper than 20 m (in most cases). The combination of these factors with the classic description of estuarine types results in the following, more precise definitions of the four filter types:

- *Internal filters* (IF) include the following types:
  - coastal lagoons (IFl)
  - rias (drowned valleys as in Britanny and Galicia) (IFr)
  - mesotidal estuaries (2 m < tides < 4 m) and macrotidal estuaries (tides > 4 m) (IFt)
  - fjords (steep and deep drowned glacial valleys) (IFf)
  - fjärds (lowland fjords with multiple channels and islands)
  - coastal archipelagoes (resulting from past glacial scour and characterized by multiple channel and sediment traps) (IFa)
- *Proximal filters* (PF) occur when river deltas protrude onto shallow and extended shelves.
- *Distal filters* (DF) are related to large and/or steep river deltas protruding on continental margins. They are often linked to the largest rivers of the world (e.g., the Niger, Indus, Brahmaputra-Ganges, and Amazon rivers). For these systems, the ultimate deposition area of river particulates is far offshore, and their river plumes may exceed the shelf width, thus directly affecting the continental margin (or shelf) and deeper ocean waters.
- *Limited filters* (LF) include systems that do not fit into any of the above categories. The category includes (1) direct river inputs to the deep ocean through undersea

canyons such as the Congo Canyon (LFc); (2) diffuse groundwater inputs, often through sandy coasts (LFs); (3) karstic resurgence of groundwater in calcareous coasts (LFk); and (4) total absence of water inputs from continents, also termed arheism (AR), conventionally limited to 3 mm y^{-1} runoff (Vörösmarty et al. 2000 a, b) as in many desert coasts.

Examples of such filter types are provided in Table 7-1 for rivers and coastlines found in all oceans and SEMSs. Many of them are clustered in specific regions of the world because of climate and lithology (fjords and fjärds), tidal range (meso- and macrotidal estuaries), coastal lithology (karst), and coastal dynamics (lagoons, deltas, and their combinations).

A qualitative description of control factors for sediment retention, organic matter processing, internal production, and nutrient retention occurring upstream from the coastline is presented in Table 7-2 for each type of filter. The typology of river catchment inputs is somewhat linked to the typology of estuarine filters, as for river sediment yields and the ratio between dissolved and particulate organic carbon (DOC/POC).

Estuarine filter functions should be based on a representative set of estuaries, which is not yet the case: Most of the estuarine literature still focuses on meso- and macrotidal estuaries (western Europe, eastern United States) and on a few large deltas (e.g., Mississippi, Amazon, Lena). When net river inputs to the coastal zone are estimated, each river basin should be assigned to a specific estuarine filter type, characterized by its retention/processing capacity for the different categories of river material (sand, silt, mud, DOC, POC, nutrients, etc.). Thanks to global databases such as those collected by the Land–Ocean Interactions in the Coastal Zone (LOICZ) program (Crossland et al. 2005; LOICZ 2008) and the availability of geographic information systems, such connections are now possible, as illustrated further for the European continent and its SEMSs (Baltic Sea, Black Sea, North Sea, and Northern Mediterranean Sea) below.

Continental shelf filters are as complex as estuarine filters. They transform, remobilize, and disperse river material that passes through estuaries on time scales of decades to 10,000 years and space scales that can extend to more than hundreds of kilometers, for example, along the Adriatic coast (Trincardi and Syvitski 2005). Quantification of these processes is the focus of sediment budgets on selected shelves using integrated approaches (STRATAFORM and EuroSTRATAFORM programs), such as for the Eel River margin in California (Nittrouer 1999), the Po River delta (Trincardi and Syvitski 2005), and the Gulf of Lions (Weaver et al. 2006). Other programs are dedicated to the fate of carbon (RiOMar; McKee 2003) and to the complex human interactions with river basins and shelves (Salomons et al. 2005). Shelf filters depend on another set of control factors, including shelf depth and width; wave, tide, and storm energy distribution; coastal currents; occurrence of undersea canyons; river plume depths; and intensity of fish trawling. The uniqueness of each shelf hinders the creation of a global-scale typology, which is needed to assess the ultimate fate of river material on shelves, on slopes, and in the deep ocean.

Table 7-1. Typology of Estuarine Filters with Regard to Net River Inputs to the Coastal Zone and Selected Examples among World Rivers and Coasts

	Internal Filters				
Ocean Basin	*Fjords*	*Rias*	*Archipelagoes and fjärds*	*Tidal estuaries*	*Lagoons*
N Atlantic	**Saguenay**	**Galicia**		**Rhine**	E USA
	E Churchill	**Britanny**		**Elbe**	**Oder**
	Norway			Tagus	**Texas bays (Trinity, Nueces rivers)**
	Labrador			St. Lawrence	
S Atlantic				Paraná **	Jacui (Los Patos Lagoon)
				Sao Francisco	Comoe
				Rio Gallegos	Oueme
W Pacific			Stikine	**N Java**	
			Skeena	**Chao Phraya**	
			Fraser	**Chanjiang (Yangtze)***	
				Zhu Jiang	
				Amur	
E Pacific	S Chile				
Indian				Shatt el Arab	Murray
Arctic	Ob-Yenisei		Back		

Rivers flowing into the SEMSs covered in this book are in **bold**.
* to **: degree of human impacts on river basin or filter

		Limited Filters			
Proximal Filters	*Distal Filters*	*Subterranean inputs*	*Karstic inputs*	*Intersecting canyon*	*Arheic*
Vistula *	**Mississippi** *	Georgia, USA	**Yucatan**	**Yesil Irmak**	Mauritania
Danube *	Niger *	**Tunisia**	**Florida**		**Lybia**
Nelson	**Magdalena**		**Dalmatia**		Namibia
Rhone					
Ebro **					
Orinoco	Orange **			Congo	
Amazon					
Mekong	Mamberano		S China	Kamchatka	
Huanghe **				Hawaii	
Song Koi *				**W Taiwan**	
Purari					
Yukon				N Chile	
				Balsas	
				Columbia **	
Zambezi *	Brahmaputra-Ganges	SE India		S Java	Nullarbor coast
	Indus **	E Africa			
	Irrawaddy				
	Godavari *				
Mackenzie					
Indigirka					
Kolyma					
Lena					

Table 7-2. Some Characteristics of Estuarine Filters for Assessing Net River Inputs to the Coastal Zone

	Internal Filters				
	Fjords	*Rias*	*Archipelagoes and fjärds*	*Tidal estuaries*	*Lagoons*
River sediment yields	Low	Low/ medium	Low/ medium	Low/ medium	Low
DOC/POC in river inputs	> or = 1	Variable	Variable	Variable	> or = 1
Average depth upstream coastline	High	Medium	Medium	Low	Very low
Turbidity	Very low	Medium	Low	High	Low/ medium
Freshwater residence time	Very high	Medium/ high	Medium	Medium	Medium
Bottom anoxia	Permanent anoxia possible	Limited		Possible at max. turbidity	
Sediment retention upstream coastline	Total	High	High	Medium	High
Autochthonous production potential	Medium	Medium	Low/ medium	Medium	High
Nutrient retention upstream coastline	High potential	Medium/ high	Medium	Medium/ high	High

NA: not applicable

Such a first-order typology does not take into account temporal variations, such as seasonal river-flow variations and flood events (return period ten to one hundred years), which can change a system from one filter type to another. For example, during large floods some macrotidal estuaries do not function as internal filters; the maximum turbidity zone moves to the outer estuary, and the estuarine filter is shifted to the distal position. Seasonal floods also control delta filters: At low and medium river stages, an important fraction of sediments is retained within estuarine deposits. During floods these accumulated sediments can be remobilized and expelled. In most deltas a large portion of sediments can be retained at the high-water stage on the flooded areas. Coastal lagoons also have seasonal dynamics: In the low-flow season their outlets to the sea can

		Limited filters			
Proximal filters	*Distal Filters*	*Subterranean inputs*	*Karstic inputs*	*Intersecting canyon*	*Arheic*
Medium/ high	Medium/ high	None	None	Low/ Very high	None
> 1 and depending on flow	< 1 and depending on flow	>> 1	> 1	Variable	NA
Low/ medium	Low/ medium	NA	NA	Low/ medium	NA
Variable	Variable	NA	NA	Low	NA
Low/ medium	Low	?	Very short	Short	NA
Seasonal anoxia may occur	Seasonal anoxia may occur	NA	None	None	NA
Limited	Limited	NA	NA	None	NA
Medium		NA	NA	Low	NA
Limited	Limited	None?	None?	Very limited	NA

be closed by littoral drift, then reopened at the high-flow season (e.g., southern Brazil and Uruguay coast).

SEMSs as Ultimate Mega-Filters of Land-to-Ocean Fluxes

Regional seas, as defined by Meybeck and colleagues (2007a), can be considered as mega-filters in the Earth system for their capacity to retain and process the dissolved and particulate material transported by fluvial systems to the open ocean. Three main types of regional seas were defined by Meybeck and colleagues (2007a) on the basis of morphological criteria and of their exchange rate with the open ocean:

- *quasi-enclosed* regional seas (e.g., Red, Baltic, and Mediterranean seas)
- *semi-enclosed* regional seas (e.g., Gulf of Mexico, South China Sea, and Sunda, Sulu, and Banda seas in Southeast Asia)
- *open* regional seas (e.g., North Sea, East China Sea, Gulf of St. Lawrence)

In addition, two other types of mega-filter were defined in relation to extended platforms (e.g., Kara and Laptev seas, Patagonia platforms) and with extended archipelagoes (e.g., British Columbia coast, Canadian Arctic Archipelago, South Chile coast). Forty mega-filters were fully characterized for their river inputs (upstream of the estuarine filters) and for their catchment characteristics, on the basis of the aggregation of 140 coastal catchments (COSCATs) that describe the world's coasts at a very coarse resolution (Meybeck et al. 2006).

There are several distinctions between the global-scale mega-filters listed by Meybeck and colleagues (2007a) and the SEMSs selected in the present volume, as follows:

- Some regional seas (Red Sea, Gulf of California, Sunda/Sulu/Banda, Arafura) are not addressed in this book.
- Siberian seas aggregated by Meybeck and colleagues (2007a) (from Barents to Chukchi) were considered too open to qualify as SEMSs and were considered extended platforms; the Kara and Laptev seas have been selected here as SEMSs.
- The Bay of Bengal was not considered by Meybeck and colleagues (2007a) as a mega-filter, as ocean dynamics were not used as criteria, but it is included here as a SEMS, as it is limited by an ocean front between the Deccan and the Andaman Islands.
- Extended archipelago coasts are not considered as SEMSs.
- The composition of regional seas and SEMSs by aggregation of COSCATs is different for some entities: The Andaman Sea is here considered part of the Bay of Bengal despite its western sills; the Gulf of Thailand is here differentiated from the East China Sea; the same holds for the Yellow Sea and the Pohai.
- The Adriatic Sea is considered here as a single entity, while Meybeck and colleagues (2007a) described the whole Mediterranean Sea in comparison to other Mediterranean entities, in the way that the Gulf of Mexico plus the Caribbean or the western Pacific seas from the Japan/East Sea to the Banda Sea could be considered together.

These differences illustrate the discrepancies in building global-scale descriptions of the Earth system and addressing a selection of Earth system entities with slightly different criteria, such as ocean physics (fronts), importance of human development, and so forth. Actually, each ocean science discipline (e.g., sedimentology, physical geography, physics, tectonics, biology) would probably define its own regional entities, although common delineation is also needed, particularly for Earth system models. Our delineation of regional entities (Meybeck et al. 2007a) is a compromise that certainly has its limits.

The SEMS concept is based on the limited exchange rates of regional entities with the world's oceans. This exchange is measured by the ratio between the cross-section of the regional sea at its interface with the neighboring ocean (SB in Figure 1.1)—generally at straits and sills, exceptionally as ocean fronts (e.g., Bay of Bengal)—and the average regional sea cross-section (SA in Figure 1.1). The SB/SA ratio is zero for landlocked seas, such as the Caspian Sea. For quasi-enclosed seas, SB/SA can be extremely low (< 1/1,000 for the Black and Mediterranean seas, between 1/100 and 1/1,000 for the Baltic Sea). Semi-enclosed regional seas are defined by SB/SA ratios between 1/100 and 1/10; open regional seas are defined by SB/SA ratios between 1/10 and 10; and for the Bay of Bengal, SB/SA is > 10.

It is hypothesized here that the efficiency of mega-filter retention or processing of land-based material in regional seas and SEMSs is inversely related to the SB/SA ratio and to the river water residence time. This control factor depends on the relative size of regional sea catchments and regional sea volumes, which is constant in time over millennia and longer; water budgets of regional sea catchments depend on climate change. According to global-scale hydrologic models, the river water residence time in regional seas is very variable, from 52,000 years for the Red Sea to 50 years for the Gulf of St. Lawrence (Meybeck et al. 2007a). It is probably also relatively short for the Bay of Bengal, which receives some of the greatest river water inputs (Brahmaputra-Ganges-Meghna and Godavari rivers) globally.

Combining Estuarine Filters and SEMSs Mega-Filters: The European Example

Estuarine and the regional sea filters are combined here for the exorheic* part of the European continent, covering 8.2 million km^2 including Iceland and limited by the Ural and north Caucasus mountains (see detailed limits in Dürr et al. 2005). We use a gridded river network and runoff field at the 30 × 30 min resolution (Vörösmarty et al. 2000a, b; Fekete et al. 2002). The river basin attributes are derived from global databases on population (Vörösmarty et al. 2000c), suspended sediments before damming (Ludwig and Probst 1998), and total nitrogen inputs in present-day conditions (including dissolved organic and particulate nitrogen; from Green et al. 2004).

The estuarine types are here determined individually for each major river and for smaller catchments at the 30 min resolution through the following steps (Dürr et al. 2008):

1. visualization of coastline on 1:1,000,000 maps and on Google Earth®
2. confirmation of fjord and fjärd coasts by the extent of ice caps at the Last Glacial Maximum (Dürr et al. 2008) based on the *Physico-Geographical World Atlas* of the Soviet Academy of Sciences (Gerasimov et al. 1964)
3. identification of sandy coasts and carbonated coasts (Dürr et al. 2005)
4. determination of tidal range (LOICZ typology; Crossland et al. 2005)
5. determination of coastal bathymetry from various atlases

*River systems that drain into water bodies connected to the ocean.

Table 7-3. Distribution of River Discharge by Estuarine Filter Types for the European Continent (8.2 Mkm^2)

	Internal filters (%)				
	Fjords	*Archipelagoes and fjärds*	*Rias*	*Macrotidal*	*Lagoons*
Arctic Ocean	29	71	—	—	—
Atlantic Ocean	20.2	3.8	42.5	22.7	0.5
Baltic Sea*	—	83.7	—	—	16.3
North Sea*	32.1	14.4	—	53.5	—
N Black Sea*	—	—	0.9	—	31.4
N Mediterranean*	—	—	—	—	5.4
Total European catchments	14.8	33.3	7.0	11.4	8.0

Caspian drainage excluded
*SEMS

After each coastal cell was attributed to an estuarine type and to a river basin, the related attributes (basin area, population) and river fluxes (water, sediments, total nitrogen) were calculated. The coastal cells were then reaggregated in coastal segments according to the delineation proposed by Meybeck and colleagues (2006) and into regional seas and oceans according to the limits suggested by Meybeck and colleagues (2007a) (Table 7-3). The Arctic Ocean is limited in Norway by the 62°N latitude and corresponds to the sum of the Norwegian Basin (COSCAT # 407) and Barents Sea (# 408). The open Atlantic Ocean includes the Hutton-Rockall Basin (# 402) and the Iberian-Biscay Plains (# 401).

Direct groundwater input can be found in Europe on sandy coastal plains (e.g., Landes coast in France, South Baltic) but could not be estimated at the 30 min resolution. European catchments are linked to the Atlantic and Arctic oceans and to Baltic, North, Black, and Mediterranean regional seas through eight major types of estuarine filters, from fjords to karsts (Table 7-3). Internal filters are the dominant types, intercepting a total of 74.5% of water runoff, with proximal filters accounting for 22.1%.

Estuarine filters are very different for each ocean drainage. Fjärds and archipelagoes are dominant in the Baltic and in the Arctic (Norwegian Basin and Barents Sea). Macrotidal estuaries are dominant in the North Sea. Proximal filters are predominant in the north Mediterranean Sea and Black Sea. Lagoons, fjords, and karsts are developed in the south Baltic and the Black Sea, the Norway coast, and the north Mediterranean Sea, respectively.

This first estimate should be refined at a finer resolution in the future in order to take into account direct groundwater inputs on sandy coasts and hydrologic variations. For

Proximal filters (%)	*Distal filters (%)*	*Limited and karst (%)*	*Total (%)*	*Discharge ($km^3 y^{-1}$)*
—	—	—	100	558
7.9	0.4	2.0	100	384
—	—	—	100	388
—	—	—	100	339
67.7	—	—	100	329
73.9	5.7	15	100	360
22.1	1.0	2.6	100	2,358

Filter types refer to Figure 7-2 and Table 7-1.
Adapted from Dürr et al. 2008

example, the Rhone River delta is a proximal filter at low river flows, while at the highest flows, the estuarine plume influences the entire shelf. Also some of the river material can transit across the shelf through the Rhone canyon.

In Europe, distal filters represent only 1% of the intercepted river flows. These rivers generally correspond to numerous small deltas along a narrow ribbon of steep coastal catchments (area < 1,000 km^2) that may have high sediment yields, as in the Ligurian, Ionian, and Aegean seas. Inland storage of river material (internal estuarine filters) and proximal storage near the coast are dominant in Europe.

The regional seas' mega-filters are exceptionally developed in Europe (Table 7-4): They intercept 67.3% of drainage area (Caspian Sea basin excluded), 60% of river runoff, 67.2% of river sediment, and 74.3% of river nitrogen. Regional sea catchments also intercept river flow related to 83.4% of the European population (Volga catchment excluded). Only 16.6% of the population, with related contaminants and nutrient loads from human activities, is directly connected to the Atlantic and Arctic oceans without being filtered by regional seas.

The Black Sea, Mediterranean Sea, and Baltic Sea can be considered as quasi-enclosed regional seas that trap all river particulates and the great majority of nutrients and organic carbon that enter them (Meybeck et al. 2007a). The North Sea is an open regional sea with lower filtering functions. When considering that internal estuarine filters (fjords, rias, macrotidal estuaries) dominate in the open Arctic and Atlantic oceans, it can be estimated that less than 1% of river particulates and a few percent of river inputs of nutrients and organic carbon originating from the European continent reach the

Table 7-4. Relative Weights of European Regional Seas River Catchments and Their Related River Fluxes in Upstream Estuarine Filters

		Open Oceans		*Semi-Enclosed Marginal Seas*				
		N. Atlantic Only	*Arctic*	*North Sea*	*Baltic Sea*	*N. Black Sea*	*N. Mediterranean*	*Europe Total*
Catchment Basin Area	10^6 km^2	1.05	1.63	0.87	1.62	2.09	0.94	8.2
	%	12.8	19.9	11.2	19.7	25.5	11.5	100
Water Volume (Discharge)	km^3 y^{-1}	384	559	339	388	328	358	2,356
	%	16.3	23.7	14.4	16.5	13.9	15.2	100
Population	10^6 persons	92	10.8	159	78	163	113	616
	%	14.9	1.7	25.8	12.7	26.4	18.3	100
Suspended Sediment	Mt y^{-1}	137	79	31	20.3	107	284	658
	%	20.8	12.0	4.7	*3.1*	16.2	**43.1**	100
Total N	Mt y^{-1}	0.99	0.475	1.75	0.64	0.90	0.95	5.7
	%	17.4	8.3	**30.6**	*11.2*	15.8	16.7	100

Population Density	Persons km^{-2}	87.6	6.6	183	48	78	120	75.1
Runoff	mm y^{-1}	366	343	390	240	157	381	287
N yield	t km^{-2} y^{-1}	0.94	0.29	2.0	0.40	0.43	1.0	0.70
Total N conc.	mg L^{-1}	2.6	0.85	5.15	1.65	2.74	2.65	2.42
Sediment Yield	t km^{-2} y^{-1}	130	48.5	35.6	*12.5*	51	**302**	80.2

Caspian drainage excluded

Bold: proportions much higher than the area weight

Italics: proportions much lower than the area weight

Adapted from Dürr et al. 2008

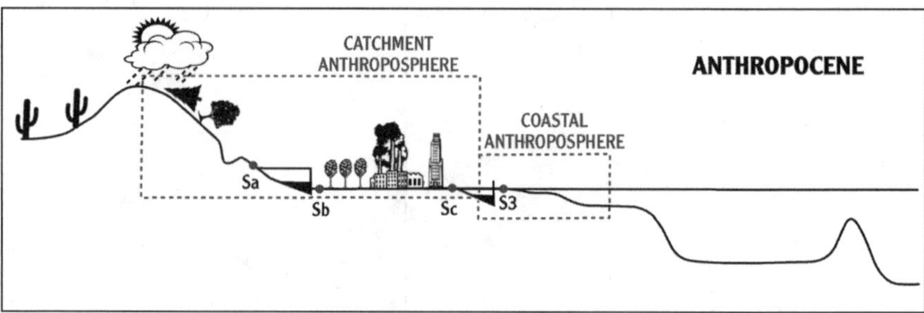

Figure 7-2. Nested filters of river material to oceans in the Anthropocene. New material sources include soil losses, deforestation, mining and extraction, atmospheric emissions, agrochemicals, and wastewaters from catchment and from coastal sources. New sinks include reservoirs, irrigated fields, and water transfers. Natural sinks such as wetlands and floodplains are also modified, both within catchments and in the estuarine zone. Contaminants presently stored in the Anthroposphere are likely to leak in the future. Positions of budget stations on river courses: Sa, Sb, Sc, and S3.

deeper waters of the open Atlantic Ocean. If such an accumulative filter function is confirmed for other oceans, then net river inputs—past, present, and future—to the ocean will be less than currently assumed, affecting regional to global ocean models.

River Inputs to SEMSs in the Anthropocene

The Anthropocene has been defined by Crutzen and Stoermer (2000) as a new state of the Earth system in which human pressures on chemical and biogeochemical fluxes match natural control factors such as climate and tectonics. This concept has been extended within the International Geosphere–Biosphere Programme and applied to river systems (Meybeck 2003; Vörösmarty and Meybeck 2004).

The Anthropocene picture of land–ocean linkages is schematically presented in Figure 7-2. In headwater streams, land-use change (such as deforestation, cultivation, and mining) generates additional sources of river-borne material. In medium and lower river courses, industrialization, urbanization, and intensive agriculture result in increased concentration of nutrients, metals, and dissolved salts; occurrence of xenobiotic substances; imbalance of oxygen levels; and eutrophication (Figure 7-2). These changes are now occurring on all continents and can be regarded as part of global change. Human activities also generate major hydrologic changes through (1) water storage in large reservoirs; (2) water diversion from one river to another, as in the Hudson–James Bay catchment; and (3) irrigated agriculture. As a result, the natural water discharge of many rivers has been greatly reduced, including some rivers draining into SEMSs, such as the Rio Grande/Bravo into the Gulf of Mexico, the Colorado River into the Gulf of California, the Huanghe (Yellow River) into the Pohai and the Nile into the Mediterranean Sea.

Human activities also greatly affect natural sinks: Wetlands are drained and filled, rivers are not allowed to inundate their floodplains, lower river courses and estuaries are channelized and dredged. The most important direct human impact to river fluxes is linked to the construction of hundreds of thousands of reservoirs, from mountainous headwaters to the lower courses of rivers. The largest reservoirs can store water for more than a year and trap all river sediments. Their expected lifetimes are 100 to 200 years (Vörösmarty et al. 2003). The amount of material from rivers, and its characteristics, can be greatly modified by reservoirs.

Depending on their location, river survey stations capture these human impacts differently (Figure 7-2). Upstream control stations (Sa) can isolate some key impacts (agriculture, mining, long-range atmospheric pollutants, deforestation). Middle-course stations (Sb) can be affected by the upper reservoirs. Lower-course stations (Sc) are affected by all pressures, including megacities and intensive agriculture. Yet these impacts may be modified if the last reservoir is located between Sc and the river mouth (S3). The majority of river surveys are recent (within the past twenty to thirty years), which is not sufficient for trend analysis; therefore, sedimentary archives in lakes, peat bogs, and estuaries are commonly used to reconstruct past riverine conditions at different time scales, from decades to millennia.

Over the past fifty years, river fluxes have greatly evolved in two opposite directions (Meybeck 2003; Meybeck and Vörösmarty 2005; Syvitski and Milliman 2007):

- Increase of fluxes has been caused by accelerated and new sources of river material (urbanization, industrialization, mining, agriculture, deforestation).
- Decrease of fluxes has been caused by reservoir trapping, river and reservoir nutrient uptake by eutrophication, water diversion, and irrigation.

The balance of these two factors varies for each watershed.

A tentative classification of "river syndromes" likely to occur on rivers flowing to SEMSs is proposed in Table 7-5, with detailed examples for European SEMSs. Flow is regulated by large reservoirs, water storage, and interbasin transfers. About 70% of all alpine rivers are now regulated, and high-water stages can be shifted from early summer to winter by reservoir operation. Series of reservoirs on a river system (reservoir cascades) are common in Spain, southwest France, Italy, and Greece and are found on some Swedish rivers and on the Vistula River. Alteration of the sediment balance occurs either when the sediment supply has greatly increased or when the river sediment flux is completely modified by reservoirs. In both cases, evolution of the coastal sediment budget is accelerated. Erosion of the Ebro River delta (85,550 km^2 catchment area, Spain) is an example of such sediment imbalance due to dozens of cascading reservoirs (Ibanez et al. 1996); the present load is estimated from 0.12 to 0.15 Mt y^{-1} compared with 3 Mt y^{-1} before damming in the 1960s. In 1907, an exceptional flood discharged 0.72 Mt of sediment in one hour; such an extreme event with marked coastal impact cannot occur today.

Table 7-5. Main Expected and/or Actual Changes in River Fluxes ("River Syndromes") Due to Global Change for Total Suspended Solids, Ionic Loads, Silica and Nutrients, Organic Carbon, and Toxic Substances in European SEMSs, Other SEMSs, and Other Mega-Filters

	Change in Sediment Flux	*Runoff Decrease*	*Contamination of River Material by Toxics*	*Local Acidification of River Basins*	*Changes of Nutrient Levels and Redfield Ratios*	*Permafrost Melting*	*Major Impacted Rivers*
European SEMSs							
Baltic Sea			++	++	++	+	Vistula, Oder, Nemanus
North Sea			+++	+	+++		Thames, Rhine, Elbe, Weser
Black Sea	+		+++		++		Danube, Dniepr, Don, Kizil Irmak, Sakarya
Adriatic Sea	++		+		++		Po, Adige, Idrija
Other SEMSs							
Okhotsk			+			++	Amur
Kara Sea			++			++	Ob
Laptev Sea						+++	Lena
Yellow Sea, Pohai Gulf	+++	++	+		+		Huanghe (Yellow)
East China Sea	++		++		+		Changjiang (Yangtze)

South China Sea			++		+		Zhu Jiang (Pearl), Hong He
Gulf of Thailand	+		+		+		Chao Phraya
Bay of Bengal			+		+		Brahmaputra-Ganges
Hudson Bay	+			+		++	Churchill, Eastmain, Great Whale
St. Lawrence Gulf			+	+			St. Lawrence
Gulf of Mexico	++	+	++		+++		Mississippi, Rio Grande/ Bravo, Colorado (Texas)
Other Mega-Filters							
Gulf of California	+++	+++			+++		Colorado (USA/Mexico)
Persian Gulf	++	+++	++		++		Shatt el Arab
Red Sea			?				None
Patagonia Platform	++	+					Negro, Colorado (Argentina), Chubut

+ to +++: relative importance of issues

In Scandinavia, acidification of river catchments resulting from atmospheric inputs of SO_2 and NO_X has accelerated heavy metal leaching. Mining activities may also result in similar evolution of river chemistry. Numerous large European rivers, such as the Rhine, Mosel, Weser, Elbe, and Vistula, have been exposed to salinization from potash, salt, and coal and iron mines, increasing some ionic contents (Na^+, K^+, Cl^-, SO_4^{2-}) by more than an order of magnitude from their natural, preindustrial, or pristine states.

Nutrient concentrations, fluxes, and their ratios (N:P and Si:N Redfield ratios) are very sensitive to human pressures, as N and P are both increasing with agriculture and urbanization. The total nitrogen concentration of European SEMS catchments can be used as an example (see Table 7-4). In relatively pristine conditions, such as the Barents Sea catchment, added nitrate and organic nitrogen concentrations do not exceed 1.0 mg N L^{-1}, with a dominance of organic nitrogen. The average total nitrogen level for North Sea catchments now exceeds 5 mg N L^{-1}, primarily as nitrate (Meybeck et al. 2006). Inputs of total nitrogen have been multiplied by 1 order of magnitude in Western European rivers in comparison with the estimated pristine levels (Green et al. 2004). A similar increase has been observed for total phosphorus in some rivers, such as the Seine (Billen et al. 2007). In many European rivers, nitrate concentrations have stabilized since 2000, while phosphorus has dramatically decreased since 1980–1990 in many basins. At the same time, the uptake of silica in major reservoirs (e.g., observed in the Iron Gates reservoir on the Danube and in some Scandinavian rivers) has decreased silica concentrations downstream (Humborg et al. 2002). The relative trajectories of the aquatic nutrients N, P, and Si in the last fifty years are therefore quite complex and cannot be generalized; they must be established for each river basin (Friedl and Wuest 2002; Turner et al. 2003; Billen et al. 2007). A similar picture is expected for micropollutants, such as heavy metals; in the Seine River the reconstruction of past levels of Cd, Cu, Hg, Pb, and Zn from floodplain cores revealed various contamination trajectories since 1935, a general decline since the 1960s, and a complex relationship with human activities (Meybeck et al. 2007b). Effects of global warming on river systems are considered in Table 7-5 only for the permafrost melting and the expected increase of dissolved organic matter, including carbon and nitrogen species.

When river control stations (Station S3 in Figures 7-1 and 7-2) are located upstream of the tidal range and salinity intrusions, the direct inputs to estuaries and to the coastline from the coastal megacities, harbors, delta agriculture, coastal industries, and oil extraction that constitute the coastal anthroposphere are not considered (Mantoura et al. 1991), or they are underestimated in the assessment of continental inputs to the coast. The greatest challenges in assessing future net river inputs to SEMSs are

- accounting for all important impacts occurring on river catchments and their fluxes
- accounting for human modification of estuarine filters, particularly for proximal filters
- assessing the direct inputs from the coastal anthroposphere to estuaries and coastal zones

- accounting for the gradual leakage of the anthropogenic material presently stored in soils (e.g., K, N, Pb, Hg), waste dumps, contaminated industrial sites, and mine tailings that is likely to occur in the next decades or centuries

In addition to global-scale models that have been developed for riverine fluxes of carbon, nitrogen, and phosphorus (Beusen et al. 2005; Seitzinger et al. 2005), local-to-regional assessments should be made, considering the great variability of both natural conditions and human impacts (Salomons et al. 2005). The most recent models can now integrate all kinds of filters, human impacts, and climate change scenarios (Garnier et al. 2006).

Conclusion

Riverine material originating from weathering, erosion, and human activities is filtered first on land, then at the land–ocean interface by estuarine and shelf filters, and finally in mega-filters such as SEMSs.

Depending on their location, river control stations may integrate differently the natural catchment filters (wetlands, lakes, floodplains) and the man-made ones (reservoirs, irrigated areas). So far, all global river flux estimates are based on the last survey stations (S3 in Figures 7-1 and 7-2) located upstream of estuaries. As these stations are located to avoid influence by tides and sea surges, they are located inland, sometimes hundreds of kilometers from the coastline; therefore, they do not take into account the effect of estuarine filters. Such river control stations also do not take into account direct inputs of local estuarine tributaries or the impacts of the coastal anthroposphere, such as sewage release, waste dumping, and leakages from harbors and agriculture. Estuarine filters can be aggregated here into four categories depending on their capacity to filter the river inputs and on their position relative to the coastline and to the shelf break:

- Internal filters (IF) are located upstream from the coastline.
- Proximal filters (PF) are located across the coastline.
- Distal filters (DF) are located across the shelf break.
- Limited filters (LF) are for direct groundwater inputs.

A first accounting of river fluxes and their estuarine filters for European SEMSs shows the dominance of internal and proximal filters on this continent, probably due to the limited size of river catchments in Europe (less than 200,000 km^2, except for the Danube).

The cascading filter approach should be applicable to the Anthropocene, with the following considerations:

- Additional sources of river material within catchments should be taken into account adequately.
- Modifications of river and estuarine filters (e.g., wetland draining, channelization, and diking of floodplains) should be considered.

- The addition of new types of filters within river basins, such as reservoirs, should be taken into account (at global scale, reservoirs are thought to already store more than 30% of riverine particulates; Vörösmarty et al. 2003).
- Direct releases of material from the coastal anthroposphere should be assessed.

These considerations can be addressed through the development of river and coastal databases, the refinement of filter typologies, and the integration of biogeochemical models from headwaters to the coastal zone. The exceptional importance of regional seas and SEMSs on net river inputs to the ocean should be considered in future ocean budgets.

References

Abril, G., M. Nogueira, H. Etcheber, G. Cabecadas, E. Lemaire, and M.J. Brogueira. 2002. Behaviour of organic carbon in nine contrasting European estuaries. *Estuarine, Coastal and Shelf Science* 54:241–262.

Alekin, O.A., and L.V. Brazhnikova. 1968. Dissolved matter discharge and mechanical and chemical erosion. *International Association of Hydrological Sciences Publication* 78:35–41.

Beusen, A.H.W., A.L.M. Dekkers, A.F. Bouwman, W. Ludwig, and J. Harrison. 2005. Estimation of global river transport of sediments and associated particulate C, N, and P. *Global Biogeochemical Cycles* 19:GS4S05, doi: 10.1029/2005GB002453.

Billen, G., J. Garnier, J. Némery, M. Sebilo, A. Sferratore, S. Barles, P. Benoit, and M. Benoît. 2007. A long-term view of nutrient transfers through the Seine river continuum. *Science of the Total Environment* 375:80–97.

Chen, C.T.A., K.K. Liu, and R. MacDonald. 2003. Continental margin exchanges. In *Ocean Biogeochemistry: A JGOFS Synthesis*, edited by M.J.K. Fasham, 53–97. Joint Global Ocean Flux Study. Berlin: Springer.

Crossland, C.J., H.H. Kremer, H.J. Lindeboom, J.I. Marshall Crossland, and M.D.A. LeTissier (eds.). 2005. *Coastal Fluxes in the Anthropocene.* Global Change—The IGBP Series. Berlin: Springer.

Crutzen, P.J., and E.F. Stoermer. 2000. The "Anthropocene." *IGBP Newsletter* 41:17–18.

Dürr, H.H., G. Laruelle, C. van Kempen, C. Slomp, H. Middelkoop, and M. Meybeck. 2008. A global typology of the proximal coastal zone for nutrient-processing studies at the land–ocean transition. Forthcoming.

Dürr, H.H., M. Meybeck, and S.H. Dürr. 2005. Lithologic composition of the earth's continental surfaces derived from a new digital map emphasizing riverine material transfer. *Global Biogeochemical Cycles* 19:GB4S10, doi: 10.1029/2005GB002515.

Fekete, B.M., C.J. Vörösmarty, and W. Grabs. 2002. High-resolution fields of global runoff combining observed river discharge and simulated water balances. *Global Biogeochemical Cycles* 16 (3):1,042, doi: 10.1029/1999GB001254.

Fournier, F. 1960. *Climat et érosion*. Paris: Presses Universitaires de France.

Friedl, G., and A. Wuest. 2002. Disrupting biogeochemical cycles, consequences of damming. *Aquatic Sciences* 64:55–65.

Garnier, J., A. Sferratore, M. Meybeck, G. Billen, and H. Dürr. 2006. Modeling silicon transfer processes in river catchments. In *The Silicon Cycle: Human Perturbations and Impacts on Aquatic Systems*, edited by V. Ittekot, D. Unger, C. Humborg, and N. Tac An, 139–162. SCOPE Series Vol. 66. Washington, DC: Island Press.

Garrels, R.M., and F.T. Mackenzie. 1971. *Evolution of Sedimentary Rocks*. New York: W.W. Norton & Co.

Gerasimov, G.N., et al. (eds.). 1964. *Physico-Geographical World Atlas*. Moscow: Scientific Academy of URSS and Cartographic and Geodesic Central Committee.

Green, P., C.J. Vörösmarty, M. Meybeck, J. Galloway, B. Peterson, and E. Boyer. 2004. Pre-industrial and contemporary fluxes of nitrogen through rivers: A global assessment based on typology. *Biogeochemistry* 68:71–105.

Howarth, R.W., G. Billen, D. Swaney, A. Townsend, N. Jaworski, K. Lajtha, J.A. Downing, R. Elmgren, N. Caraco, T. Jordan, F. Berendse, J. Freney, V. Kudeyarov, P. Murdoch, and Z. Zhao-Liang. 1996. Regional nitrogen budgets and riverine N & P fluxes for the drainages to the North Atlantic Ocean: Natural and human influences. *Biogeochemistry* 35 (1):75–139.

Humborg, C., S. Blomqvist, E. Avsan, Y. Bergensund, and E. Smedberg. 2002. Hydrological alterations with river damming in northern Sweden: Implications for weathering and river biogeochemistry. *Global Biogeochemical Cycles* 16 (3):1–13.

Ibanez, C., N. Prat, and A. Canicio. 1996. Changes in the hydrology and sediment transport produced by large dams on the lower Ebro River and its estuary. *Regulated Rivers* 12:51–62.

Livingstone, D.A. 1963. Chemical composition of rivers and lakes. Data of geochemistry chapter G. *U.S. Geological Survey Professional Paper* 440-G, G1–G64. http://pubs.er.usgs.gov/usgspubs/pp/pp440G.

LOICZ. 2008. Land–Ocean Interactions in the Coastal Zone. www.loicz.org.

Ludwig, W., and J.-L. Probst. 1998. River sediment discharge to the oceans: Present-day controls and global budgets. *American Journal of Science* 298:265–295.

Mantoura, R.F.C., J.M. Martin, and R. Wollast. 1991. *Ocean Margin Processes in Global Change*. Dahlem Workshop Reports. Chichester, UK: John Wiley & Sons.

McKee, B.A. 2003. *RiOMar: The Transport, Transformation and Fate of Carbon in River-Dominated Ocean Margins*. Report of the RiOMar Workshop 1–3 November 2001, Tulane University, New Orleans, LA.

Meybeck, M. 1979. Major elements contents of river waters and dissolved inputs to the oceans. *Revue de Geologie Dynamique et de Geographie Physique* 21:215–246.

Meybeck, M. 1982. Carbon, nitrogen, and phosphorus transport by world rivers. *American Journal of Science* 282:401–450.

Meybeck, M. 2003. Global analysis of river systems: From Earth system controls to Anthropocene syndromes. *Philosophical Transactions of the Royal Society B* 358 (1440):1,935–1,955.

Meybeck, M., H.H. Dürr, S. Roussennac, and W. Ludwig. 2007a. Regional seas and their interception of riverine fluxes to oceans. *Marine Chemistry* 106:301–325.

Meybeck, M., H.H. Dürr, and C.J. Vörösmarty. 2006. Global coastal segmentation and its river catchment contributors: A new look at land–ocean linkage. *Global Biogeochemical Cycles* 20:GB1S90, doi: 10.1029/2005GB002540.

Meybeck, M., P. Green, and C. Vörösmarty. 2001. A new typology for mountains and other relief classes. *Mountain Research and Development* 21:34–45.

Meybeck, M., L. Lestel, P. Bonté, R. Moilleron, J.-L. Colin, O. Rousselot, D. Hervé, C. de Pontevès, C. Grosbois, and D.R. Thévenot. 2007b. Historical perspective of heavy metals contamination (Cd, Cr, Cu, Hg, Pb, Zn) in the Seine River basin (France) following a DPSIR approach (1950–2005). *Science of the Total Environment* 375:204–231.

Meybeck, M., and A. Ragu. 1995. *GEMS/Water Contribution to the Global Register of River Inputs (GLORI).* Provisional Final Report. Geneva: UNEP Global Environment Monitoring System/WHO/UNESCO.

Meybeck, M., and C. Vörösmarty. 2005. Fluvial filtering of land-to-ocean fluxes: From natural Holocene variations to Anthropocene. *Comptes Rendus Geosciences* 337:107–123.

Milliman, J.D., E. Rutkowski, and M. Meybeck. 1995. *Global Register of Rivers (GLORI).* IGBP-LOICZ (Land–Ocean Interactions in the Coastal Zone) Series. Texel, The Netherlands: LOICZ International Project Office.

Nittrouer, C.A. (ed.). 1999. The formation of continental margin strata. *Marine Geology* 154:1–420.

Perillo, G.M.E. (ed.). 1995. *Geomorphology and Sedimentology of Estuaries.* Paris: Elsevier.

Perillo, G.M.E., M.C. Piccolo, and M. Pino Quivira. 1999. What do we know about the geomorphology and physical oceanography of South American estuaries? In *Estuaries of South America: Their Geomorphology and Dynamics*, edited by G.M.E. Perillo, M.C. Piccolo, and M. Pino Quivira, 1–13. Berlin: Springer.

Salomons, W., H. Kremer, and R.K. Turner. 2005. The catchment to coast continuum. In *Coastal Fluxes in the Anthropocene*, edited by C.J. Crossland, H.H. Kremer, H.J. Lindeboom, J.I. Marshall Crossland, and M.D.A. LeTissier, 145–200. Global Change—The IGBP Series. Berlin: Springer.

Seitzinger, S.P., J.A. Harrison, E. Dumont, A.H.W. Beusen, and A.F. Bouwman. 2005. Sources and delivery of carbon, nitrogen, and phosphorus to the coastal zone: An overview of global Nutrient Export from Watersheds (NEWS) models and their application. *Global Biogeochemical Cycles* 19:GB4S01, doi: 10.1029/2005GB002606.

Stallard, R.F. 1998. Terrestrial sedimentation and the carbon cycle: Coupling weathering and erosion to carbon burial. *Global Biogeochemical Cycles* 12 (2):231–257.

Syvitski, J.M.P., N. Harvey, E. Wolanski, W.C. Burnett, G.M.E. Perillo, V. Gornitz, H. Bokuniewicz, M. Huettel, W.S. Moore, Y. Saito, M. Taniguchi, P. Hesp, W.W.-S. Yim, J. Salisbury, J. Campbell, M. Snoussi, S. Haida, R. Arthurton, and S. Gao. 2005. Dynamics of the coastal zone. In *Coastal Fluxes in the Anthropocene*, edited by C.J. Crossland, H.H. Kremer, H.J. Lindeboom, J.I. Marshall Crossland, and M.D.A. LeTissier, 39–94. Global Change—The IGBP Series. Berlin: Springer.

Syvitski, J.M.P., and J.D. Milliman. 2007. Geology, geography and humans battle for dominance over the delivery of fluvial sediments to the coastal ocean. *Journal of Geology* 115:1–19.

Trincardi, F., and J.P.M. Syvitski (eds.). 2005. Mediterranean prodelta systems. *Marine Geology* 222/223:1–514.

Turner, R.E., N.N. Rabalais, D. Justic, and Q. Dortch. 2003. Global patterns of dissolved N, P and Si in large rivers. *Biogeochemistry* 64:297–317.

UN GEMS/Water Programme. 2008. Global Environment Monitoring System of UNEP. www.gemswater.org.

Van Bennekom, A.J., and W. Salomons. 1981. Pathways of organic nutrients and organic matter from land to oceans through rivers. In *River Input to the Ocean System*, edited by J.M. Martin, J.D. Burton, and D. Eisma, 33–51. UNESCO–UNEP–SCOR Workshop, Rome 1979.

Vörösmarty, C.J., B.M. Fekete, M. Meybeck, and R.B. Lammers. 2000a. Geomorphometric attributes of the global system of rivers at 30-minute spatial resolution. *Journal of Hydrology* 237:17–39.

Vörösmarty, C.J., B.M. Fekete, M. Meybeck, and R.B. Lammers. 2000b. Global system of rivers: Its role in organizing continental land mass and defining land-to-ocean linkages. *Global Biogeochemical Cycles* 14 (2):599–621, doi: 10.1029/1999GB900092.

Vörösmarty, C.J., P. Green, J. Salisbury, and R.B. Lammers. 2000c. Global water resources: Vulnerability from climate change and population growth. *Science* 289 (5477):284–288.

Vörösmarty, C.J., and M. Meybeck. 2004. Responses of continental aquatic systems at the global scale: New paradigms, new methods. In *Vegetation, Water, Humans and the Climate*, edited by P. Kabat, M. Claussen, P.A. Dirmeyer, J.H.C. Gash, L. Bravo de Guenni, M. Meybeck, R.A. Pielke, C.J. Vörösmarty, R.W.A. Hutjes, and S. Lütkemeier, 375–413. Berlin: Springer.

Vörösmarty, C.J., M. Meybeck, B. Fekete, K. Sharma, P. Green, and J. Syvitski. 2003. Anthropogenic sediment retention: Major global-scale impact from the population of registered impoundments. *Global and Planetary Change* 39:169–190.

Walling, D.E. 1983. The sediment delivery problem. *Journal of Hydrology* 65:209–237.

Weaver, P.P.E., M. Canals, and F. Trincardi (eds.). 2006. Sources to sinks sedimentation on the European margin. *Marine Geology* 234:1–292.

8

Fluxes of Nutrients and Selected Organic Pollutants Carried by Rivers

Kon-Kee Liu, Sybil Seitzinger, Emilio Mayorga, John Harrison, and Venugopalan Ittekkot

It is estimated that the global load of nitrate and ammonia carried by rivers doubled and that of dissolved phosphate tripled from the 1970s to the 1990s as human population increased by more than 1.6 billion. Not only did nutrient loads increase, but a suite of man-made toxins also infiltrated the environment and will eventually be transported to the coastal zone. It is conceivable that these chemical loadings will affect semi-enclosed marine systems (SEMSs) most strongly. In this chapter, we provide estimates of the loads of nutrients (N and P) and two hazardous man-made chemicals—polychlorinated biphenyls (PCBs) and polycyclic aromatic hydrocarbons (PAHs)—carried by rivers into thirteen SEMSs (Plate 5; Table 8-1) and assess the potential impacts of these loadings on coastal environments.

Introduction

The global human population increased by 1%–2% annually during the last three decades of the twentieth century, reaching a current population of about 6.6 billion. While the global average population density is about 50 individuals km^{-2}, the population density in the coastal zones is almost double this, exceeding 90 individuals km^{-2} (Talaue-McManus 2009). Accompanying the population growth, human activities have quickly altered the landscapes of catchment basins around the world, mostly via deforestation, urbanization, and agriculture. Many rivers are experiencing alarmingly rapid rises in loadings of nutrients and other man-made chemicals, most of which are eventually discharged to coastal seas (Meybeck 2003; Smith et al. 2003; Turner et al. 2003; Seitzinger et al. 2005).

The total freshwater runoff discharged to coastal seas is estimated at 39,000 km^3 y^{-1}

Table 8-1. SEMSs for Discussion: Geographic Information on the Marine Systems and the Drainage Basins of Rivers Emptying to Them

System	*Abbreviations*	*Whole Area (10^6 km^2)*	*Shelf (10^6 km^2)*	*Basin (10^6 km^2)*	*Marine Climate Zone*	*Drainage Area (10^6 km^2)*	*Discharge (km^3 y^{-1})*	*Pop. Density (km^{-2})*
Baltic Sea	BA	0.39	0.38		Temperate–subarctic	1.74	291	46.6
Bay of Bengal	BB	0.61	0.19	0.42	Tropic	3.52	2,064	237.5
Black Sea	BL	0.47	0.15	0.33	Temperate	2.40	266	77.0
East China Sea	EC	1.21	0.98	0.23	Subtropic–temperate	4.07	747	251.8
Gulf of Mexico	GM	1.54	0.56	0.98	Tropic–subtropic	5.42	667	26.3
Gulf of St. Lawrence	GS	0.22	0.15	0.07	Temperate–subarctic	1.55	764	29.5
Gulf of Thailand	GT	0.28	0.28		Tropic	0.35	264	105.4
Hudson Bay	HB	0.84	0.82	0.02	Subarctic–arctic	3.31	587	1.7
Kara Sea	KS	0.78	0.78		Arctic	6.73	833	6.1
Laptev Sea	LS	0.49	0.49		Arctic	3.69	411	0.4
Northern Adriatic Sea	NA	0.07	0.07		Subtropic–temperate	0.18	102	164.0
North Sea	NS	0.60	0.56	0.04	Temperate–subarctic	0.87	338	184.5
Sea of Okhotsk	SO	1.57	0.68	0.89	Temperate–subarctic	2.51	388	25.2

The shelf represents the area within the 200 m isobath.

Only those systems with a basin area that is more than 5% of the whole are divided into the shelf and basin areas.

Areas of the marine systems are integrated according to the ETOPO2v2 Database (NGDC 2006) and the boundaries shown in Plate 5.

Watershed Climate Zone	Note
Temperate–subarctic	The Denmark straits serve as the outer boundary.
Tropic–subtropic	Salinity front and high ocean-color region define the domain, including the northern Andaman Sea.
Temperate	The Bosporus Strait serves as the outer boundary; the Sea of Azov is included.
Subtropic–temperate	The Okinawa Island Arc, the Taiwan Strait, and the Korea/Tsushima Strait serve as the outer boundary; the Yellow Sea and Bohai Sea are included.
Subtropic–temperate	The line from the Yucatan Peninsula to the Florida Keys defines the outer boundary.
Temperate–subarctic	The line from Cape Ray, Newfoundland, to Cape North, Nova Scotia, defines the outer boundary.
Tropic–subtropic	The line from the southern tip of Vietnam to Kota Baharu, Malaysia, defines the outer boundary.
Subarctic–arctic	The Foxe Basin and the Hudson Strait are not included in the system. Southampton Island and the eastward extension of its NE coastline serve as the outer boundary. This portion is likely the most impacted by riverine loads.
Subarctic–arctic	The 200 m isobath on the shelf serves as the outer boundary.
Subarctic–arctic	The 200 m isobath on the shelf serves as the outer boundary.
Temperate	The meridian of 16°E defines the outer boundary as suggested by the salinity front and high-ocean-color region.
Temperate–subarctic	The Shetland Islands and the shortest lines connecting them with Orkney Islands and Norway serve as the outer boundary.
Temperate–subarctic	The Kuril Islands serve as the outer boundary.

The drainage basin information is provided by the NEWS model (Seitzinger et al. 2005), where river discharge is from Fekete et al. 2002.

The climate zones are defined according to the following rough latitudinal classification: tropic, 0–23.5°N; subtropic, 23.5–33°N; temperate, 33–47°N; subarctic, 47–66.6°N; arctic, > 66.6°N.

(Fekete et al. 2000), and the total loading of terrigenous materials in both dissolved and particulate forms is about 14.6 Pg y^{-1} (Syvitski 2003; Syvitski et al. 2005). Humans have increased the amount of sediment transported by rivers by 2.3 ± 0.6 Pg y^{-1} via soil erosion (Syvitski et al. 2005). In this chapter we estimate discharges of nutrients and some hazardous man-made chemicals by rivers to a suite of SEMSs exhibiting a wide range of conditions, such as population density and climate in the adjacent catchment basins. Our purpose is to estimate the loadings of biogeochemically active materials in the Anthropocene (Meybeck 2003) and to assess their potential impacts on different SEMSs.

Rivers discharge many different forms of nutrient elements, including organic and inorganic species in dissolved and particulate form. The species most readily available for biological uptake by primary producers such as phytoplankton and other photosynthetic organisms are dissolved inorganic nitrogen and dissolved inorganic phosphorus (DIN and DIP, respectively). Dissolved organic nitrogen and dissolved organic phosphorus (DON and DOP, respectively) are also available to primary producers once they are converted into their inorganic forms. Particulate organic and inorganic forms of nitrogen and phosphorus are not easily distinguishable by regular chemical analysis, and for environmental purposes they are collectively called particulate nitrogen and particulate phosphorus (PN and PP, respectively). It is noted that for PN the *organic* form is usually more important, whereas the *inorganic* form of PP is more important.

The first global estimates of discharge of various nutrient elements by rivers in the 1970s were presented by Meybeck (1982). In recent years, several attempts have been made to update these global estimates. Most notable are those of the Land–Ocean Interactions in the Coastal Zone (LOICZ) project and the Nutrient Export from Watersheds (NEWS) project; the latter provided estimates for the 1990s (Seitzinger et al. 2005). Meybeck's estimates of global riverine loads of DIN and DIP were 821 Gmol N y^{-1} and 26 Gmol P y^{-1}, respectively, whereas those from the NEWS project were 1,771 Gmol N y^{-1} (Dumont et al. 2005) and 78 Gmol P y^{-1} (Harrison et al. 2005b), respectively. Despite the uncertainties pertaining to these estimates, it is conceivable that significant increases in riverine loads of nutrients occurred in the twenty years between the two assessments. Some forms of nutrients (i.e., DIN and DIP) are more likely to be augmented by human activities (Smith et al. 2003; Seitzinger et al. 2005). The human contributions have doubled for DIN and quadrupled for DIP. It is worth noting that the natural contributions also appear to have doubled for both DIN and DIP. Whether these changes are real or an artifact originating from uncertainties of the assessments warrants further investigation.

As far as the nutrient budget of the coastal ocean is concerned, the supply primarily comes from the open ocean (Walsh 1991; Liu et al. 2009a). However, nutrient fluxes carried by rivers often produce special effects in the coastal ecosystem. Increased nutrient loading from rivers may enhance primary productivity and fish yields (Nixon 1982; Nixon et al. 1986) or even enable drawdown of atmospheric CO_2 (Green et al. 2006), but negative effects often prevail. The impacts are undoubtedly strongest in SEMSs with

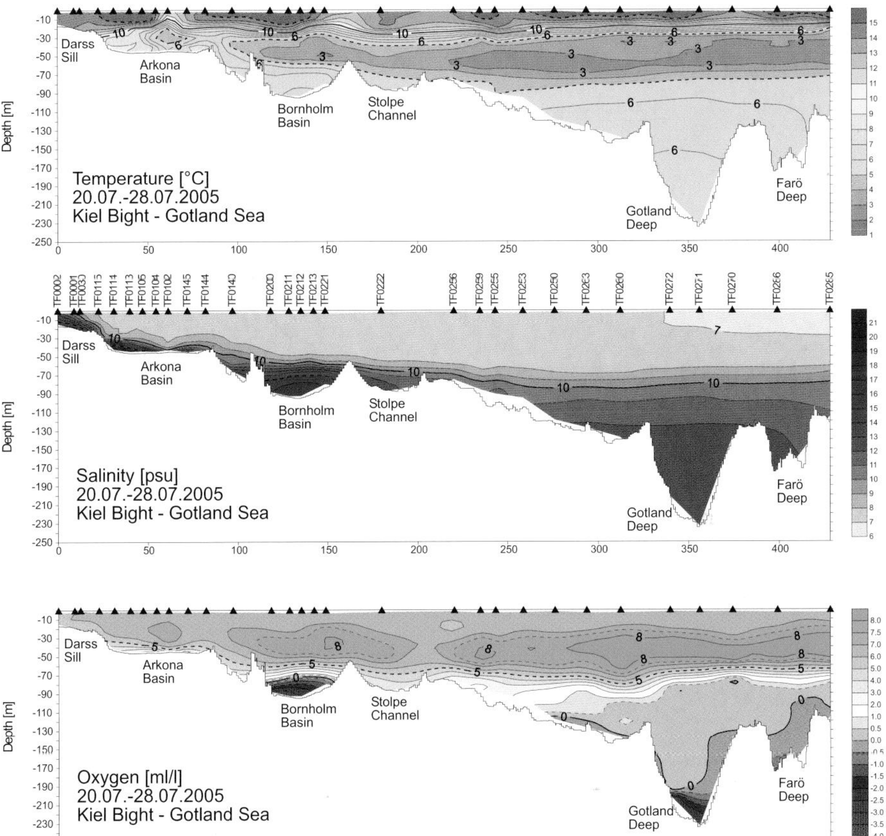

Plate 1. Horizontal–vertical sections of temperature (top), salinity (middle), and oxygen (bottom) as recorded during a Baltic monitoring cruise. Negative oxygen is used to indicate hydrogen sulfide. Observations were supported by the Bundesamt für Seeschifffahrt und Hydrographie (BSH).

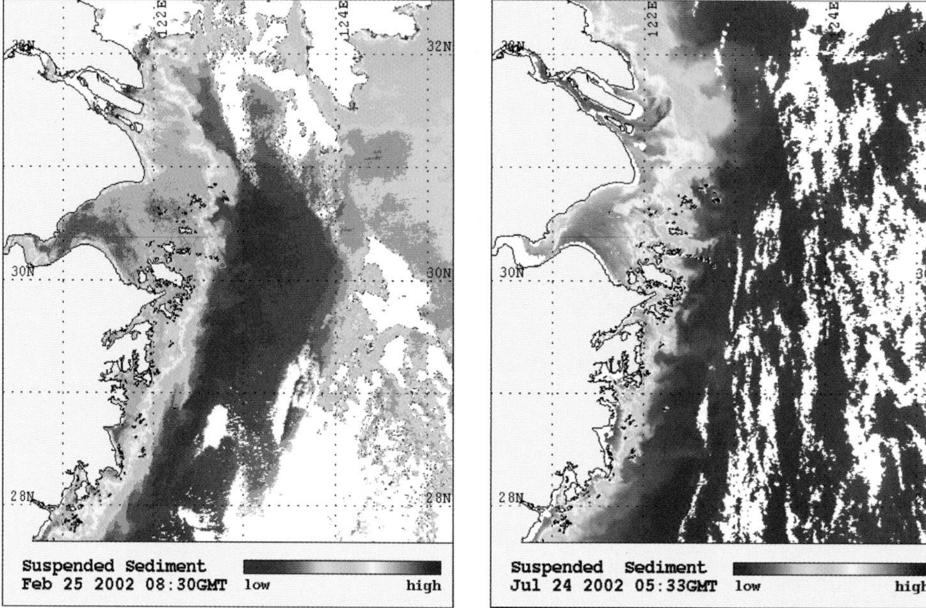

Plate 2. Coastal front in East China Sea under winter northeast monsoon (left) and summer southwest monsoon (right). SeaWiFS imagery courtesy of GeoEye, via the Hong Kong University of Science and Technology.

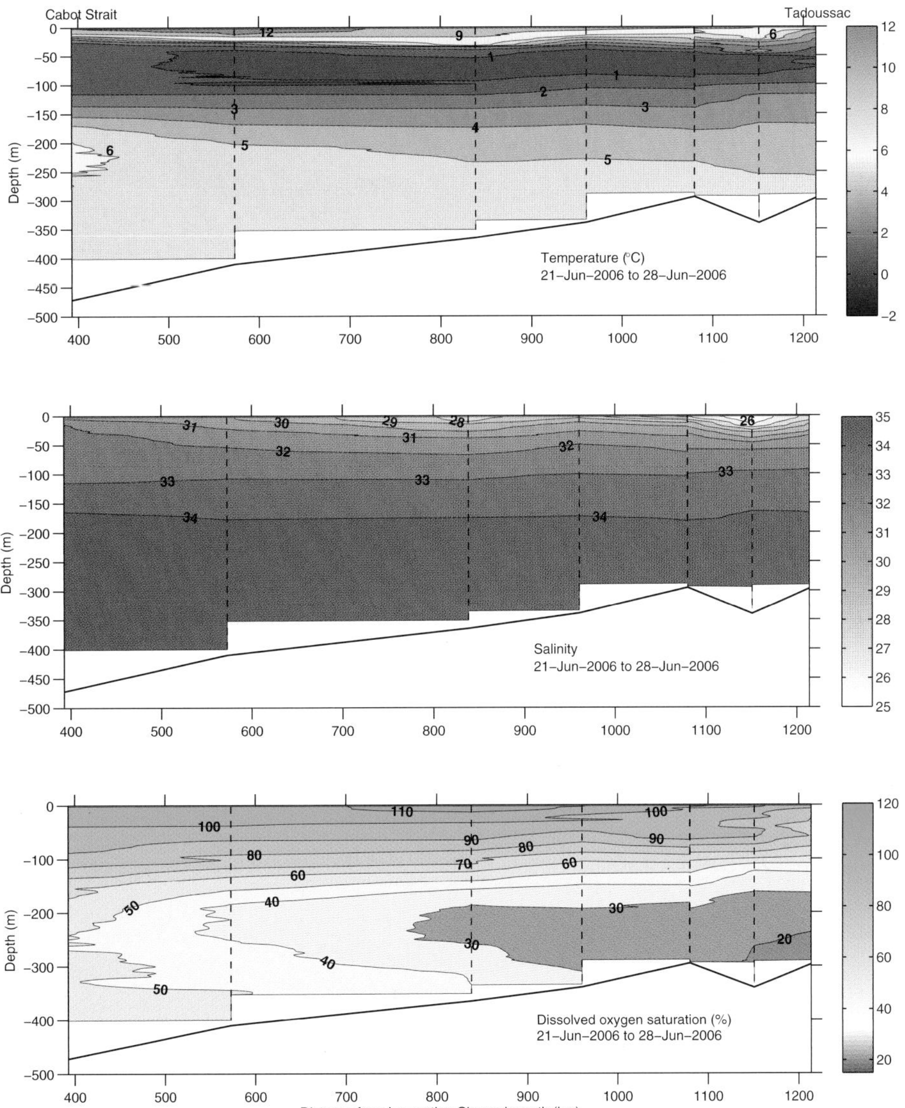

Plate 3. Horizontal–vertical sections of temperature (Panel A), salinity (Panel B), and oxygen (Panel C) along the deepest part of the Laurentian Channel, in the Gulf of St. Lawrence.

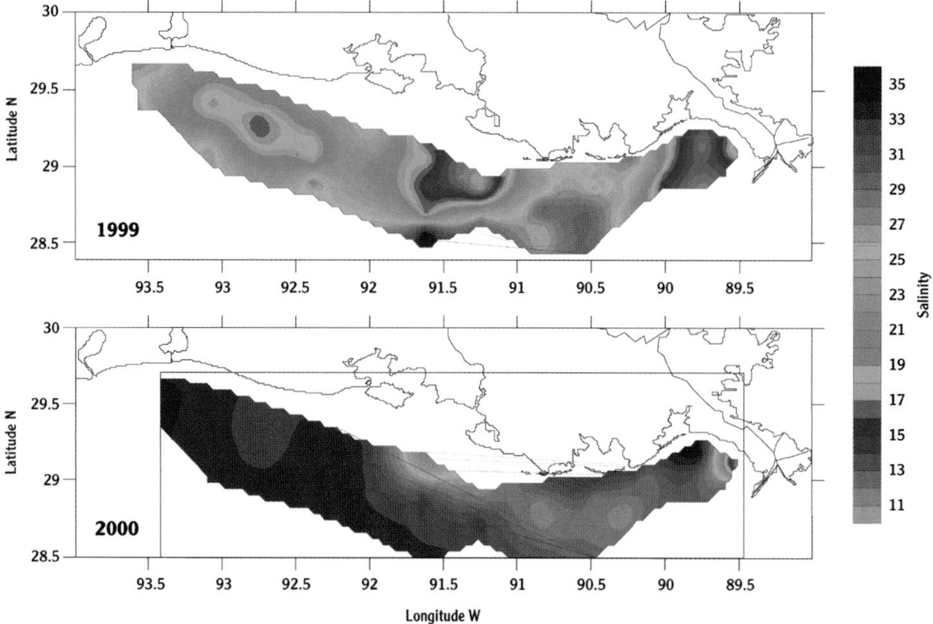

Plate 4. Surface salinity on the Louisiana–Texas (LATEX) continental shelf during July of 1999 and 2000, under conditions of high and low freshwater discharges, respectively. Figure provided by Nazan Atilla and Nancy Rabalais, Louisiana Universities Marine Consortium (unpublished data).

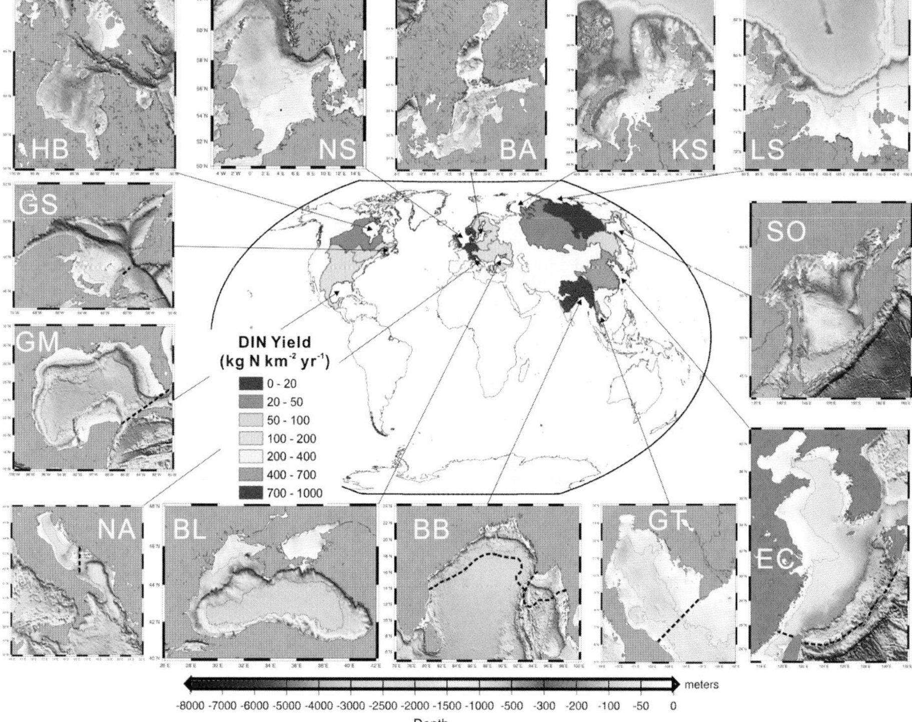

Plate 5. Map of the catchment basins connected to the SEMSs analyzed in this study. The color code indicates the yield of dissolved inorganic nitrogen, a major riverine nutrient. The bathymetry map of each system is shown with the boundaries of the system shown as blue or black dashed curves. The abbreviation of the name of each system is shown. (Please refer to Table 8-1 for the full names of the systems.) From the Continental Margins Task Team, http://cmtt.pangaea.de/0_List_of_maps.htm.

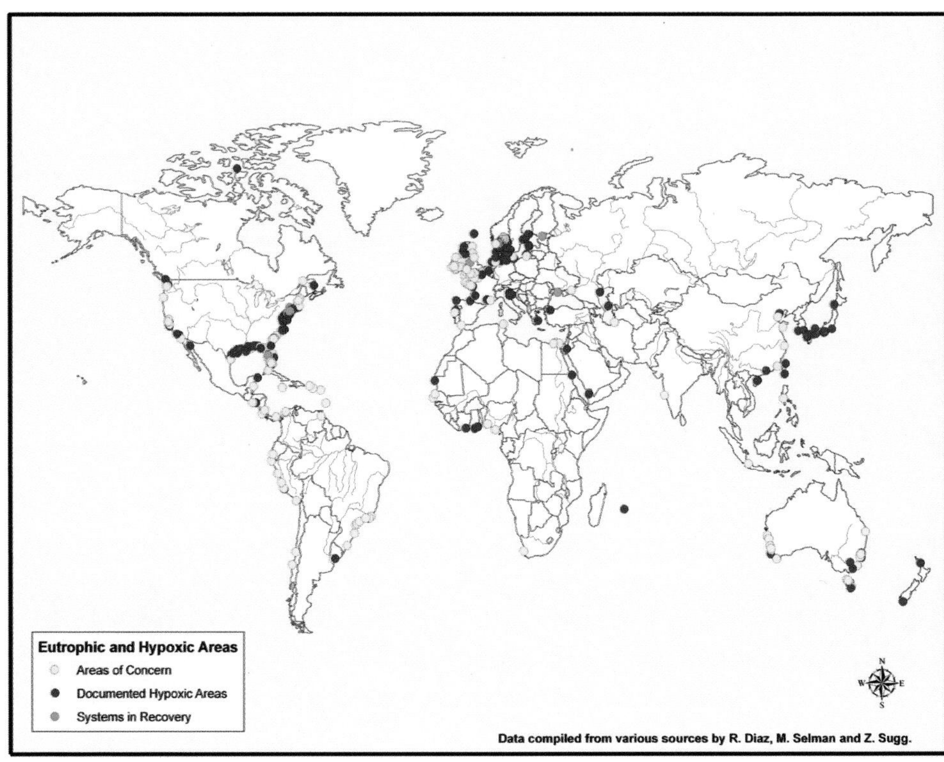

Plate 6. Distribution of areas with coastal hypoxic bottom waters on a worldwide basis. Original data in Diaz and Rosenberg 1995; modified by R.J. Diaz, Virginia Institute of Marine Science, based on a compilation by R.J. Diaz for Selman et al. 2008. Used with permission from the World Resources Institute.

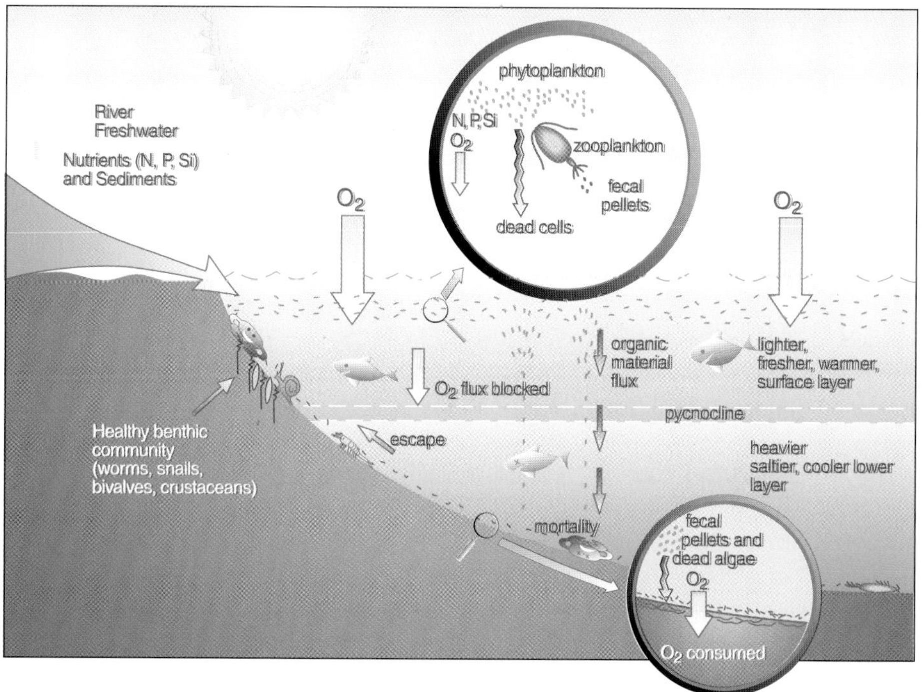

Plate 7. Processes involved in the development and maintenance of hypoxia on the northern Gulf of Mexico continental shelf where there is sustained and high freshwater discharge from the Mississippi River, year-round haline and seasonal thermal stratification, nutrient-enhanced primary production, and accumulation and decomposition of organic matter (CAST 1999, p. 5). Used with permission from the Council for Agricultural Science and Technology (CAST).

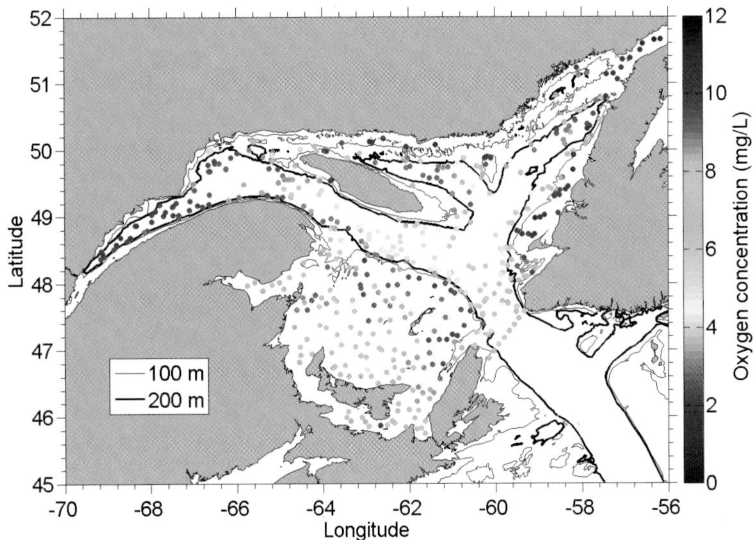

Plate 8. Near-bottom oxygen concentration (mg L^{-1}) measured during the 2004 (southern and northern Gulf of St. Lawrence) and 2005 (only the northern part of the Gulf of St. Lawrence) fish stock assessment surveys. Modified from data published in Gilbert et al. 2007. Reproduced with permission from Société Provancher.

significant freshwater discharges, which may cause strong stratification in river plumes, which in turn may reduce vertical mixing. Sluggish mixing in the water column often results in hypoxia or anoxia beneath the river plume. Restricted exchanges with the open ocean prolong the residence time of the freshwater and the nutrient loads, enhancing their effects.

Aside from nutrients, the loading of man-made chemicals in rivers is increasing. Most notable is the loading of hydrophobic organic contaminants (HOCs), which are easily taken up by particulate organic matter and transported in the solid phase (Ko and Baker 2004). There are many kinds of HOCs; most notable are organohalogens and polycyclic aromatic hydrocarbons (PAHs). The former include pesticides, such as dichlorodiphenyl trichloroethane (DDT); industrial chemicals, such as polychlorinated biphenyls (PCBs); and unintentionally produced pollutants, such as dioxins. Many of these organochlorines are persistent organic pollutants (POPs) due to their resistance to degradation in the environment (U.S. EPA 2002). Even though production of POPs has been banned in many countries since the adoption of the Stockholm Convention in 2001, POPs still exist throughout the environment due to accumulation from past use and slow decomposition. POPs are soluble in fat, so they tend to concentrate in fatty tissues of organisms, including those found in SEMSs; thus they can reach humans at the top of the food chain. People who consume large amounts of fish and shellfish are at high risk of POP exposure. Accumulation of POPs in the human body may cause reproductive, developmental, neurological, and immunological problems (ATSDR 1995). Especially vulnerable are fetuses and breast-fed babies of women who are exposed to POPs, because POPs can pass through the placenta and enter breast milk.

All PAHs have chemical structures of two or more fused benzene rings and occur in thousands of varieties (Eisler 1987). They occur naturally in coal and petroleum but can also be produced during combustion of fossil fuel (Ko and Baker 2004). Many of the four- to seven-ring PAHs are carcinogenic. They also cause harmful effects on skin and immune systems of animals. Contrary to POPs, PAHs are not bio-amplified in food chains, even though they tend to dissolve in fats. This is probably because most PAHs can be metabolized. The transport of PCBs and PAHs by rivers is assessed in this chapter.

Semi-Enclosed Marine Systems for Discussion

Thirteen SEMSs are examined in this chapter and other chapters of this book (Table 8-1; Plate 5). They include basins with a wide range of physical characteristics, from the tropics to the Arctic. All of them receive a sizable discharge of freshwater, ranging from 102 to 2,064 $km^3 y^{-1}$, which may drive a positive estuarine circulation (Cameron and Pritchard 1963; Simpson 1997). The outflow of river plumes, mainly controlled by Coriolis force, tides, and wind (Chao and Boicourt 1986), tends to disperse the riverine loads of terrigenous materials, whereas the landward intrusion of salt wedges tends to bring back particulate matter that sinks out of the surface plume. This positive estuarine circulation pattern may retain riverine loads near the coastal zone, where detrimental conditions, such as eutrophication and contaminated food chains, may arise due to the cumulative effects of riverine loads.

The domains of the SEMSs are briefly described in Table 8-1. Some of the systems are more closed than others because of their surrounding landmasses and islands. The extent of the marine portion of each basin is defined such that the more restricted water exchange with external water bodies may result in more discernible effects of the riverine loads of nutrients and man-made chemicals. Systems that have narrow channels or straits connecting with the external water body, such as the Black and Baltic seas, are defined by the outer edges and the extensions of the landmasses forming the geographic features. Systems that have sills, ridges, or shelf breaks on the opening side connecting to the external water body, such as the East China Sea or Kara Sea, are defined by these submarine topographic features. Those that do not have restricting geographic features, such as the Bay of Bengal, are defined by the front separating the low salinity and high ocean-color coastal water from the open-ocean water.

The areas of the marine systems considered in this study vary considerably from 6.8×10^4 km^2 (Northern Adriatic Sea) to 1.57×10^6 km^2 (Sea of Okhotsk). Some of the systems have rather deep seafloors, as deep as 4,160 m in the Gulf of Mexico, where the influences of riverine loads are probably limited, except for downslope transports of sediments and associated particulate materials. To assess the more direct impacts exerted upon the shallower parts of the SEMSs, we divide some of the systems into the shelf (shallower than 200 m) and the basin parts (Table 8-1). This was done for systems in which the deeper parts are off the shelf, instead of depressions within the shelf, and occupy more than 5% of the total area of the SEMS.

The drainage basins of the rivers that empty into the marine systems also vary in size, from 1.77×10^5 km^2 (Northern Adriatic Sea) to 6.73×10^6 km^2 (Kara Sea), and the population densities vary from as low as 0.4 persons km^{-2} (Laptev Sea) to as high as 237.5 persons km^{-2} (Bay of Bengal). The ratios between the drainage areas and the marine systems vary from 1.26 for the Gulf of Thailand to 8.63 for the Kara Sea, but if only the shelf region is concerned, the maxima go as high as 16.2 and 18.6 for the Black Sea and the Bay of Bengal, respectively. Because the latter two drainage basins have high population densities, human stresses on these marine systems can be very high. Annual river runoffs, if evenly distributed in the marine systems, represent a layer of freshwater with thickness varying from 0.25 m for the Sea of Okhotsk to 3.47 m for the Gulf of St. Lawrence. If only the shelves are considered, the shelf of the Bay of Bengal receives almost 11 m of freshwater annually.

Approaches

Although nutrient analyses are rather common for river water quality monitoring, they are not conducted for all rivers on a regular basis. Therefore, some scaling techniques must be employed to obtain the total nutrient loads of all rivers discharged to a basin. For this study, we adopted the most recently developed model from the NEWS project as the primary source of information on riverine exports of nutrient elements. For comparison, we rely on two approaches: (1) estimates from a different model, and (2) data from an existing compilation of riverine exports of nutrients to the coastal seas.

For the NEWS project, a spatially explicit model consisting of multiple submodels was developed for various forms of bioactive elements, namely, C, N, and P (Beusen et al. 2005; Dumont et al. 2005; Harrison et al. 2005a, b; Seitzinger et al. 2005). The mean annual yield of each form of N, P, and C in a river basin is predicted on the basis of a consistent data set of environmental and anthropogenic factors (such as water runoff, population density, fertilizer input, land use, atmospheric N deposition, sewage treatment, and natural nitrogen fixation rate, among others) and mechanistic components (such as removal of nutrients in dammed reservoirs and by crop harvest). The NEWS model is based on the STN30-p (version 6) global river system data set (Vörösmarty et al. 2000) and runoff and discharge data from Fekete and colleagues (2002).

For comparison purposes we use the LOICZ model, a multiple regression model developed to predict the DIN and DIP production rate in catchment basins as a function of population density and runoff (Smith et al. 2003). For the second comparison, we use the estimates of nutrient fluxes to various coastal seas from the regional syntheses of a large set of continental margins produced by the Continental Margins Task Team (CMTT) cosponsored by the LOICZ project and the Joint Global Ocean Flux Study (JGOFS) (Liu et al. 2009b). The CMTT estimates were based on individual contributors' own compilations rather than a single unified algorithm.

For PCBs and PAHs, it is more difficult than for nutrients to obtain quantitative estimates, because of much less data availability and the lack of modeling approaches to generalize the limited observations. Because of the high affinities of PCBs and PAHs for particulate organic matter, we estimated their loadings using their relationships with organic carbon and the POC loads predicted by the NEWS model.

Riverine Loads of Nutrients and Other Man-Made Chemicals

Riverine Fluxes of Nutrients

Riverine exports of each nutrient form to the marine systems predicted by the NEWS model are listed in Table 8-2. The dissolved inorganic forms are the most readily available to marine primary producers, including mostly phytoplankton but also macro-algae, sea grasses, and mangroves. A significant fraction of the riverine PP is likely to be released as DIP, when PP reaches the saline environment (Froelich 1988; Van der Zee et al. 2007). Some fractions of organic forms of nutrients may be readily decomposable by heterotrophic microbes or by enzymes released by autotrophs, such as alkaline phosphatase, and become available. The proportion of DON from land-based sources that is bioavail-

Table 8-2. Riverine Exports (Gmol y^{-1}) of Different Forms of Bioactive Elements and Two Types of Hazardous Organic Chemicals (PCBs and PAHs in ton y^{-1}) to SEMSs

	NEWS						
System	*DIN*	*DIP*	*DON*	*DOP*	*PN*	*PP*	*POC*
Baltic Sea	19.8	0.68	7.5	0.22	4.7	0.86	49
Bay of Bengal	222.5	3.54	51.8	1.58	99.0	24.37	1,396
Black Sea	23.6	1.47	10.5	0.38	8.8	1.75	100
East China Sea	122.7	2.55	25.3	1.03	71.6	13.72	783
Gulf of Mexico	77.2	1.28	17.1	0.54	29.7	5.96	340
Gulf of St. Lawrence	15.7	0.29	17.5	0.47	5.3	0.96	55
Gulf of Thailand	10.4	0.22	6.1	0.17	8.9	1.74	100
Hudson Bay	8.3	0.14	11.8	0.27	6.7	1.21	69
Kara Sea	16.0	0.47	17.1	0.42	18.7	3.51	200
Laptev Sea	3.4	0.09	7.8	0.17	10.6	1.98	113
Northern Adriatic Sea	11.9	0.50	3.3	0.10	4.5	1.06	61
North Sea	58.4	2.44	13.8	0.47	7.4	1.40	80
Sea of Okhotsk	19.9	0.58	8.5	0.21	8.7	1.63	93

Different estimates of riverine exports (Gmol y^{-1}) of dissolved inorganic N and P to SEMSs are compared.
NEWS represents output for the 1990s from the model described in Seitzinger et al. 2005.
LOICZ represents the output from the model described in Smith et al. 2003.
CMTT represents the compiled results in Liu et al. 2009a.

able varies considerably. For example, the percent of DON that was bioavailable in runoff from urban/suburban land, agricultural pastures, and forests was 59% ± 11%, 30% ± 14%, and 23% ± 19%, respectively, based on ten- to twelve-day incubation experiments (Seitzinger et al. 2002b). Algal uptake of DOP can be as efficient as uptake of DIP (Cotner and Wetzel 1992), if reactive DOP is available at a relatively high concentration. However, significant fractions of the organic and particulate forms are decomposed or solubilized into dissolved inorganic forms gradually, so they become available to primary producers at a longer time scale. Temperature probably plays an important role in setting the mineralization rate of nutrients in organic and particulate form; as temperature increases, rates increase. Sedimentary bacteria are especially important in regeneration of nutrients in tropical environments (Alongi 1994).

A comparison of the NEWS estimates of DIN and DIP river exports with the LOICZ and CMTT estimates is shown in Table 8-2 and Figure 8-1. While there are several differences for both DIN and DIP, overall there is reasonably good correlation between the

LOICZ		*CMTT*		*This study*	
DIN	*DIP*	*DIN*	*DIP*	*PCBs*	*PAHs*
16.8	0.89	14.2	0.55	0.8–8	8–37
156.4	8.73	35.7		23–235	235–1,055
20.6	1.08	40.7		1.7–17	17–75
77.2	4.17	110.0	1.00	13–132	132–592
34.6	1.80	70.0[a]	1.35[a]	6–57	57–257
29.1	1.58			0.9–9	9–41
14.2	0.79	3.2[b]	0.04[b]	1.7–17	17–75
10.7	0.55			1.2–12	12–52
25.8	1.33	59.0	1.90	3.4–34	34–151
5.0	0.25	4.1	0.16	1.9–19	19–85
6.8	0.38	8.56[c]	0.23[c]	1.0–10	10–46
26.0	1.43	43.6	0.88	1.3–13	13–60
18.7	0.98	10.0[d]	1.00[d]	1.6–16	16–70

[a] Only Mississippi-Atchafalaya Rivers are included.
[b] Only Chao Phraya River is included.
[c] Only Po River is included.
[d] These are upper limits of the estimated fluxes.

estimates from the NEWS and the LOICZ models, with R^2 values 0.94 for DIN and 0.70 for DIP. However, there are systematic differences between the NEWS and the LOICZ estimates for high loading values. In most cases, the LOICZ model predicts lower values for DIN, with a regression slope of 0.70 relative to the NEWS estimates, and predicts higher values for DIP, with a regression slope of 1.71. On the other hand, the data points representing the relationship between the NEWS and CMTT estimates are mostly scattered about the 1:1 line, with a few outliers. For DIN, the two outliers are those for the Bay of Bengal and the Kara Sea. For the Bay of Bengal, the NEWS estimate (222.5 Gmol y^{-1}) is much higher than the CMTT estimate (35.7 Gmol y^{-1}) provided by Naqvi and colleagues (2009). However, the estimates of total nitrogen loading, 373 Gmol y^{-1} from NEWS and 286 Gmol y^{-1} from Naqvi and colleagues, are much closer. This reflects the difficulties in partitioning the different forms of a nutrient element when estimating their respective loads. For the Kara Sea, the NEWS estimate (16 Gmol y^{-1}) is considerably lower than the CMTT estimate (59 Gmol y^{-1}) provided by Macdon-

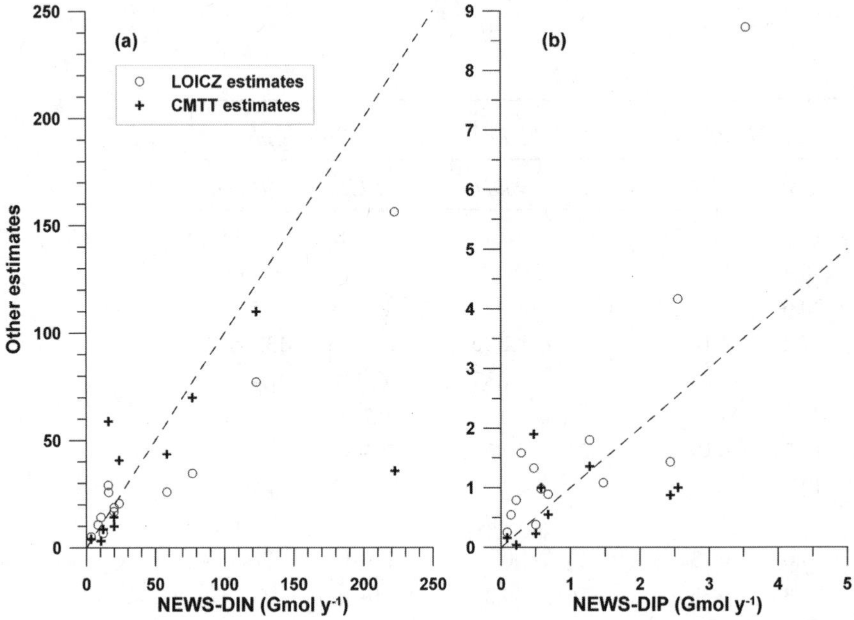

Figure 8-1. Comparison of estimates of nutrient loads predicted by the NEWS model, the LOICZ project, and the CMTT. (A) Dissolved inorganic nitrogen (DIN); (B) dissolved inorganic phosphorus (DIP). The dashed line is the 1:1 line.

ald and colleagues (2009). The estimates of the total dissolved nitrogen loading, 33 Gmol y^{-1} from NEWS and 72 Gmol y^{-1} from Macdonald and colleagues, are closer. For DIP, the three outliers belong to the Kara Sea, East China Sea, and North Sea. The NEWS estimate is lower for the Kara Sea, as is the situation for DIN, but higher for the other two systems. In general, the NEWS model generates nutrient loads in reasonably good agreement with other estimates, aside from the Bay of Bengal, where the large tropical rivers warrant further investigation of their biogeochemical behaviors.

The results of the NEWS model indicate that river runoff and relief of the watershed (natural factors) and population density, anthropogenic input, and alterations to hydrology (human factors) in the watershed are the most important factors controlling nutrient loads (Seitzinger et al. 2005). The LOICZ model also confirms the importance of runoff and population density in controlling the delivery of nutrients (Smith et al. 2003). Both models predict that human sources account for two-thirds or more of dissolved inorganic nutrients delivered by rivers. Smith and colleagues (2003) suggest that the natural or pristine conditions for nutrient export from rivers may exist in watersheds with population densities below 1 individual km^{-2}.

Biogeochemical Characteristics of Amounts and Forms of Nutrients from Land

Adopting the NEWS estimates for the riverine nutrient loads, we explore the characteristics of the nutrients. First we investigate the relative importance of the three forms of nutrients, namely the dissolved inorganic form, the dissolved organic form, and the particulate form, using ternary diagrams (Figure 8-2). This type of figure shows the relative contributions of three components—in this case, the three forms of a nutrient element. The contribution of a component in a riverine load represented by any point plotted on the diagram can be determined as the intersection of the percentage line with the axis representing the component. An example is given for the riverine nitrogen loads to the Laptev Sea (LS) in Figure 8-2A, which shows the contributions are 16%, 36%, and 48% for DIN, DON, and PN, respectively.

From the ternary diagrams we can see that nitrogen and phosphorus have quite different partitioning among the three forms. For nitrogen, DIN is the dominant form in all but four (GS, HB, KS and LS) of the thirteen SEMSs (Figure 8-2A). By contrast, PP is the dominant form of phosphorus (Figure 8-2B) in all but one system (NS). PN is dominant in the two mainly arctic watersheds, those associated with the Kara and Laptev seas; DON is dominant in the drainage basins of two mainly subarctic watersheds, those associated with the Gulf of St. Lawrence and the Hudson Bay. Both areas are sparsely populated. The watershed associated with the Sea of Okhotsk is also partly in the subarctic zone, but its nutrient characteristics do not resemble the group of the other four high-latitude watersheds. DIP is the dominant form only in the runoff to the North Sea (Figure 8-2B), which also has the highest fraction in DIN (Figure 8-2A). In fact, all European drainage basins, including those of the Baltic, the Black, and the Northern Adriatic seas, have relatively high contribution fractions in dissolved inorganic forms of nutrients (Figure 8-2).

The predicted riverine loads of N and P discharged to the thirteen SEMSs fall in wide ranges over 3 orders of magnitude (Table 8-2). The riverine exports to the Bay of Bengal are the largest in all categories, apparently due to the very high water discharge, which is 2.5 to 20 times as high as other SEMS in this analysis. The total population, which is the second highest in all drainage areas, must also contribute significantly to the high riverine chemical loads. The river runoffs to the Laptev Sea carry the lowest loads of dissolved inorganic nutrients, although the water discharge is the median one. Its drainage basin has extremely low population density (Table 8-1). The Northern Adriatic Sea receives the smallest loads of DON, DOP, and PN, while the Baltic Sea receives the lowest loads of POC and PP.

The predicted mean concentrations of the DIN and DIP in river waters vary widely across the study basins, from 8 μM and 0.2 μM (Laptev Sea) to 173 μM and 7 μM (North Sea) for DIN and DIP, respectively (Figure 8-3A). The predicted mean concentrations of

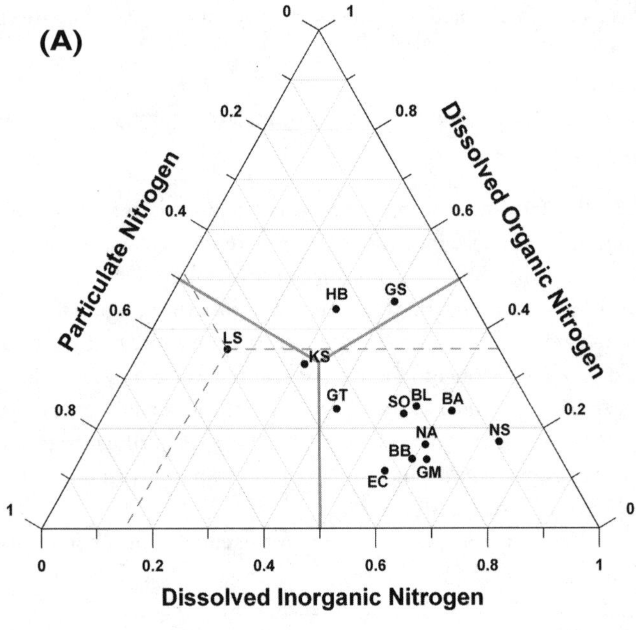

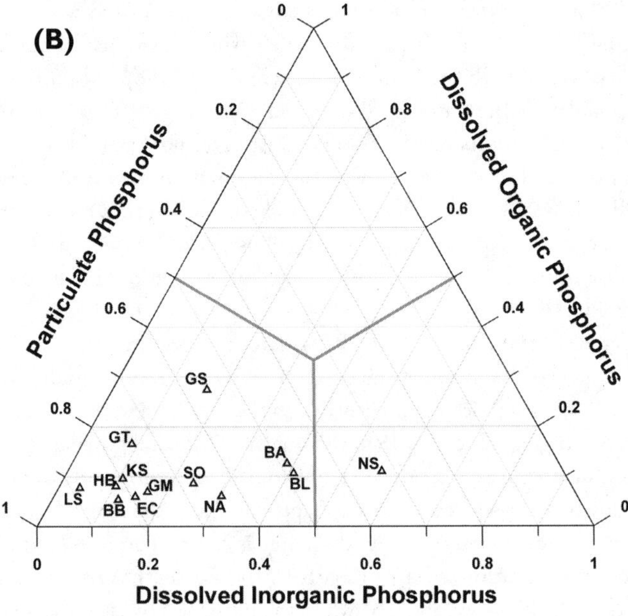

Figure 8-2. Ternary diagram illustrating the relative importance of the three forms of nutrients—the dissolved inorganic form, the dissolved organic form, and the particulate form—for nitrogen (A) and phosphorus (B). The abbreviation at each data point indicates its associated SEMS (see Table 8-1).

the particulate form of organic carbon (POC) and of PN and PP also vary considerably. The highest concentrations—1,048 μM, 96 μM, and 18 μM for PC, PN, and PP, respectively—occur in the water runoffs to the Bay of Bengal, while the lowest ones—71 μM, 6.9 μM, and 1.3 μM—occur in the runoffs to the Gulf of St. Lawrence. By contrast, the concentrations of the dissolved organic forms of N and P are relatively uniform—19–41 μM for DON and 0.42–1.44 μM for DOP.

Several additional factors must be taken into account to extrapolate from loads, concentrations, or nutrient ratios to eutrophication effects or nutrient limitation. First of all, the growth of algae in marine environments is often limited by the availability of either N or P nutrients, depending on which of the two most closely approaches the critical minimum concentration to sustain algal growth (Smith 1984). Because the algal uptake of N and P nutrients during photosynthesis has a reasonably constant atomic ratio of 16:1 (Redfield et al. 1963), it is not the absolute concentration of the two nutrients that determines which one may become the limiting nutrient during the course of algal growth, but rather the atomic (or molar) ratio between them. Moreover, the distribution and composition of the discharged nutrients change with time in coastal waters, depending on physical factors such as circulation, water residence time, dispersion, dilution, and water depth of the receiving system, and depending on biological and chemical transformations of N and P, which are usually temperature dependent. These transformations can affect nutrient ratios, forms, and amounts. For example, when PP encounters salt water, strong ions such as Cl^- compete with PO_4^{3-} for ion exchange sites on particulates. The resulting release of phosphate from particulates, both suspended and deposited, to seawater has been shown to increase P availability by as much as five times (Froelich 1988). Conversely, humic-complexed iron can bind to, flocculate, and remove PP from suspension at low salinities in estuaries (Howarth et al. 1995). N can also be affected greatly by coastal processes. For example, rapid denitrification in estuaries may significantly reduce the amount of N available for use by marine primary producers. In addition, preferential regeneration of nutrient elements may further complicate the N–P relationship for the net community production, which is the balance between primary production and total community respiration that results in net accumulation of organic carbon in SEMSs. Elevated C:N ratios (up to 7 times the Redfield ratio) and C:P ratios (up to 4.2 times the Redfield ratio) in elemental uptake were observed for the net community production in the North Sea (Bozec et al. 2006).

Although riverine nutrient ratios cannot be used directly to infer nutrient limitation in receiving water bodies, because of rapid transformations, dilution, and so on in coastal zones, it is still instructive to examine nutrient ratios in riverine N and P loads. The DIN:DIP ratios of riverine loads to SEMSs, in all SEMSs examined except the Black Sea, are considerably higher than the Redfield N:P ratio (16 by atoms). The highest DIN:DIP ratio is as high as 69. The regression slope gives a DIN:DIP ratio of 30. The dissolved organic form is less important relative to the dissolved inorganic form except in the group of the four high-latitude watersheds mentioned above. The addition of the

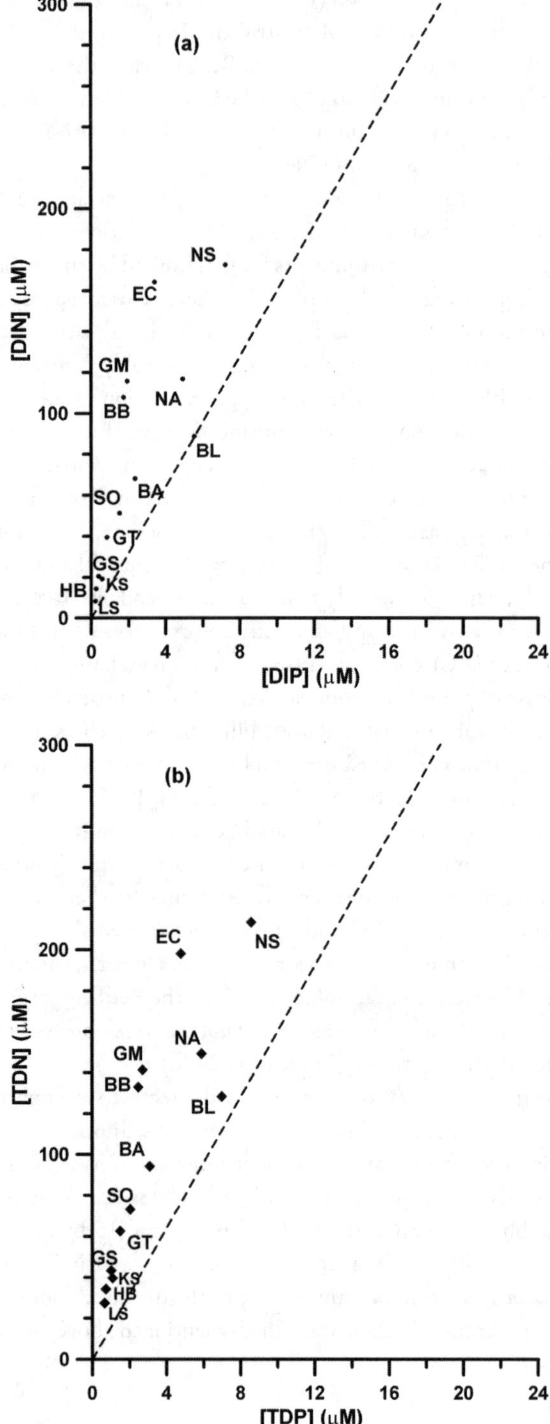

(a)
[DIN] (μM)
[DIP] (μM)
NS
EC
GM
NA
BB
BL
BA
SO
GT
GS
KS
HB
LS
(b)
[TDN] (μM)
[TDP] (μM)
EC
NS
NA
GM
BB
BL
BA
SO
GT
GS
KS
HB
LS

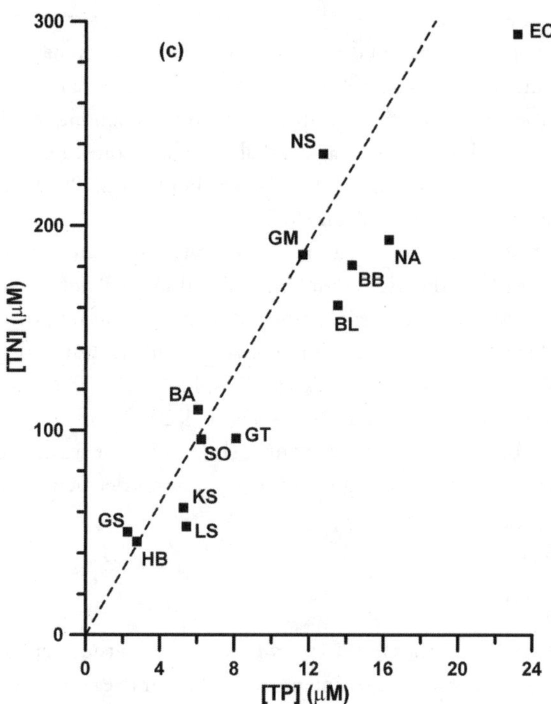

Figure 8-3. *(left and above)* The N–P relationship of the riverine nutrient loads in different forms discharged to the SEMSs: (A) dissolved inorganic nutrients, (B) total dissolved nutrients, (C) total nutrients. The dashed line in each plot shows the Redfield ratio of N:P = 16, which is the average N:P atomic ratio in marine plankton. Compared with the Redfield ratio, the dissolved nutrient loads mostly show nitrogen excesses in Panels A and B, whereas the total nutrient loads, which include the particulate forms, show N:P relationships more compatible with the Redfield ratio in Panel C.

dissolved organic form did not change the N:P ratio much (Figure 8-3B). However, the addition of the particulate form changes the N–P relationship significantly, such that the N:P ratios of the total nutrient loads are all close to, or even less than, the Redfield ratio (Figure 8-3C).

The very high river nutrient concentrations in certain basins may make river plumes very enriched in nutrients, and eutrophication may occur. However, the mismatch of the N and P concentrations may reduce the full impact of the high nutrient concentrations in areas such as the East China Sea. It has been demonstrated by nutrient manipulation experiments that P limitation occurs to microplankton on the periphery of the Changjiang plume (Chang and Liu 2007) and other river plumes on the Chinese coast (Harrison et al. 1990). Widespread occurrences of excess nitrate have been observed in the surface water of the East China Sea (Wong et al. 1998), suggesting that the high N:P ratio (48.2) in the riverine load of dissolved inorganic nutrients may play an important role in determining the limiting nutrient in the river plume. Because the N:P ratio of the riverine load to the East China Sea is only the fifth highest among the thirteen systems, the P limitation in the East China Sea is probably not a unique case. A similar situation has also been observed in the Sagami Bay, Japan (Fujiki et al. 2004) and the Gironde plume in the Bay of Biscay (Labry et al. 2002).

On a longer time scale, the situation may become quite different. Once the PP is deposited on the seafloor, the iron-bound phosphorus may be mobilized in the reducing environment because of dissolution of its host minerals, the ferric hydroxides (Canfield 1993). The benthic regenerated DIP may compensate for the relative deficiency of phosphorus in the dissolved form. In fact, excess phosphate has been observed in the inner shelf bottom water of the East China Sea (Wong et al. 1998) and may be partially attributed to the release of DIP from the sediments. Alternatively, intrusion of the intermediate water onto the shelf may also augment the phosphorus deficiency (Gong et al. 1996; Liu et al. 2000).

PCBs and PAHs

PCBs and PAHs are ubiquitous in the environment, even in the Arctic Ocean. The total PCB and PAH concentrations in sediments from the Kara Sea (Sericano et al. 2001) are 0.41 ± 0.34 ppb and 130 ppb, respectively, while those in sediments from the Mississippi River delta (Santschi et al. 2001) are not much higher—7 ppb and 660 ppb, respectively. It is noteworthy that fish from the Kara Sea, as compared with sediments, have much higher PCB and PAH concentrations, 290 ± 200 ppb and 590 ± 170 ppb, respectively. The extremely high enrichment of PCB in fish indicates the bio-amplification effect of PCBs.

The relatively high levels of HOCs in the Kara Sea sediments illustrate the fact that the transport of HOCs is not confined to catchment basins but is transboundary because of their tendency to travel with air masses, either in the gas or aerosol phase (Baker et al. 1991). Airborne HOCs may be washed out by rain or removed by snow during cold condensation processes (Baker et al. 1991; Melnikov et al. 2003).

The concentrations of HOCs in river waters vary over huge ranges—0.11 to 403 ng L^{-1} for total PCB (Zhang et al. 2007) and 4–26,920 ng L^{-1} for total PAH (Guo et al. 2007). However, many of the extremely high values are attributed to local sources of pollution, such as the urban watershed or disposal of petroleum wastes (Hartmann et al. 2004). For relatively large watersheds that have different types of land uses, the HOC concentrations normalized to POC concentrations fall in relatively narrow ranges. Thus, if one can estimate POC loads, then one should be able to estimate PAH loads. The mean ratio between the total PCB and POC concentrations was found to be 14 $\mu g\ g^{-1}$ for both the Mississippi River (Zhang et al. 2007) and the Ebro River in France (Gómez-

Gutiérrez et al. 2006). A low value of 1.4 μg g^{-1} was found for the Susquehanna River in the United States (Ko and Baker 2004). The ratios found in other water bodies on land, including suspended particulate matter from two lakes in The Netherlands (Koelmans et al. 1997), primary sewage in Thessaloniki, Greece (Katsoyiannis and Samara 2004), and sediments from the Danshuei River in Taiwan (Hung et al. 2006), mostly fall between the two extremes.

The mean total PAH:POC ratio in rivers was found to be 14 ± 2 μg g^{-1} for the York River in the United States (Countway et al. 2003) and 63 ± 40 μg g^{-1} for the Xijiang in China (Deng et al. 2006). The ratios for Lake Superior (Baker et al. 1991), the Susquehanna River (Ko and Baker 2004), the Mississippi River (Zhang et al. 2007), and the Mackenzie River in Canada (Yunker et al. 2002) all fall within this range. Some rivers with catchment basins in coal and petroleum production areas or refinery areas have extremely high PAH concentrations. For example, total PAH concentrations as high as 13.4 μg L^{-1} have been observed in the Daliao River in China (Guo et al. 2007), and the total PAH:POC ratio may reach as high as 280. However, the PAH:POC ratio for sediments from the Daliao River was 27 ± 5, which is within the range mentioned above. This seems to suggest that the excessive PAHs from fossil fuel sources may be either degraded quickly or lost from water bodies by volatilization, whereas the more inert fraction of PAHs exported to SEMSs is likely within the above-mentioned range.

Thus, we estimated the ranges of the PCB and PAH loads using the upper and lower bounds of their ratios against POC and the POC loads predicted by the NEWS model. The results are listed in Table 8-2. The estimated PCB loading for the East China Sea is 13–132 t y^{-1}, which is reasonably consistent with the estimate of the total organochlorine flux, 286 t y^{-1} (Wu et al. 1999), taking into consideration the very preliminary nature of this assessment.

Assessment of Potential Impacts

The potential impacts of the riverine loads of nutrients and man-made chemicals on SEMSs depend on many factors, including the amount and composition of the riverine loads; the bioavailable fractions of organic and particulate nutrients; the seasonal variation of discharge rates; and the geometry, circulation pattern, mixing processes, and ecosystem structure of the receiving marine systems. In addition, biological activities, especially the recycling rate, are strongly dependent on the ambient temperature (e.g., Harrison 1983; Shiah et al. 2000). Hence, the climate zone of the system is probably also a key factor controlling responses to the loadings. Here we only provide preliminary assessments of the potential impacts based on N and P loading estimates and the area of the receiving body.

The immediate impact of the riverine loads on the coastal environment is via the river plume, which is controlled mainly by tides, winds, and buoyancy forces (Chao and Boicourt 1986). Although the complex phosphorus dynamics discussed previously make it difficult to estimate the relative availability of P versus N in river plumes, it is instructive to make educated guesses about nutrient status based on some assumptions. In the few systems for which data are available, less than 20% of PP is labile (Froelich 1988). If this is true for the SEMSs examined in this study, then ten out of thirteen systems would be P-limited in the river plume, barring significant contribution from the marine nutrient sources. If this is the case, the river plumes in the North Sea, the Black Sea, the Northern Adriatic Sea, and the East China Sea would receive the strongest immediate impacts from riverine nutrient loads because of their high DIP concentrations (Figure 8-3A) and the potential release of DIP from PP (Figure 8-3C). The highest average DIN and DIP concentrations occur in the runoffs to the North Sea (Figure 8-3A), apparently resulting from the very high nutrient yields in the catchment basin (see Plate 5 for DIN yield).

On a longer time scale, the riverine loads of nutrients would be dispersed throughout the SEMSs. Then the total nutrient load and the area of the system would determine the potential impact; the higher the total load and the smaller the area, the stronger would be the impact. In all but four systems, the shelf region occupies 70% or more of the whole area. In a system with a deep basin, such as the Gulf of Mexico, the shelf is often effectively separated from the basin by currents along the shelf break (Huthnance 1995) and, thus, experiences a stronger impact from the riverine loads. In order to indicate the maximal potential impacts, we calculated the ratios between the chemical loads and the shelf areas (Figure 8-4A–C). The potential impacts show a general decreasing trend with increasing latitude, but quite a few exceptions exist. The rather low potential impacts exerted upon the Gulf of Thailand are attributable to its very limited catchment basin, which is the smallest in proportion to the sea area among all SEMSs examined.

The deficiency of DIP in the runoff may be compensated by DIP from the subsurface layer. It has been demonstrated that P limitation could be alleviated in certain parts of the Changjiang River plume, where the vertical supply of subsurface nutrients was active (Gong et al. 1996). If the vertical P supply is substantial, then DIN loading should determine the impact of riverine nutrient loading in the short run. For this, the ratio between the DIN load and the shelf area indicates the potential impact in each SEMS (Figure 8-4A). Total nitrogen (TN) loading should have a significant impact in the longer term; the ratio between the TN load and the shelf area indicates the potential impact in each SEMS (Figure 8-4B).

In both categories (Figures 8-4A and 8-4B), the Bay of Bengal sustains the strongest potential impacts from riverine nutrient loads, due to the very high nutrient yields in its catchment basins (see Plate 5 for DIN yield). However, it is noted that the Bay of Bengal is the most open of all the systems examined, with no physical barriers separating its coastal zone from the open ocean. The boundary (Plate 5) is indicated by the mean position of the front separating the coastal water from the open sea. In this case, the rate of exchange between coastal and open-ocean water masses will dictate the system's response to the riverine loads.

Next to the Bay of Bengal, in terms of potential short-term impact of nutrient loads

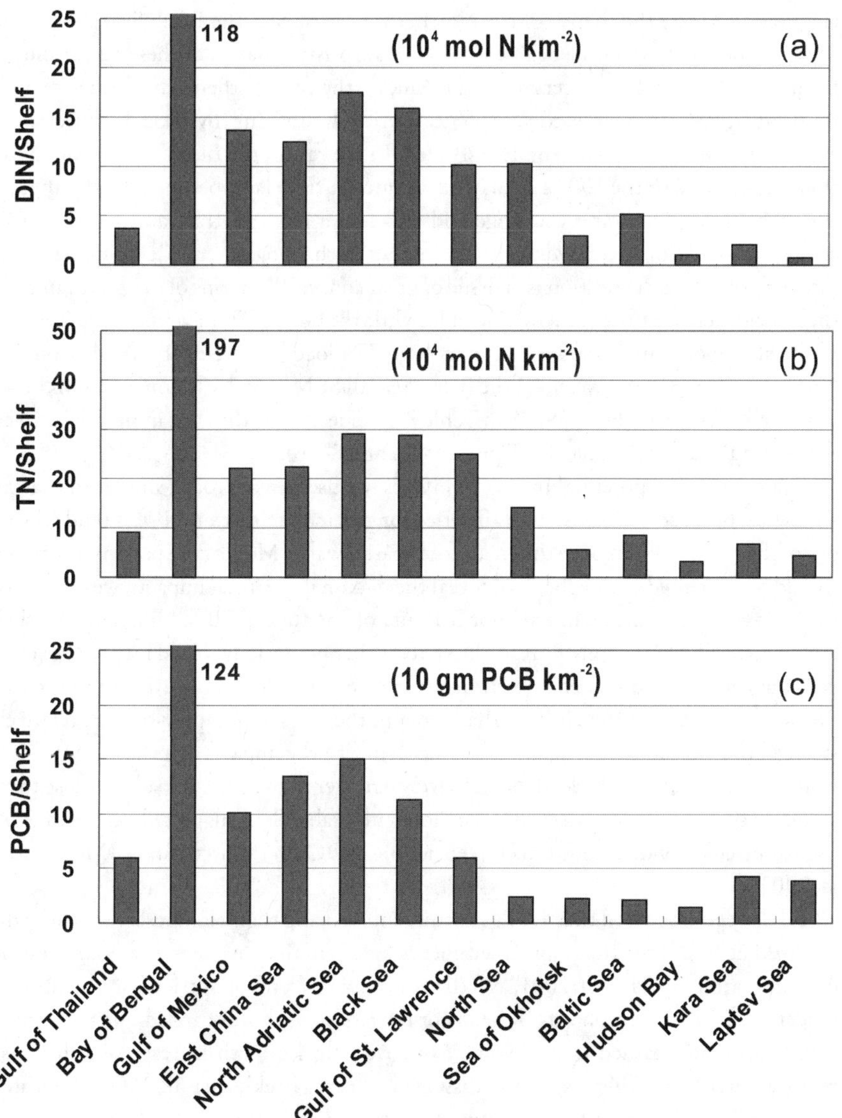

Figure 8-4. Impact assessment plots showing ratios between the chemical loadings and the continental shelf areas of the SEMSs, which are arranged according to midlatitude of catchment basin: (A) for dissolved inorganic nitrogen (DIN), in units of 10^4 mol N km^{-2}; (B) for total nitrogen (TN), in units of 10^4 mol N km^{-2}; (C) for total PCB, in units of 10 gm PCB km^{-2}. (See text for more detail.) The numbers in the plots represent the values for the Bay of Bengal.

(Figure 8-4A), are the shelves of the Northern Adriatic Sea, the Black Sea, the Gulf of Mexico, and the East China Sea. It is then not surprising that all of these systems suffer from hypoxia or anoxia in certain parts. Among the four of them, the former two are probably more vulnerable to eutrophication and resulting hypoxia because water exchange with the open sea is more restricted. In the case of the Black Sea, anoxic conditions occur beneath the 100–200 m of oxygenated surface layer because of its configuration, which restricts water exchange and makes it a nutrient trap, and because of its strong stratification (Richards 1965). Whether anthropogenic nutrients may further intensify the anoxic condition is an issue of great concern. In terms of long-term impact, the same systems are also rather vulnerable, while the Gulf of St. Lawrence surpasses two of the aforementioned systems due to its high TN load (Figure 8-4B). Aside from the more densely populated system of the Baltic Sea, all SEMSs with catchment basins in the arctic-subarctic zone (HB, KS, LS in Table 8-1) appear to be the least impacted by riverine N and P loadings (Table 8-1, Figures 8-4A and 8-4B).

Concerning the potential impact of HOCs, we used an approach similar to that for nutrients. Because of their strong affinities for particulate matter, HOCs tend to sink out of the water column once they reach the coastal water. Most of the particulate matter should be deposited on the shelf. To reveal the maximal potential impacts, we calculated the ratios between the estimated upper limits of the total PCB loading and the shelf areas for the SEMSs (Figure 8-4C). The systems that potentially would receive the strongest impacts are those with the strongest nutrient impacts. However, the potential impacts of HOCs on the Gulf of Thailand and the Kara and Laptev seas are relatively stronger than the nutrient impacts, as compared with the impacts on other SEMSs. This is simply because the POC loads are relatively stronger than nutrient loads in those three systems. The two arctic systems may be more vulnerable than suggested here because of the strong cold condensation effect in enriching HOCs in the arctic snow (Melnikov et al. 2003).

It has been estimated that 1.4 Pg of sediments is being trapped in reservoirs annually (Syvitski et al. 2005). The trapped sediments may contain considerable amounts of soil organic matter (Smith et al. 2001) as well as particulate forms of nutrients. As mentioned earlier, particulate phosphorus is the major form of phosphorus carried by rivers, and it could easily be retained in reservoirs. Moreover, algal growth in reservoirs may also remove nutrient loading, especially dissolved silicate (Ittekkot et al. 2006). Repeated hydrographic surveys off the Changjiang River mouth indicated the depletion of dissolved silicate, and therefore diatoms, following the first-stage water storage in the Three Gorges Dam in 2003 (Gong et al. 2006). This observation is consistent with current understanding, but further observation is required to confirm the link between damming and silicate depletion. Although particulate nitrogen is also trapped, it could be remobilized because most of it is in organic form, which may be decomposed into ammonium. (See Biogeochemical Characteristics of Amounts and Forms of Nutrients from Land, above, for more description of nitrogen cycling.) Since nitrogen is a relatively mobile element and its particulate form is less important, the withholding of particulate nitrogen by reservoirs is probably not as significant as that of Si or P. The different trapping efficiencies of reservoirs for nitrogen, phosphorus, and silicon mean that reservoirs can substantially alter nutrient ratios. It is noteworthy that the hypoxic or anoxic bottom water in reservoirs may enhance production of nitrous oxide (N_2O), which is a potent greenhouse gas. It is also noted that the trapped particulate organic matter may serve as a carrier of organic pollutants because of their strong affinity for organic matter. The fate of the trapped pollutants warrants further investigation.

Summary and Concluding Remarks

Riverine nutrient (N and P) loads are divided into three fractions, namely, the dissolved inorganic form, the dissolved organic form, and the particulate form. DIN is the dominant form of nitrogen, whereas PP is the dominant form of phosphorus. The N:P ratios in the dissolved inorganic form are all above the Redfield ratio for the SEMSs studied, except for the runoffs to the Black Sea. Since a finite fraction of the riverine PP may be readily released as DIP when it reaches the saline environment, it may make up for the DIP deficiency. On a longer time scale, significant proportions of nutrients in organic and particulate form may become bioavailable. Nevertheless, P limitation still occurs in some river plumes.

The river plumes in the North Sea, Black Sea, and Baltic Sea are potentially most strongly influenced by N and P loadings. For the entire shelf region beyond river plumes, the Bay of Bengal is strongly impacted due to its huge riverine nutrient loads, surpassing all other systems. In the second tier are the shelves of the Northern Adriatic Sea, the Black Sea, the Gulf of Mexico, and the East China Sea. The same systems would potentially also sustain the strongest impacts of hazardous man-made chemicals (PCBs and PAHs). However, two arctic systems, namely, the Kara and Laptev seas, would be rather strongly influenced by the PCBs and PAHs because of their high riverine POC fluxes, which are the main carriers of PCBs and PAHs, and because of the effective cold condensation in scavenging airborne PCBs and PAHs by the arctic snow.

Though we have provided state-of-the-art estimates of C, N, P, and organic contaminant loads to SEMSs, clearly many questions remain. More work is certainly needed to address many important issues concerning the riverine chemical loads—the bioavailability of nutrients in organic and particulate forms and the rates at which they become bioavailable; the changing nutrient composition in runoff, especially the deficiency of silicate, which may change ecosystem structure in the coastal zone (e.g., Ittekkot et al. 2006); the future trend of chemical loads in runoff; and the seasonal variation of the chemical loads, which may assert different impacts in different seasons, to name a few. It is very likely that the riverine loads of nutrients and man-made organic pollutants will increase with time because of increasing population, fertilizer use, and industrial productivity worldwide (Seitzinger et al. 2002a, 2007). How climate change will alter the riverine export of nutrients is an important topic for future study. The Bay of Bengal and the East China Sea are probably among the areas of most rapid growth in C, N, P, and organic contaminant loads due to the very fast economic development in their catchment basins. By contrast, the catchment basin of the Laptev Sea, where the population density is less than 1 individual km^{-2} (Table 8-1), may represent the pristine condition for nutrient export, but this does not hold for loads of organic contaminants.

It is also noteworthy that marine sources provide more nutrients to many coastal seas, such as the East China Sea (Chen and Wang 1999), than do riverine loads, by various physical processes (such as tidal mixing, coastal upwelling, slope-water intrusion). The estimated marine DIN and DIP fluxes to continental margins globally are larger than the total riverine loads by a factor of 6–9 for N and by 2–3 for P (Walsh 1991; Liu et al. 2009a), but the oceanic supply is probably more uniform and gradual. Like the riverine nutrient loads, the oceanic supply is subject to alteration due to climate change, which may change ocean circulation (e.g., Vecchi and Soden 2007) and, consequently, the nutrient fluxes carried by ocean currents. Despite the continual growth in riverine loads of nutrients and the potential change in oceanic nutrient supply, the marine contributions will probably remain the dominant nutrient source in the foreseeable future. It is quite important to understand how the infusion of marine nutrients may alter the impacts of the riverine loads in SEMSs. The marine input may alternately reinforce, supplement, or homogenize the effects of the riverine loads.

Aside from scientific significance, some issues are of importance to watershed management and may have critical impacts on the socioeconomic aspects of SEMSs. A good example of such an issue is the debate over what controls the N or P limitation in the receiving water body of riverine discharges. In addition, this work also has strong implications for watershed management because different processes and costs are involved in the control of N or P discharges to the aquatic system. It is important to understand these critical processes better. It is highly desirable to develop watershed models to predict future trends of chemical loads and to construct coupled physical–biogeochemical models to explore their impacts on SEMSs. It is also intended that the qualitative assessment presented here will serve as the starting point for more quantitative future work.

References

Alongi, D.M. 1994. The role of bacteria in nutrient recycling in tropical mangrove and other coastal benthic ecosystems. *Hydrobiologia* 285:19–32.

ATSDR (Agency for Toxic Substances and Disease Registry). 1995. *Public health statement: Polycyclic aromatic hydrocarbons (PAHs).* Atlanta, GA: U.S. Department of Health and Human Services, Public Health Service.

Baker, J.E., S.J. Eisenreich, and B.J. Eadie. 1991. Sediment trap fluxes and benthic recycling of organic carbon, polycyclic aromatic hydrocarbons, and polychlorobiphenyl congeners in Lake Superior. *Environmental Science & Technology* 25:500–509.

Beusen, A.H.W., A.L.M. Dekkers, A.F. Bouwman, W. Ludwig, and J. Harrison. 2005.

Estimation of global river transport of sediments and associated particulate C, N, and P. *Global Biogeochemical Cycles* 19:GB4S05, doi: 10.1029/2005GB002453.

Bozec, Y., H. Thomas, L.-S. Schiettecatte, A.V. Borges, K. Elkalay, and H.J.W. de Baar. 2006. Assessment of the processes controlling seasonal variations of dissolved inorganic carbon in the North Sea. *Limnology and Oceanography* 51:2,746–2,762.

Cameron, W.M., and D.W. Pritchard. 1963. Estuaries. In *The Sea*, edited by M.N. Hill, 306–324. New York: John Wiley & Sons.

Canfield, D.E. 1993. Organic matter oxidation in marine sediments. In *Interaction of C, N, P and S Biogeochemical Cycles*, edited by R. Wollast, F.T. Mackenzie, and L. Chou, 333–363. Berlin: Springer.

Chang, J., and H.-C. Liu. 2007. Relationships between phosphorus limitation among microplankton and Changjiang runoff in the East China Sea. Taiwan Geoscience Assembly, Lungtang, Taoyuan 15–18 May 2007.

Chao, S.-Y., and W.C. Boicourt. 1986. Onset of estuarine plumes. *Journal of Physical Oceanography* 16:2,137–2,149.

Chen, C.T.A., and S.L. Wang. 1999. Carbon, alkalinity and nutrient budgets on the East China Sea continental shelf. *Journal of Geophysical Research—Oceans* 104:20,675–20,686.

Cotner, J.B., and R.G. Wetzel. 1992. Uptake of dissolved inorganic and organic phosphorus compounds by phytoplankton and bacterioplankton. *Limnology and Oceanography* 37:232–243.

Countway, R.E., R.M. Dickhut, and E.A. Canuel. 2003. Polycyclic aromatic hydrocarbon (PAH) distributions and associations with organic matter in surface waters of the York River, VA estuary. *Organic Geochemistry* 34:209–224.

Deng, H.M., P.A. Peng, W.L. Huang, and H.Z. Song. 2006. Distribution and loadings of polycyclic aromatic hydrocarbons in the Xijiang River in Guangdong, south China. *Chemosphere* 64:1,401–1,411.

Dumont, E., J.A. Harrison, C. Kroeze, E.J. Bakker, and S.P. Seitzinger. 2005. Global distribution and sources of dissolved inorganic nitrogen export to the coastal zone: Results from a spatially explicit, global model. *Global Biogeochemical Cycles* 19:GB4S02, doi: 10.1029/2005GB002488.

Eisler, R. 1987. *Polycyclic Aromatic Hydrocarbon Hazards to Fish, Wildlife, and Invertebrates: A Synoptic Review.* Biological Report 85 (1.11). Laurel, MD: U.S. Fish and Wildlife Service.

Fekete, B.M., C.J. Vörösmarty, and W. Grabs. 2002. High-resolution fields of global runoff combining observed river discharge and simulated water balances. *Global Biogeochemical Cycles* 16 (3):1,042, doi: 10.1029/1999GB001254.

Froelich, P.N. 1988. Kinetic control of dissolved phosphate in natural rivers and estuaries: A primer on the phosphate buffer mechanism. *Limnology and Oceanography* 33:649–668.

Fujiki, T., T. Toda, T. Kikuchi, H. Aono, and S. Taguchi. 2004. Phosphorus limitation of primary productivity during the spring–summer blooms in Sagami Bay, Japan. *Marine Ecology Progress Series* 283:29–38.

Gómez-Gutiérrez, A.I., E. Jover, L. Bodineau, J. Albaigés, and J.M. Bayona. 2006. Organic contaminant loads into the western Mediterranean Sea: Estimate of Ebro River inputs. *Chemosphere* 65:224–236.

Gong, G.-C., J. Chang, K.-P. Chiang, T.-M. Hsiung, C.-C. Hung, S.-W. Duan, and L.A.

Codispoti. 2006. Reduction of primary production and changing of nutrient ratio in the East China Sea: Effect of the Three Gorges Dam? *Geophysical Research Letters* 33:L07610, doi: 10.1029/2006GL025800.

Gong, G.C., Y.L.L. Chen, and K.K. Liu. 1996. Chemical hydrography and chlorophyll *a* distribution in the East China Sea in summer: Implications in nutrient dynamics. *Continental Shelf Research* 16:1,561–1,590.

Green, R.E., T.S. Bianchi, M.J. Dagg, N.D. Walker, and G.A. Breed. 2006. An organic carbon budget for the Mississippi River turbidity plume, and plume contributions to air–sea CO_2 fluxes and bottom-water hypoxia. *Estuaries and Coasts* 29:579–597.

Guo, W., M.C. He, Z.F. Yang, C.Y. Lin, X.C. Quan, and H.Z. Wang. 2007. Distribution of polycyclic aromatic hydrocarbons in water, suspended particulate matter and sediment from Daliao River watershed, China. *Chemosphere* 68:93–104.

Harrison, J.A., N. Caraco, and S.P. Seitzinger. 2005a. Global patterns and sources of dissolved organic matter export to the coastal zone: Results from a spatially explicit, global model. *Global Biogeochemical Cycles* 19: GB4S04.

Harrison, J.A., S.P. Seitzinger, A.F. Bouwman, N.F. Caraco, A.H.W. Beusen, and C.J. Vörösmarty. 2005b. Dissolved inorganic phosphorus export to the coastal zone: Results from a spatially explicit, global model. *Global Biogeochemical Cycles* 19:GB4S03.

Harrison, P.J., M.H. Hu, Y.P. Yang, and X. Lu. 1990. Phosphate limitation in estuarine and coastal waters of China. *Journal of Experimental Marine Biology and Ecology* 140:79–87.

Harrison, W.G. 1983. The time-course of uptake of inorganic and organic nitrogen compounds by phytoplankton from the eastern Canadian Arctic: A comparison with temperate and tropical populations. *Limnology and Oceanography* 28:1,231–1,237.

Hartmann, P.C., J.G. Quinn, R.W. Cairns, and J.W. King. 2004. The distribution and sources of polycyclic aromatic hydrocarbons in Narragansett Bay surface sediments. *Marine Pollution Bulletin* 48:351–358.

Howarth, R., H. Jensen, R. Marino, and H. Postma. 1995. Transport to and processing of phosphorus in near-shore and oceanic waters. In *Phosphorus in the Global Environment*, SCOPE Series Vol. 54, edited by H. Tiessen, 323–345. Chichester, UK: John Wiley & Sons.

Hung, C.C., G.C. Gong, K.T. Jiann, K.M. Yeager, P.H. Santschi, T.L. Wade, J.L. Sericano, and H.L. Hsieh. 2006. Relationship between carbonaceous materials and polychlorinated biphenyls (PCBs) in the sediments of the Danshui River and adjacent coastal areas, Taiwan. *Chemosphere* 65:1,452–1,461.

Huthnance, J.M. 1995. Circulation, exchange and water masses at the ocean margin: The role of physical processes at the shelf edge. *Progress in Oceanography* 35:353–431.

Ittekkot, V., D. Unger, C. Humborg, and N.Tac An (eds.). 2006. *The Silicon Cycle: Human Perturbations and Impacts on Aquatic Systems.* SCOPE Series Vol. 66. Washington, DC: Island Press.

Katsoyiannis, A., and C. Samara. 2004. Persistent organic pollutants (POPs) in the sewage treatment plant of Thessaloniki, northern Greece: Occurrence and removal. *Water Research* 38:2,685–2,698.

Ko, F.C., and J.E. Baker. 2004. Seasonal and annual loads of hydrophobic organic contaminants from the Susquehanna River basin to the Chesapeake Bay. *Marine Pollution Bulletin* 48:840–851.

Koelmans, A.A., F. Gillissen, W. Makatita, and M. van den Berg. 1997. Organic carbon normalisation of PCB, PAH and pesticide concentrations in suspended solids. *Water Research* 31:461–470.

Labry, C., A. Herbland, and D. Delmas. 2002. The role of phosphorus on planktonic production of the Gironde plume waters in the Bay of Biscay. *Journal of Plankton Research* 24:97–117.

Liu, K.-K., L. Atkinson, R. Quiñones, and L. Talaue-McManus. 2009a. Biogeochemistry of continental margins in a global context. In *Carbon and Nutrient Fluxes in Continental Margins: A Global Synthesis*, edited by K.-K. Liu, L. Atkinson, R. Quiñones, and L. Talaue-McManus. Global Change—The IGBP Series. Berlin: Springer.

Liu, K.-K., L. Atkinson, R. Quiñones, and L. Talaue-McManus. 2009b. *Carbon and Nutrient Fluxes in Continental Margins: A Global Synthesis*. Global Change—The IGBP Series. Berlin: Springer.

Liu, K.-K., T.Y. Tang, G.C. Gong, L.Y. Chen, and F.K. Shiah. 2000. Cross-shelf and along-shelf nutrient fluxes derived from flow fields and chemical hydrography observed in the southern East China Sea off northern Taiwan. *Continental Shelf Research* 20:493–523.

Macdonald, R.W., L.G. Anderson, J.P. Christensen, L.A. Miller, I.P. Semiletov, and R. Stein. 2009. The Arctic Ocean. In *Carbon and Nutrient Fluxes in Continental Margins: A Global Synthesis*, edited by K.-K. Liu, L. Atkinson, R. Quiñones, and L. Talaue-McManus. Global Change—The IGBP Series. Berlin: Springer.

Melnikov, S., J. Carroll, A. Gorshkov, S. Vlasov, and S. Dahle. 2003. Snow and ice concentrations of selected persistent pollutants in the Ob–Yenisey River watershed. *Science of the Total Environment* 306:27–37.

Meybeck, M. 1982. Carbon, nitrogen and phosphorus transport by world rivers. *American Journal of Science* 282:401–450.

Meybeck, M. 2003. Global analysis of river systems: From Earth system controls to Anthropocene syndromes. *Philosophical Transactions of the Royal Society B* 358 (1440):1,935–1,955.

Naqvi, S.W.A., H. Naik, W.D. Souza, P.V. Narvekar, A.L. Paropkari, and H.W. Bange. 2009. North Indian Ocean margins. In *Carbon and Nutrient Fluxes in Continental Margins: A Global Synthesis*, edited by K.-K. Liu, L. Atkinson, R. Quiñones, and L. Talaue-McManus. Global Change—The IGBP Series. Berlin: Springer.

Nixon, S.W. 1982. Nutrient dynamics, primary production and fisheries yields of lagoons. *Oceanologica Acta* 5:357–371.

Nixon, S., C. Oviatt, J. Frithsen, and B. Sullivan. 1986. Nutrients and the productivity of estuarine and coastal marine ecosystems. *Journal of Limnology Society of South Africa* 12:43–71.

Redfield, A.C., B.H. Ketchum, and F.A. Richards. 1963. The influence of organisms on the composition of seawater. In Vol. 2 of *The Sea*, edited by M.N. Hill, 26–77. New York: John Wiley & Sons.

Richards, F.A. 1965. Anoxic basins and fjords. In Vol. 1 of *Chemical Oceanography*, edited by J.P. Riley and G. Skirrow, 611–645. New York: Academic Press.

Santschi, P.H., B.J. Presley, T.L. Wade, B. Garcia-Romero, and M. Baskaran. 2001. Historical contamination of PAHs, PCBs, DDTs, and heavy metals in Mississippi River delta, Galveston Bay and Tampa Bay sediment cores. *Marine Environmental Research* 52:51–79.

Seitzinger, S.P., J.A. Harrison, E. Dumont, A.H.W. Beusen, and A.F. Bouwman. 2005.

Sources and delivery of carbon, nitrogen, and phosphorus to the coastal zone: An overview of global Nutrient Export from Watersheds (NEWS) models and their application. *Global Biogeochemical Cycles* 19:GB4S01, doi: 10.1029/2005GB002606.

Seitzinger, S.P., C. Kroeze, A.F. Bouwman, N. Caraco, F. Dentener, and R.V. Styles. 2002a. Global patterns of dissolved inorganic and particulate nitrogen inputs to coastal systems: Recent conditions and future projections. *Estuaries* 25:640–655.

Seitzinger, S., E. Mayorga, A. Beusen, A. Bouwman, E. Dumont, J. Harrison, C. Kroeze, D. Wisser, B. Fekete, and C. Vörösmarty. 2007. Past, Current and Future Trajectories of Watershed Nutrient Sources, Forms and Exports: A Global NEWS Application to the Millennium Ecosystem Assessment Scenarios. American Geophysical Union fall meeting, San Francisco, CA 10–14 December 2007.

Seitzinger, S.P., R.W. Sanders, and R. Styles. 2002b. Bioavailability of DON from natural and anthropogenic sources to estuarine plankton. *Limnology and Oceanography* 47:353–366.

Sericano, J.L., J.M. Brooks, M.A. Champ, M.C. Kennicutt, and V.V. Makeyev. 2001. Trace contaminant concentrations in the Kara Sea and its adjacent rivers, Russia. *Marine Pollution Bulletin* 42:1,017–1,030.

Shiah, F.K., G.C. Gong, T.Y. Chen, and C.C. Chen. 2000. Temperature dependence of bacterial specific growth rates on the continental shelf of the East China Sea and its potential application in estimating bacterial production. *Aquatic Microbial Ecology* 22:155–162.

Simpson, J.H. 1997. Physical processes in the ROFI regime. *Journal of Marine Systems* 12:3–15.

Smith, S.V. 1984. Phosphorus versus nitrogen limitation in the marine environment. *Limnology and Oceanography* 29:1,149–1,160.

Smith, S.V., W.H. Renwick, R.W. Buddemeier, and C.J. Crossland. 2001. Budgets of soil erosion and deposition for sediments and sedimentary organic carbon across the conterminous United States. *Global Biogeochemical Cycles* 15:697–707.

Smith, S.V., D.P. Swaney, L. Talaue-McManus, J.D. Bartley, P.T. Sandhei, C.J. McLaughlin, V.C. Dupra, C.J. Crossland, R.W. Buddemeier, B.A. Maxwell, and F. Wulff. 2003. Humans, hydrology, and the distribution of inorganic nutrient loading to the ocean. *Bioscience* 53:235–245.

Syvitski, J.P.M. 2003. Sediment fluxes and rates of sedimentation. In *Encyclopedia of Sediments and Sedimentary Rocks*, edited by G.V. Middleton, 600–606. Dordrecht, The Netherlands: Kluwer Academic Publishers.

Syvitski, J.P.M., C.J. Vörösmarty, A.J. Kettner, and P. Green. 2005. Impact of humans on the flux of terrestrial sediment to the global coastal ocean. *Science* 308:376–380.

Talaue-McManus, L. 2009. Examining human impacts on global biogeochemical cycling via the coastal zone and ocean margins. In *Carbon and Nutrient Fluxes in Continental Margins: A Global Synthesis*, edited by K.-K. Liu, L. Atkinson, R. Quiñones, and L. Talaue-McManus. Global Change—The IGBP Series. Berlin: Springer.

Turner, R.E., N.N. Rabalais, D. Justic, and Q. Dortch. 2003. Global patterns of dissolved N, P and Si in large rivers. *Biogeochemistry* 64:297–317.

U.S. EPA. 2002. *Persistent Organic Pollutants: A Global Issue, a Global Response*. EPA 160-F-02-001 (2610R). Washington, DC: U.S. EPA Office of International Programs. http://www.epa.gov/oia/toxics/pop.htm.

Van der Zee, C., N. Roevros, and L. Chou. 2007. Phosphorus speciation, transformation and retention in the Scheldt estuary (Belgium/The Netherlands) from the freshwater tidal limits to the North Sea. *Marine Chemistry* 106:76–91.

Vecchi, G.A., and B.J. Soden. 2007. Global warming and the weakening of the tropical circulation. *Journal of Climate* 20:4,316–4,340.

Vörösmarty, C., B. Fekete, M. Meybeck, and R. Lammers. 2000. Global system of rivers: Its role in organizing continental land mass and defining land-to-ocean linkages. *Global Biogeochemical Cycles* 14 (2):599–621, doi: 10.1029/1999GB900092.

Walsh, J.J. 1991. Importance of continental margins in the marine biogeochemical cycling of carbon and nitrogen. *Nature* 350:53–55.

Wong, G.T.F., G.C. Gong, K.K. Liu, and S.C. Pai. 1998. "Excess nitrate" in the East China Sea. *Estuarine Coastal and Shelf Science* 46:411–418.

Wu, Y., J. Zhang, and Q. Zhou. 1999. Persistent organochlorine residues in sediments from Chinese river/estuary systems. *Environmental Pollution* 105:143–150.

Yunker, M.B., S.M. Backus, E. Graf Pannatier, D.S. Jeffries, and R.W. Macdonald. 2002. Sources and significance of alkane and PAH hydrocarbons in Canadian arctic rivers. *Estuarine Coastal and Shelf Science* 55:1–31.

Zhang, S.Y., Q.A. Zhang, S. Darisaw, O. Ehie, and G.D. Wang. 2007. Simultaneous quantification of polycyclic aromatic hydrocarbons (PAHs), polychlorinated biphenyls (PCBs), and pharmaceuticals and personal care products (PPCPs) in Mississippi River water, in New Orleans, Louisiana, USA. *Chemosphere* 66:1,057–1,069.

9

Biogeochemical Cycling in Semi-Enclosed Marine Systems and Continental Margins

Helmuth Thomas, Daniela Unger, Jing Zhang, Kon-Kee Liu, and Elizabeth H. Shadwick

Introduction

The cycling of bioactive elements in semi-enclosed marine systems (SEMSs) is controlled by a variety of physical, biological, chemical, and anthropogenic factors. Bottom topography and exchange of water with the continent (via estuaries) and the ocean are the physical drivers of this cycling. External factors, such as temperature, precipitation, and wind may be active on time scales from days to months, producing seasonal or permanent stratification or mixing. Exchange of matter between SEMSs and interfacing reservoirs (i.e., the atmosphere, land and rivers, ocean and sediments) governs the biogeochemistry and ecosystem structure of SEMSs, which are coupled through feedback loops. Many of the processes relevant to biogeochemical cycling have been affected by human activities either directly or through climate change. The extent to which these processes control system biogeochemistry varies by region; thus it is these processes that characterize a system.

This overview will outline selected fundamental processes that contribute to the complex patterns of biogeochemical cycling in SEMSs. First, we briefly illustrate the main biogeochemical processes relevant for the carbon, nitrogen, and phosphorus cycles. We then discuss, by way of selected examples (Figure 9-1), processes at the interfaces that link SEMSs to adjacent reservoirs. We close with a brief perspective on the carbon cycle in SEMSs and continental margins, and on the role these systems play in the global carbon cycle.

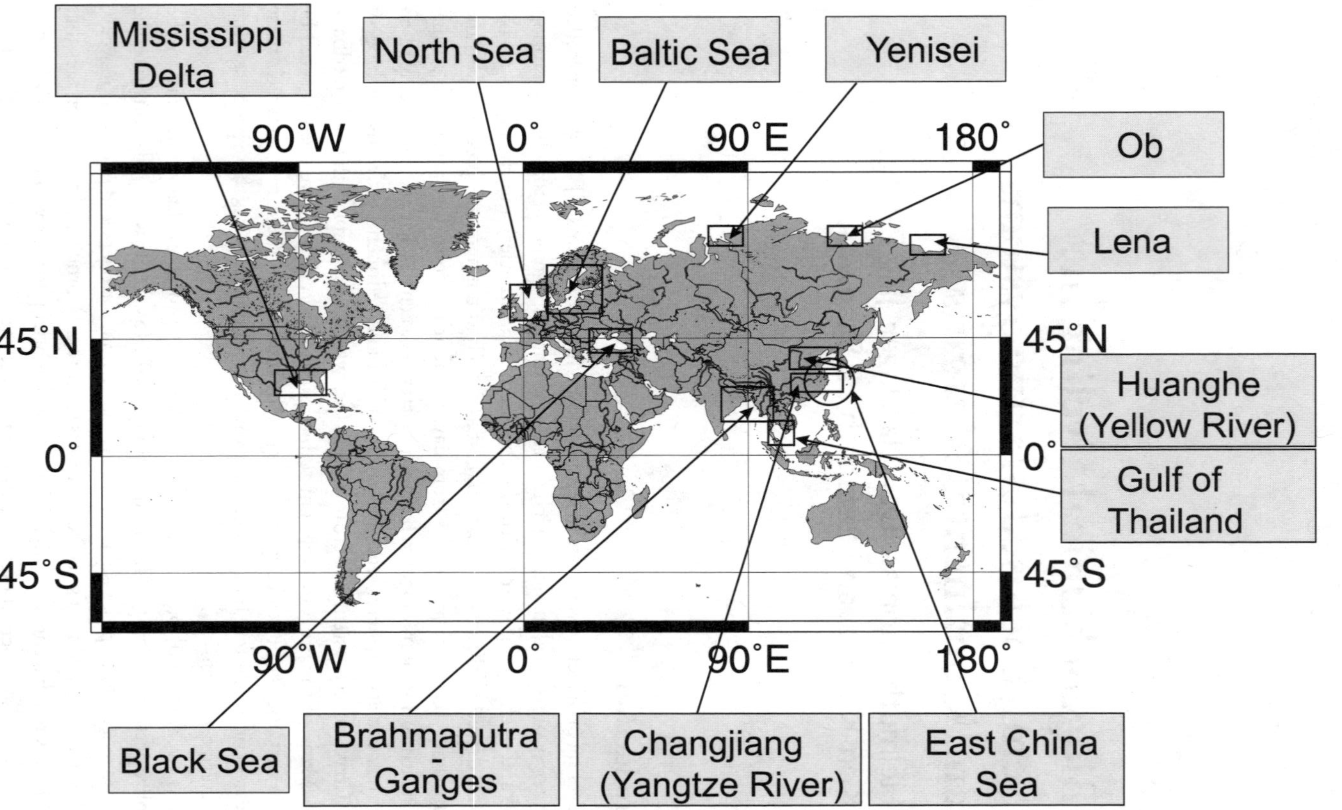

Figure 9-1. The SEMSs and river systems addressed in Chapter 9.

Fundamental Processes of the Key Elements Carbon, Nitrogen, and Phosphorus

SEMSs constitute highly diverse ecosystems (e.g., Gattuso et al. 1998), which may assimilate, regenerate, transform, and/or remove carbon and nutrient elements to varying extents, depending on local characteristics. Carbon, nitrogen, and phosphorus cycles are subject to biogeochemical processes involving both reduction–oxidation (redox) and multiphase reactions.

The Carbon Cycle

The biogeochemical cycle of carbon is very complex; only those pathways involving the major compounds (Garrels et al. 1975; Morel and Hering 1993) in the earth surface environment will be discussed (these are depicted in Figure 9-2). Photosynthesis is carried out primarily by phytoplankton, although macro-algae, sea grass, and benthic micro-algae may also make significant contributions to ecosystem production in SEMSs. Mangroves and symbiotic algae in corals play a particularly important role in tropical margins. The direct product of photosynthesis—glucose—provides essential energy to organisms and is also used in building other biochemical compounds. A major fraction of photosynthetic production is eventually degraded and respired back to CO_2 and water by marine organisms. In contrast to the strong seasonality of primary production, respiration shows a less pronounced, but still significant, seasonal variability.

Often coupled to photosynthesis is the formation by organisms of calcium carbonate shells and skeletons. Calcareous plankton includes both the autotrophic coccolithophores and the heterotrophic pteropods and foraminifera, which are the major producers of calcium carbonate shells in the water column. Another form of calcium carbonate formation takes place in coral reefs, where coral polyps and coralline macro-algae produce massive carbonate deposits. Calcium carbonate exists in two major forms: calcite and aragonite. Of the two, aragonite is more soluble and may be transformed into calcite. As particulate organic matter deposits on the seafloor in continental margins, a major fraction is recycled in the sediment, resulting in a strong benthic–pelagic coupling (Middelburg and Soetaert 2004; Liu et al. 2007). In shallower parts of the continental shelf and in semi-enclosed seas, 20%–50% of primary production may be consumed on the seafloor (Rabouille et al. 2001). Without sufficient oxygen or other compounds functioning as oxidants, organic matter may be split into compounds with higher and lower oxidation states during fermentation reactions by anaerobic microbes (Figure 9-2). The end products of these fermentation reactions are CO_2 and methane. Carbon dioxide is very soluble in seawater, but methane is less so and therefore tends to escape to the atmosphere as gas.

The distributions of nitrogen and phosphorus in marine environments are often closely related; however, their biogeochemical pathways are quite different.

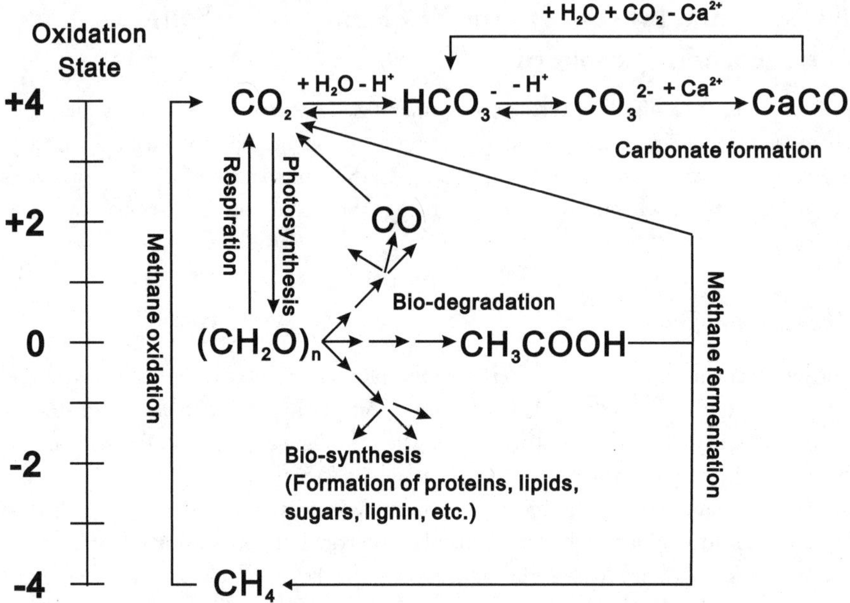

Figure 9-2. Biogeochemical pathways of the carbon cycle. Pathways of biosynthesis and biodegradation are shown as schematics to represent the complex biochemical pathways. Only the most common compounds are shown. Redrawn from Liu et al. 2009 and used by permission of Springer.

The Nitrogen Cycle

Nitrogen compounds span oxidation states from -3 to +5 and are therefore sensitive to redox conditions. The redox pathways of the nitrogen cycle are depicted in Figure 9-3. In the center of the diagram is organic nitrogen, indicated by RNH_2, which represents the nitrogenous organic matter that is produced by organisms and also the organic nitrogen compounds that are discharged from human activities.

- Phytoplankton, bacteria, and archaea compete for ammonium (NH_4^+). In a mainly light-dependent process, phytoplankton, bacteria, and archaea assimilate ammonium to form organic matter. During nitrification, bacteria and archaea oxidize ammonium to nitrite (NO_2^-) and in a second step to nitrate (NO_3^-).
- Under low-oxygen or anoxic conditions, ammonium may be lost from the environment due to its conversion to nitrogen gas (N_2) during anaerobic oxidation of ammonium, also known as anammox (Jetten et al. 2003; Kuypers et al. 2003).

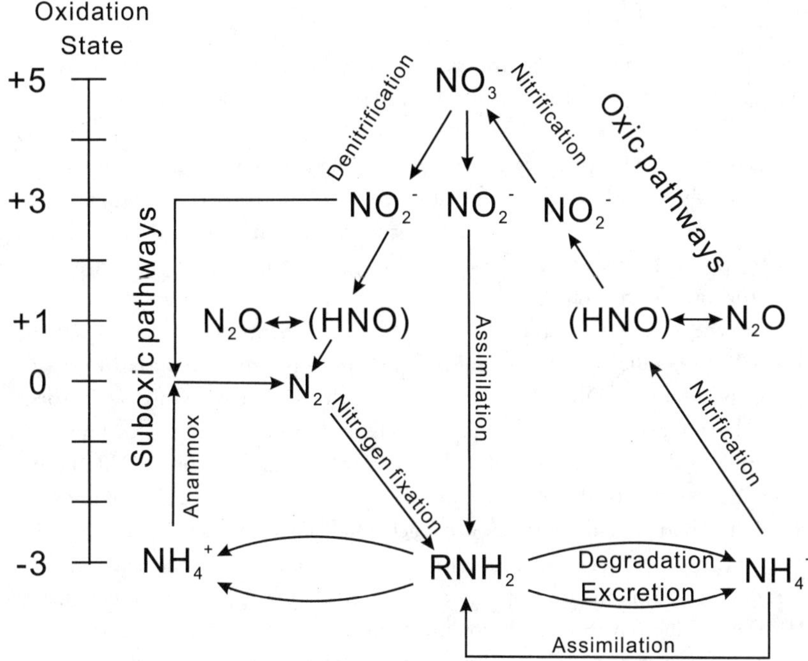

Figure 9-3. Biogeochemical pathways of the nitrogen cycle (modified after Liu and Kaplan 1982 and Codispoti et al. 2001). Redrawn from Liu et al. 2009 and used by permission of Springer.

- Under low-oxygen conditions, nitrate may be consumed during nitrate respiration (also called denitrification) by heterotrophic bacteria (Delwiche and Bryan 1976). Nitrite is an intermediate product of nitrate respiration that can be further reduced to nitrous oxide (N_2O) or dinitrogen gas during denitrification.
- Nitrogen fixation can serve as a source of bioavailable nitrogen in SEMSs (see The Semi-Enclosed Marine System–Atmosphere Interface below) and counteracts the loss of fixed nitrogen from the environment through the above-mentioned processes.
- Conversion of fixed nitrogen to nitrogen gas, which will then escape to the atmosphere, is a major pathway for removal of fixed nitrogen in the coastal zone. Nitrous oxide (N_2O) may also escape from the coastal environment to the atmosphere. N_2O is a potent greenhouse gas that makes a potentially significant contribution to global warming (Forster et al. 2007). The conversion of dissolved inorganic nitrogen (DIN) to N_2 during denitrification, or anammox, in SEMSs and continental margins represents the most important nitrogen sink in the ocean. Consequently, this sink may regulate the global carbon cycle (Altabet et al. 2002).

The Phosphorus Cycle

In contrast to nitrogen, phosphorus remains constantly at an oxidation state of +5. However, its speciation and availability are still sensitive to redox conditions because of its chemical affiliates. Particulate phosphorus is primarily introduced to the marine environment by rivers, with primary production playing a minor role in the supply of particulate phosphorus (Meybeck 1982; Beusen et al. 2005). As particulate organic matter sinks to the seafloor (Figure 9-4A), a major fraction of it may be decomposed. This decomposition consumes oxygen, resulting in significantly reduced oxygen concentrations in the lower water column.

Within the oxygenated surface layer of seafloor sediments, particulate organic phosphorus (POP) is efficiently remineralized (Figure 9-4C) and converted to dissolved inorganic phosphorus (DIP). A major fraction of this DIP may diffuse out of the sediments and reenter the water column phosphorus cycle. On the other hand, phosphorus-containing minerals may be formed authigenically from the buildup of DIP in the pore waters of the sediments. Burial of these minerals represents a major sink of phosphorus in the ocean (Berner et al. 1993). Low-oxygen conditions may occur near the sediment surface on the seafloor, or even in the bottom water in contact with the seafloor, favoring the reduction and dissolution of phosphorus-containing ferric hydroxides (Canfield 1993). These reduction and dissolution processes act to release DIP to the pore water. If the oxygenated layer is thin, the DIP released in this layer may escape back into the water column. Conversely, if the oxic layer is thick, most of the DIP released will form authigenic minerals, as mentioned above.

The Semi-Enclosed Marine System–River Interface

SEMSs are very often influenced by enhanced river inputs. This river input establishes the linkage of the marine system to natural processes and human activities occurring on land. Rivers supply not only large amounts of water, but also particulate and dissolved constituents to the marine system. As a result, riverine influence affects both the baseline conditions and the temporal variability of SEMSs. As these dissolved and particulate constituents are discharged to the coastal zone, they pass through the estuarine system, undergoing significant modifications due to changes in both salinity and pH. Hydrodynamics ultimately determine whether, and where, this riverine sediment is deposited.

Natural variability of river input on monthly to interannual time scales (e.g., caused by the variability of precipitation), as well as variability stimulated by human activities, including climate change, will affect the biogeochemical characteristics of SEMSs on interannual and decadal time scales (Rabalais et al. 2002; Vicchi et al. 2004; Gypens and Lancelot 2006). The more closed the system is, or the more river input it receives, the higher the potential impact of altered riverine inputs on a SEMS. Thus, systems receiving large riverine inputs are more vulnerable to human forcing. Unfortunately, interannual variability of river runoff may hamper the identification of long-term trends during the

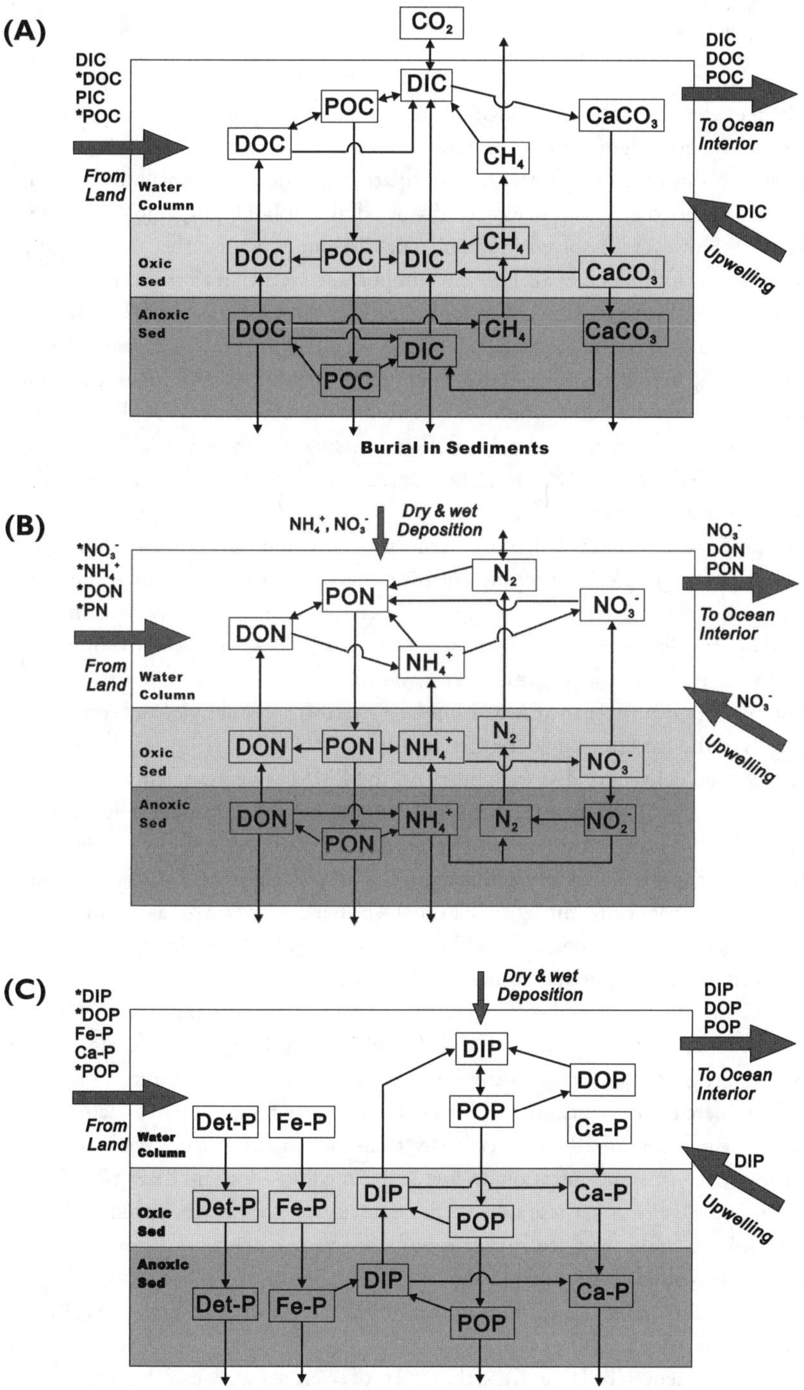

(A)
CO_2
DIC
*DOC
PIC
*POC
From Land
DIC
DOC
POC
To Ocean Interior
DIC
Upwelling
Water Column
Oxic Sed
Anoxic Sed
DOC
POC
DIC
$CaCO_3$
CH_4
Burial in Sediments
(B)
NH_4^+, NO_3^-
Dry & wet Deposition
$*NO_3^-$
$*NH_4^+$
*DON
*PN
From Land
NO_3^-
DON
PON
To Ocean Interior
NO_3^-
Upwelling
Water Column
Oxic Sed
Anoxic Sed
N_2
PON
NO_3^-
DON
NH_4^+
NO_2^-
(C)
Dry & wet Deposition
*DIP
*DOP
Fe-P
Ca-P
*POP
From Land
DIP
DOP
POP
To Ocean Interior
DIP
Upwelling
Water Column
Oxic Sed
Anoxic Sed
DIP
DOP
POP
Det-P
Fe-P
Ca-P

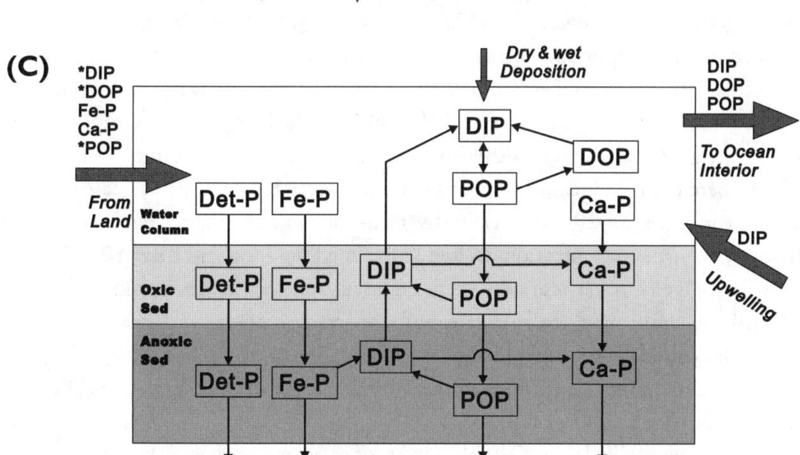

Figure 9-4. Biogeochemical cycles in SEMSs and shallow marine environments. The compounds marked with asterisks are significantly influenced by human activities. (A) Carbon cycle; (B) nitrogen cycle; (C) phosphorus cycle. Redrawn from Liu et al. 2009 and used by permission of Springer.

early phase of change, making it difficult to recognize and respond to alterations before ecosystems have been significantly affected.

Abiotic processes such as aggregation, flocculation, and desorption play an important role in the sedimentation and release of dissolved and colloidal organic constituents (Figure 9-4). However, dissolved inorganic constituents remain rather unaffected by these processes (Hollibaugh et al. 1991). An important exception to this is the interaction of trace metals with particles or colloids. Particles can serve as sources of trace metals through desorption or dissolution (e.g., Watson et al. 2000). However, particularly in estuarine mixing zones, particles and colloids can also act as a sink for trace metals, removing them from the water column (e.g., Sañudo-Wilhelmy et al. 1996; Pan and Liss 1998). Photochemical reactions can break down dissolved organic matter (Amon and Benner 1996), making otherwise resistant constituents available for further microbial consumption (Bushaw et al. 1996).

Biotic processes, on the other hand, determine the distribution of dissolved inorganic constituents. For example, autotrophic organisms may assimilate materials at the river–ocean interface, while heterotrophic organisms play a role in the regeneration and consumption of organic matter at this interface (Dagg et al. 2004). Autotrophic organisms fuel food webs by producing organic matter and releasing oxygen. In a reverse process, the supply of organic matter to SEMSs stimulates heterotrophic activity and respiration, which leads to the consumption of oxygen.

The increased delivery of dissolved nutrients to SEMSs as a consequence of fertilizer use, municipal sewage, or manure disposal results in enhanced productivity in these regions. This can be considered one of the most obvious threats to SEMSs over recent decades (Turner and Rabalais 1991; Humborg et al. 1997; Vicchi et al. 2004; Seitzinger et al. 2005). Anthropogenic nitrogen (Figure 9-4B) may enter SEMSs as organic nitrogen and inorganic fixed nitrogen, namely as ammonium, nitrate, or nitrite.

Enhanced nutrient input and eutrophication—which results in the internal production of "excess" organic matter—have either led to, or exacerbated, hypoxic or anoxic conditions in many ocean regions. The Gulf of Mexico is the best-known example of this phenomenon (e.g., Rabalais et al. 2000; Justic et al. 2002). Another example is the East China Sea, where the riverine nutrient load sustains the high productivity occurring on the shelf. In the region of 20–50 m isobaths, near-bottom hypoxia has been observed, with dissolved oxygen concentrations as low as 1–2 mg L^{-1} (Tong and Zhang 2007).

Effective management in these areas, such as wastewater treatment and more-efficient use of fertilizer, could work to mitigate human pressure on coastal ocean regions. Recently implemented policies designed to reduce riverine nutrient loads have helped to prevent the onset of anoxic conditions in the southern bight of the North Sea (OSPAR Commission 2000).

In SEMSs influenced by large rivers, the effects of riverine inputs can extend appreciably into the basin. The summer plume from the Changjiang (Yangtze) River, for example, spreads as far as 300–400 km from the river mouth. The Bay of Bengal is almost entirely influenced by freshwater input, as evident from reduced salinities over the whole region. The area of a SEMS affected by river plumes, however, can shrink drastically as water discharge is reduced. For example, in the Bohai (North China), the annual freshwater influx from the Huanghe (Yellow) River and other rivers has been dramatically decreased (i.e., 50%–60%) as a result of human activities (e.g., irrigation) in the catchment areas over the last two decades (Zhang et al. 2004). Associated changes in circulation, vertical mixing, and nutrient supply have the potential to alter the entire system.

Many processes in SEMSs are related to the input of particulate matter. Drainage basin characteristics (e.g., relief, geology, erosion, and runoff) determine the sediment yield for a region. This may result in significant differences in suspended matter concentration in rivers with high relief and erosion rates, such as the Brahmaputra-Ganges and the Huanghe in Asia, versus low relief and erosion in high-latitude rivers such as the Ob, Yenisei, and Lena in Russia (Milliman and Meade 1983). Suspended matter plays a major role in biogeochemical processes because particulates are important sites for both microbial activity (Herman and Heip 1999) and enhanced adsorption of colloidal matter. Unless particles settle or are removed from the water column by suspension feeders (e.g., Dagg et al. 1996), turbidity (at suspended particulate matter concentration > 10 mg L^{-1}) may inhibit the autotrophic uptake of nutrients due to light limitation (DeMaster and Pope 1996). However, suspended matter of riverine origin represents a major source of bioavailable nutrients. Nitrogen from particulate matter, for example, can make a significant contribution to total nitrogen, especially in turbid river systems. Mayer and colleagues (1998) estimated the contribution from particulate nitrogen at between 40% and 90% for rivers such as the Mississippi and the Huanghe.

The majority of the phosphorus delivered to most SEMSs by rivers is also in particulate form (Chapter 8, this volume). River-borne particulate organic phosphorus may originate from terrestrial ecosystems as well as from waste from human activities. However, of greater significance is the particulate inorganic phosphorus, which prevails as iron-bound phosphorus (Fe-P) (Figure 9-4C), and is often associated with coatings on mineral grains (Berner et al. 1993). Reduced sediment input from rivers because of damming affects the total amount of P delivered to the estuary (Dagg et al. 2004). This is significant, for example, in the Huanghe and the Bohai, where the sediment load has decreased by 60%–80% during recent decades due to damming (Yang et al. 1998; Wang et al. 2007). Another factor worth noting is that phosphorus, like dissolved silica (Humborg et al. 1997), can be retained in reservoirs upstream, while the watersheds downstream supply extra nitrogen to the river as a result of agricultural activities. Depending on the regional conditions, a reduction of phosphorus input may counteract overall eutrophication, but this reduction may also cause P limitation, enhancing the obvious shifts in nutrient ratios (e.g., N:P) and consequently contributing to environmental problems.

The Semi-Enclosed Marine System–Open-Ocean Interface

The open ocean exerts further control on nutrient conditions in SEMSs. In open shelf seas that are well ventilated, ocean waters constitute a major and, in some cases, dominant source for nutrients (Figure 9-4). For example, oceanic nutrients dominate in the open shelves of the Mid-Atlantic Bight (Fennel et al. 2006). The competition between riverine and oceanic inputs can be exemplified in the North Sea. Its southern part receives most of the freshwater and thus riverine nutrients, whereas its northern part is primarily under oceanic control, with riverine nutrients playing a negligible role (Thomas et al. 2005; Bozec et al. 2006). The enormous input of nutrients from the Atlantic Ocean (Thomas et al. 2008) makes the North Sea one of the most productive ocean areas in the world. This productivity is further enhanced by riverine nutrient inputs in the southern regions.

The East China Sea provides another example of the importance of oceanic circulation on the nutrient dynamics of SEMSs. The seasonal flow pattern of the Kuroshio Current and its branches (e.g., Taiwan Warm Current) over the East China Sea Shelf has an important effect on circulation in this region (Zhang and Su 2006; Chapter 6, this volume). In summer, effluents from the Changjiang River disperse over the broad East China Sea, inducing a strong gradient from eutrophic coastal waters to offshore oligotrophic waters (Gong et al. 1996); Kuroshio waters can be distinguished at water depths of 50–100 m. In winter, the river plumes spread southward, while the Taiwan Warm Current water occupies the broader shelf. The surface waters of the Kuroshio are devoid of nutrients, whereas nutrient concentrations in its deep water are higher than in shelf waters of the East China Sea. Furthermore, river and oceanic waters have distinct nutrient compositions. Open-ocean waters act as the primary source for phosphorus, and to a lesser extent for nitrogen, while river input is phosphorus-deficient relative to nitrogen (Chen and Wang 1999). Given the short residence time (a few months) for waters on the East China Sea Shelf, circulation is of great importance in the biogeochemical cycling of nutrients there (Zhang et al. 2007a).

The Semi-Enclosed Marine System–Atmosphere Interface

The atmosphere constitutes another significant source of nutrients to SEMSs, particularly for the delivery of nitrogen to systems surrounded by densely populated areas. Atmospheric nutrient input can exceed riverine input and can thus make a significant contribution to the annual nutrient availability and related productivity of coastal areas (Thomas et al. 1999, 2003; Bozec et al. 2006). In the Yellow Sea and East China Sea, atmospheric deposition accounts for 40%–50% of total DIN input from land sources (i.e., river plus atmosphere), 10%–40% of total phosphate, and 1%–5% of dissolved silicate (Zhang et al. 2007b). In the Bay of Bengal, despite the enormous riverine input received, atmospheric N deposition is estimated to be three times greater than the DIN input by rivers (Naqvi et al. 2006).

In some areas, nitrogen fixation serves as a further source of reactive nitrogen. Fixation of atmospheric nitrogen appears to supply significant amounts of bioavailable nitrogen to oligotrophic subtropical areas but can also play a role in higher-latitude systems. For example, in the Baltic Sea, input by nitrogen fixation is of comparable magnitude to input by rivers. It has been reported that some river plumes provide favorable conditions for nitrogen fixation, possibly due to the physical stabilization of the water column (Ohlendiek et al. 2000; Voss et al. 2006). On the other hand, the atmosphere serves as a sink for elemental nitrogen and nitrous oxide produced by denitrification in anoxic water columns or surface sediments (Fennel et al. 2006).

SEMSs play a significant role in exchanging CO_2 between the ocean and atmosphere (Thomas et al. 2004; Borges 2005; Borges et al. 2005). Particularly relevant for carbon cycling and air–sea exchange in SEMSs is the river supply of dissolved inorganic carbon (DIC) and alkalinity (A_T). The characteristics of river DIC inputs are established by the soil conditions of the drainage area; rivers draining lime-rich soil carry higher inorganic carbon loads than rivers draining siliceous soils (e.g., Thomas and Schneider 1999; Thomas et al. 2003). Land-use changes and water management activities such as damming and irrigation also affect the riverine alkalinity concentrations (Raymond and Cole 2003). On geologic time scales, the enhanced weathering caused by rising atmospheric CO_2 conditions will eventually affect alkalinity in SEMSs and the open ocean.

The DIC/A_T ratio influences air–sea fluxes of carbon in the estuarine mixing zone. In the case of large rivers, this ratio may also influence fluxes downstream in the plume (Körtzinger 2003). DIC/A_T ratios set the conditions for CO_2 super- or undersaturation, thus defining thermodynamically driven CO_2 fluxes, which then are modulated by biological processes. Inorganic nutrients stimulate uptake of atmospheric CO_2 from the atmosphere to replenish biologically fixed carbon, while the input of organic nutrients supports heterotrophic activity, which causes a net CO_2 release to the atmosphere. Estuarine zones appear to act predominately as CO_2 sources (Frankignoulle et al. 1998; Borges 2005), but CO_2 uptake in river plumes is observed in some regions, for example, in the East China Sea (Chen and Wang 1999) and in the Bay of Bengal (Kumar et al. 1996). Some semi-enclosed basins are thought to act as "continental shelf pumps" that transfer carbon from the atmosphere into the deeper ocean (Tsunogai et al. 1999; Thomas et al. 2004). Biological and physical mechanisms have been inferred; however, the individual physical, chemical, and biological characteristics of specific SEMSs require case-by-case studies, and up-scaling tools have yet to be developed (Borges 2005).

The Semi-Enclosed Marine System–Sediment Interface

Sediments in the shallow waters of SEMSs play an important role in mediating the concentrations of nutrients (Figure 9-4) and A_T in bottom waters. On the seasonal time scale, sedimentation delays water column remineralization of organic matter and consequently supplies recycled nutrients to the pelagic environment. Sedimentary particulate organic phosphorus, for example, is efficiently remineralized and converted to DIP within the oxygenated surface layer of the sediments. A major fraction of DIP may then diffuse out of the sediments and reenter the phosphorus cycle in the water column. If sediments accumulate over time, they buffer eutrophication by temporarily storing nutrients. In this way, sediments also "buffer" mitigation strategies for nutrient reduction by slowly releasing temporarily stored nutrients back to the water column, effectively delaying the response of SEMSs to these measures.

Nutrient retention has been examined using numerical simulations, for example, in the Gulf of Thailand (Liu et al. 2007). The gulf receives nutrients mainly from the Chao Phraya River, which discharges an estimated 1.8 Gmol of DIN annually into the northern end of the gulf. Decreasing nutrient concentrations in the southern gulf have been reported and effectively simulated by a numerical model of this area (Liu et al. 2007). However, the modest influx of nutrients from the Chao Phraya River can account for less than 1% of the total primary production, which is estimated to be 399 mg C m^{-2} d^{-1} (Liu et al. 2007). Additional input of nutrients may come from the nearby Mekong River and from upwelling of nutrient-rich water from the deep South China Sea. However, numerical simulations have demonstrated that the total input of nutrients can account for only a quarter of the estimated primary production, while the other three-quarters must be sustained by nutrients regenerated from the sediments. It has been further demonstrated that benthic regeneration provides nutrients not only to the overlying water column in the Gulf of Thailand, but also to the euphotic zone in the South China Sea proper, enhancing primary production there by 15% (Liu et al. 2007).

Based on numerical simulations of the South China Sea, our assumption is that particulate organic matter reaching the seafloor on the shelves represents 21% of the primary production, which compares well with the estimated global averages of 15%–36% (Smith and Hollibaugh 1993; Wollast 1998; Mackenzie et al. 2004). Sedimentary denitrification may remove 14% of the particulate organic nitrogen reaching the seafloor, which corresponds to an average nitrogen removal of 67 mmol N m^{-2} y^{-1}. This estimate is considerably smaller than that for the Mid-Atlantic Bight (400 mmol N m^{-2} y^{-1}) (Fennel et al. 2006) but closer to the global average of 167–245 mmol N m^{-2} y^{-1} (Wollast 1993). In the Yellow Sea, Liu and colleagues (2003) found that nutrient inputs through the sediment–water interface can be five times higher than river inputs for nitrate and dissolved silicate, but they represent only 10% of the riverine input of phosphorus. In the East China Sea, the burial of organic matter is 7.4×10^6 t C y^{-1}, which is about an order of magnitude higher than the outflow of particulate organic carbon (POC) (i.e., 0.25×10^6 t C y^{-1}) into the northwest Pacific Ocean (Deng et al. 2006).

Furthermore, bottom sediments in SEMSs affect nutrient ratios in the water column through sediment–pore-water–water column exchange (Figure 9-4B, C), because burial efficiencies and regeneration rates differ among different nutrients. Redox conditions play a major role in releasing and retaining phosphorus in sediments. In the case of nitrogen, sediments can be a major site for denitrification, removing combined nitrogen at a rate close to, or exceeding, the rate of riverine or oceanic nitrogen supply (Vicchi et al. 2004; Fennel et al. 2006).

Seepage of groundwater through unconsolidated sediments constitutes a significant source of nutrients and freshwater to SEMSs, particularly in nearshore areas. Estimates of nutrient and water inputs via groundwater cover a wide range, and these inputs may be comparable to, or even exceed, those from rivers. Accordingly, groundwater-borne nutrients likely exert significant control on the productivity of SEMSs. Nutrients in groundwater originate from point sources such as septic devices or larger-scale urban or agricultural sources. Similar to riverine inputs, the effects of groundwater nutrient inputs on the ecosystems of SEMSs are complex and depend strongly on the particular setting of the system. For example, denitrification may significantly decrease nitrogen inputs when groundwater is released into a denitrifying layer. Moreover, nutrients may be intensively exchanged across the groundwater–SEMS interface in both directions, which makes the quantification of this "new" nutrient input difficult. There is, however, much evidence to suggest that groundwater inputs play a major, and likely underestimated, role in the ecosystems of SEMSs (e.g., Giblin and Gaines 1990; Valiela et al. 1990; Moore 1996; Crusius et al. 2005; Niencheski et al. 2007). Future research should further improve our understanding of the role of groundwater discharge in coastal ecosystems.

A Global Perspective

SEMSs and continental margins provide an important link between the land and the ocean interior and also between the atmosphere and the deep ocean (Liu et al. 2000). Here we summarize the role of SEMSs and continental margins as both a major conduit and a supplementary carbon repository in global biogeochemical cycles (Figure 9-5).

According to the latest assessment from the Intergovernmental Panel on Climate Change (IPCC) (Denman et al. 2007), fossil fuel burning and deforestation contributed 384 Pg C and 140 Pg C, respectively, to anthropogenic CO_2 in the atmosphere (up until 1994). Of this 524 Pg C contribution, 165 Pg C remains in the atmosphere. Rising atmospheric CO_2 concentrations have led to an undersaturation (with respect to CO_2) of ocean surface waters at the global scale, rendering the ocean a net sink for CO_2 and increasing the carbon content of the ocean over time.

The current production rate of anthropogenic CO_2 is 8 Pg C y^{-1}, of which 6.4 Pg C y^{-1} is from fossil fuel burning and the remainder from deforestation. Forty percent of the emitted anthropogenic CO_2 remains in the atmosphere, and 27.5% is taken up by the ocean, while the balance is presumably taken up by the land biota, likely by temperate forests. However, the net rate of oceanic uptake of atmospheric CO_2 has been reduced to 1.6 Pg C y^{-1} due to the outgassing of river-borne carbon. On the other hand, the net carbon flux into the open ocean is offset by 0.6 Pg C y^{-1} introduced from land. This yields an increase in the rate of oceanic carbon uptake of 2.2 Pg C y^{-1}. The surface ocean stores

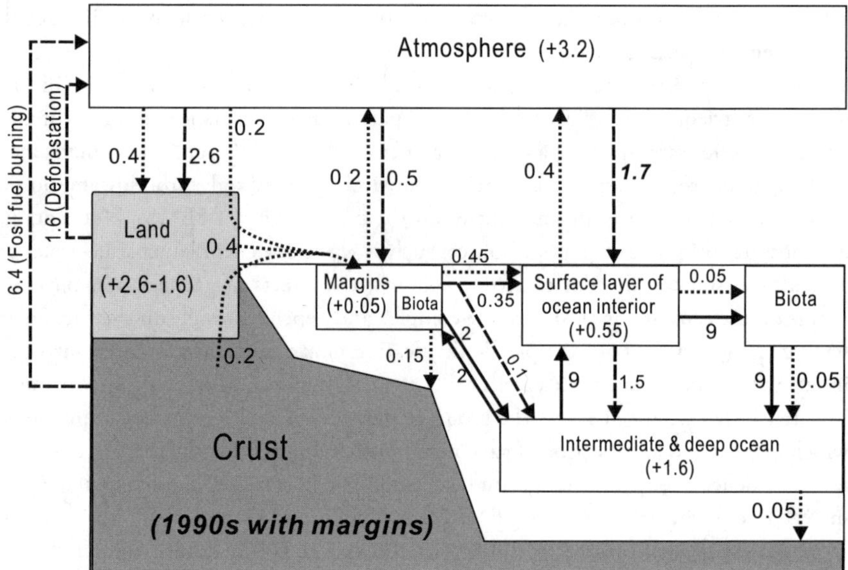

Figure 9-5. The present-day global carbon cycle (after Denman et al. 2007) and shallow marine environments in comparison. The inventories are in units of Pg C; the fluxes are in units of Pg C y^{-1}. The dotted arrows indicate terrigenous carbon flux discharged to the ocean and its fate. The dashed curves represent anthropogenic carbon fluxes. The values in parentheses indicate the increase rate (Pg C y^{-1}) of carbon inventory. For clarity, only net transboundary fluxes are shown. Redrawn from Liu et al. 2009 and used by permission of Springer.

0.6 Pg C y^{-1}, while the intermediate and deep waters store 1.6 Pg C y^{-1} of this "extra" carbon.

Based on the IPCC 2007 report (Denman et al. 2007) and a recent synthesis of the biogeochemistry of semi-enclosed seas and continental margins (Liu et al. 2009), we will illustrate the role of continental margin carbon fluxes in the global ocean carbon budget (Figure 9-5). First, the fate of terrigenous carbon, which is discharged by rivers to SEMSs and continental margins, is described (Table 9-1). Present-day release of terrestrial carbon from continental margins and semi-enclosed seas to the atmosphere is within the range of estimated values of preindustrial CO_2 release (0.14–0.36 Pg C y^{-1}) (Mackenzie et al. 2004). The partitioning of (terrestrial) carbon deposition between the margins (0.15 Pg y^{-1}) and the interior (0.05 Pg y^{-1}) approximates the proportions suggested by Sabine and colleagues (2004). Of riverine carbon inputs, 55% are exported to the open ocean, which is similar to the proportion suggested by Smith and Hollibaugh (1993).

Second, the contribution of export production from SEMSs and marginal seas to global ocean export production is considered (Table 9-2). We assume here that the export

Table 9-1. Fate of Terrestrial Carbon Inputs in SEMSs and Marginal Seas

Riverine carbon inputs	0.8 Pg C y^{-1}	
Burial as organic or carbonate carbon	0.15 Pg C y^{-1}	
Release to atmosphere	0.2 Pg C y^{-1}	
Export to open ocean:	0.45 Pg C y^{-1}	
Deep sea burial	→	0.05 Pg C y^{-1}
Release to atmosphere in open ocean	→	0.4 Pg C y^{-1}

Table 9-2. Export Production from SEMSs and Marginal Seas

World ocean:		
Primary production	50 Pg C y^{1}	
Export production	→	11 Pg C y^{-1} (22%)
SEMSs and marginal seas:		
Primary production	9.7 Pg C y^{-1}	
Export production	→	2.1 Pg C y^{-1} (22%)
POC burial on continental slopes	0.19 Pg C y^{-1}	

efficiency of the open oceans applies to the semi-enclosed and marginal seas as well. This assumption is based on the idea that shelf-dominated margins may have lower export efficiency but that this is partially compensated for by the higher export efficiency of slope-dominated margins. Therefore, it is reasonable to assume that the export production of margins is about 2 Pg C y^{-1}. The estimated flux of (marine) POC of 0.19 Pg C y^{-1}, deposited on the continental slope and rise, thus reflects 9.5% of the export production of semi-enclosed and marginal seas. This delivery efficiency agrees well with the recent observations during the European Ocean Margin Exchange (OMEX) project that suggest that 8%–10% of POC exported from the continental margin reaches the seafloor (Wollast and Chou 2001). Similar burial rates have been obtained for SEMSs and marginal seas, by Duarte and colleagues (2005) and Dunne and colleagues (2007) for example. In particular, Duarte and colleagues provide an exhaustive discussion on the strengths and limitations of the above estimates, consideration of which appears to be crucial when discussing the carbon budget of SEMSs and marginal seas at a global scale.

Third, the fate of anthropogenic CO_2 taken up by semi-enclosed seas and continental margins is assessed (Table 9-3). As anthropogenic CO_2 builds up in the atmosphere,

Table 9-3. Anthropogenic CO_2 Uptake in SEMSs and Marginal Seas

Net CO_2 uptake from the atmosphere	0.29–0.36 Pg C y^{-1}	
Release to the atmosphere (terrestrial carbon)	0.2 Pg C y^{-1}	
Uptake of anthropogenic CO_2:	0.5 Pg C y^{-1}	
Accumulation in SEMSs and marginal seas	→	0.05 Pg C y^{-1}
Export to the open ocean	→	0.45 Pg C y^{-1}

there is additional uptake by ocean surface waters. This uptake is counteracted by the release of CO_2 from terrestrial origins to the atmosphere, yielding an estimated net CO_2 uptake by the ocean of approximately 0.3 Pg C y^{-1}.

Validation of the proposed scenario requires a quantitative understanding of the function of the continental shelf pump, which has yet to be achieved. This exercise is intended to demonstrate that carbon fluxes in continental margins have important implications for the ocean carbon cycle at the global scale. If the proposed scenario is valid, the uptake of anthropogenic CO_2 in semi-enclosed and marginal seas contributes 0.5 Pg C y^{-1} of the total oceanic uptake of 2.2 Pg C y^{-1}. This would attribute an uptake of 1.7 Pg C y^{-1} of anthropogenic CO_2 to the open ocean. This line of argument is based on the assumption that the estimated open-ocean uptake of 2.2 Pg C y^{-1} is representative of the world ocean's uptake of anthropogenic CO_2.

Alternatively, it has been argued that the uptake of anthropogenic CO_2 in semi-enclosed and marginal seas constitutes a complement, rather than a fraction, of the open-ocean uptake (Tsunogai et al. 1999; Thomas et al 2004). In this case a fraction of the 0.5 Pg C y^{-1} taken up by SEMSs and marginal seas would complement the total open ocean uptake of 2.2 Pg C y^{-1}. It is clear that a proper assessment and understanding of the continental shelf pump (Tsunogai et al. 1999; Thomas et al. 2004) and its sensitivity to climate change is thus of critical relevance to reliably assessing the global ocean uptake of anthropogenic CO_2.

References

Altabet, M.A., M.J. Higginson, and D.W. Murray. 2002. The effect of millennial-scale changes in Arabian Sea denitrification on atmospheric CO_2. *Nature* 415:159–162.

Amon, R.M.W., and R. Benner. 1996. Photochemical and microbial consumption of dissolved organic carbon and dissolved oxygen in the Amazon River system. *Geochimica et Cosmochimica Acta* 60:1,783–1,792.

Berner, R.A., K.C. Ruttenburg, E.D. Ingall, and J.-L. Rao. 1993. The nature of phosphorus burial in modern marine sediments. In *Interaction of C, N, P and S Biogeochemical Cycles*, edited by R. Wollast, F.T. Mackenzie, and L. Chou, 365–378. Berlin: Springer.

Beusen, A.H.W., A.L.M. Dekkers, A.F. Bouwman, W. Ludwig, and J.A. Harrison. 2005. Estimation of global river transport of sediments and associated particulate C, N, and P. *Global Biogeochemical Cycles* 19 (4):GB4S05, doi: 10 1029/2005GB002453 (2005).

Borges, A.V. 2005. Do we have enough pieces of the jigsaw to integrate CO_2 fluxes in the Coastal Ocean? *Estuaries* 28 (1):3–27.

Borges, A.V., B. Delille, and M. Frankignoulle. 2005. Budgeting sinks and sources of CO_2 in the coastal ocean: Diversity of ecosystems counts. *Geophysical Research Letters* 32:L14601, doi: 10 1029/2005GL023053.

Bozec, Y., H. Thomas, L.-S. Schiettecatte, A.V. Borges, K. Elkalay, and H.J.W. de Baar. 2006. Assessment of the processes controlling the seasonal variations of dissolved inorganic carbon in the North Sea. *Limnology and Oceanography* 51:2,746–2,762.

Bushaw, K.L., R.G. Zepp, M.A. Tarr, D. Schulz-Jander, W.L. Miller, D.A. Bronk, and M.A. Moran. 1996. Photochemical release of biologically available nitrogen from aquatic dissolved organic matter. *Nature* 381:404–407.

Canfield, D.E. 1993. Organic matter oxidation in marine sediments. In *Interaction of C, N, P and S Biogeochemical Cycles*, edited by R. Wollast, F.T. Mackenzie, and L. Chou, 333–363. Berlin, Springer.

Chen, T.A.C., and S.L. Wang. 1999. Carbon, alkalinity and nutrient budgets on the East China Sea continental shelf. *Journal of Geophysical Research* 104:20,675–20,686.

Codispoti, L.A., J.A. Brandes, J.P. Christensen, A.H. Devol, S.W.A. Naqvi, H.W. Paerl, and T. Yoshinari. 2001. The oceanic fixed nitrogen and nitrous oxide budgets: Moving targets as we enter the Anthropocene? *Scientia Marina* 65:85–105.

Crusius, J., D. Koopmans, J.F. Bratton, M.A. Charette, K.D. Kroeger, P. Henderson, L. Ryckman, K. Halloran, and J.A. Colman. 2005. Submarine groundwater discharge to a small estuary estimated from radon and salinity measurements and a box model. *Biogeosciences* 2:141–157.

Dagg, M., R. Benner, S. Lohrenz, and D. Lawrence. 2004. Transformation of dissolved and particulate materials on continental shelves influenced by large rivers: Plume processes. *Continental Shelf Research* 24:833–858.

Dagg, M.J., E.P. Green, B.A. McKee, and P.B. Ortner. 1996. Biological removal of fine-grained lithogenic particles from a large river plume. *Journal of Marine Research* 54:149–160.

Delwiche, C.C., and B.A. Bryan. 1976. Denitrification. *Annual Review of Microbiology* 30:241–262.

DeMaster, D., and R.H. Pope. 1996. Nutrient dynamics in the Amazon shelf waters: Results from AMASSEDS (Amazon Shelf Sediment Study). *Continental Shelf Research* 16:263–289.

Deng, B., J. Zhang, and Y. Wu. 2006. Recent sediment accumulation and carbon burial in the East China Sea. *Global Biogeochemical Cycles* 20:GB3014, doi: 10 1029/2005GB002559.

Denman, K.L., G. Brasseur, A. Chidthaisong, P. Ciais, P.M. Cox, R.E. Dickinson, D. Hauglustaine, C. Heinze, E. Holland, D. Jacob, U. Lohmann, S. Ramachandran, P.L.d.S. Dias, S.C. Wofsy, and X. Zhang. 2007. Couplings between changes in the climate system and biogeochemistry. In *Climate Change 2007: The Physical Science Basis,* 499–588. Working Group I Contribution to the Fourth Assessment Report of the IPCC (Intergovernmental Panel on Climate Change). Cambridge, UK: Cambridge University Press.

Duarte, C.M., J.J. Middelburg, and N. Caraco. 2005. Major role of marine vegetation on the oceanic carbon cycle. *Biogeosciences* 2:1–8.

Dunne, J.P., J.L. Sarmiento, and A. Gnanadesikan. 2007. A synthesis of global particle export from the surface ocean and cycling through the ocean interior and on the seafloor. *Global Biogeochemical Cycles* 21:GB4006, doi: 10 1029/2006GB002907.

Fennel, K., J. Wilkin, J. Levin, J. Moisan, J. O'Reilly, and D. Haidvogel. 2006. Nitrogen cycling in the Middle Atlantic Bight: Results from a three-dimensional model and implications for the North Atlantic nitrogen budget. *Global Biogeochemical Cycles* 20:GB3007, doi: 10 1029/2005GB002456.

Forster, P., V. Ramaswamy, P. Artaxo, T. Berntsen, R. Betts, D.W. Fahey, J. Haywood, J. Lean, D.C. Lowe, G. Myhre, J. Nganga, R. Prinn, G. Raga, M. Schulz, and R.V. Dorland. 2007. Changes in atmospheric constituents and in radiative forcing. In *Climate Change 2007: The Physical Science Basis,* 129–234. Working Group I Contribution to the Fourth Assessment Report of the IPCC (Intergovernmental Panel on Climate Change). Cambridge, UK: Cambridge University Press.

Frankignoulle, M., G. Abril, A. Borges, I. Bourge, C. Canon, B. Delille, E. Libert, and J.-M. Théate. 1998. Carbon dioxide emission from European estuaries. *Science* 282:434–436.

Garrels, R.M., F.T. Mackenzie, and C. Hunt. 1975. *Chemical Cycles and the Global Environment.* Los Altos, CA: William Kaufmann.

Gattuso, J.-P., M. Frankignoulle, and R. Wollast. 1998. Carbon and carbonate metabolism in coastal aquatic ecosystems. *Annual Reviews of Ecological Systems* 29:405–434.

Giblin, A.E., and A.G. Gaines. 1990. Nitrogen inputs to a marine embayment: The importance of groundwater. *Biogeochemistry* 10:309–328.

Gong, G.C., Y.L.L. Chen, and K.-K. Liu. 1996. Chemical hydrography and chlorophyll *a* distribution in the East China Sea in summer: Implications in nutrient dynamics. *Continental Shelf Research* 16:1,561–1,590.

Gypens, N., and C. Lancelot. 2006. Response of the Belgian coastal zone (southern North Sea) to increased atmospheric CO_2 and nutrient loads: From pristine to 2015. CARBO-OCEAN 2nd annual meeting. Gran Canaria, Spain, 4–8 December 2006.

Herman, P.M.J., and C.H.R. Heip. 1999. Biogeochemistry of the MAximum TURbidity zone of Estuaries (MATURE): Some conclusions. *Journal of Marine Systems* 22: 89–104.

Hollibaugh, J.T., R.W. Buddemeier, and S.V. Smith. 1991. Contributions of colloidal and high molecular weight dissolved material to alkalinity and nutrient concentrations in shallow marine and estuarine systems. *Marine Chemistry* 34:1–27.

Humborg, C., V. Ittekkot, A. Cociasu, and B. von Bodungen. 1997. Effect of Danube River dam on Black Sea biogeochemistry and ecosystem structure. *Nature* 386:385–388.

Jetten, M.S.M., O. Sliekers, M. Kuypers, T. Dalsgaard, L. van Niftrik, I. Cirpus, K. van de Pas-Schoonen, G. Lavik, B. Thamdrup, D. Le Paslier, H.J.M. Op den Camp, S. Hulth, L.P. Nielsen, W. Abma, K. Third, P. Engström, J.G. Kuenen, B.B. Jørgensen, D.E. Canfield, J.S. Sinninghe Damsté, N.P. Revsbech, J. Fuerst, J. Weissenbach, M. Wagner, I. Schmidt, M. Schmid, and M. Strous. 2003. Anaerobic ammonium oxidation by marine and freshwater planctomycete-like bacteria. *Journal of Applied Microbiology and Biotechnology* 63:107–114.

Justic, D., N.N. Rabalais, and R.E. Turner. 2002. Modeling the impacts of decadal changes in riverine nutrient fluxes on coastal eutrophication near the Mississippi River delta. *Ecological Modelling* 152:33–46.

Körtzinger, A. 2003. A significant CO_2 sink in the tropical Atlantic Ocean associated with the Amazon River plume. *Geophysical Research Letters* 30 (24):2,287, doi: 10 1029/2003GL018841.

Kumar, M.D., S.W.A. Naqvi, M.D. George, and D.A. Jayakumar. 1996. A sink for atmospheric carbon dioxide in the northeast Indian Ocean. *Journal of Geophysical Research* 101:18,121–18,125.

Kuypers, M.M.M., A.O. Sliekers, G. Lavik, M. Schmid, B.B. Jorgensen, J.G. Kuenen, J.S.S. Damste, M. Strous, and M.S.M. Jetten. 2003. Anaerobic ammonium oxidation by anammox bacteria in the Black Sea. *Nature* 422:608–611.

Liu, K.-K., L. Atkinson, R. Quiñones, and L. Talaue-McManus. 2009. Biogeochemistry of continental margins in a global context. In *Carbon and Nutrient Fluxes in Continental Margins: A Global Synthesis*, edited by K.-K. Liu, L. Atkinson, R. Quiñones, and L. Talaue-McManus. Global Change—The IGBP Series. Berlin: Springer.

Liu, K.-K., Y.-J. Chen, C.-M. Tseng, I.-I. Lin, H. Liu, and A. Snidvongs. 2007. The significance of phytoplankton photo-adaptation and benthic–pelagic coupling to primary-production in the South China Sea: Observations and numerical investigations. *Deep-Sea Research II* 54:1,546–1,574.

Liu, K.-K., K. Iseki, and S.-Y. Chao. 2000. Continental margin carbon fluxes. In *The Changing Ocean Carbon Cycle: A Midterm Synthesis of the Joint Global Ocean Flux Study*, edited by R.B. Hanson, H.W. Ducklow, and J.G. Field 187–239. Cambridge, UK: Cambridge University Press.

Liu, K.-K., and I.R. Kaplan. 1982. Nitrous oxide in the sea off Southern California. In *The Environment of the Deep Sea*, edited by W.G. Ernst and J.G. Morin, 73–92. Upper Saddle River, NJ: Prentice Hall/Pearson.

Liu, S.M., J. Zhang, S.Z. Chen, H.T. Chen, G.H. Hong, H. Wei, and Q.M. Wu. 2003. Inventory of nutrient compounds in the Yellow Sea. *Continental Shelf Research* 23:1,161–1,174.

Mackenzie, F.T., A. Andersson, A. Lerman, and L.M. Ver. 2004. Boundary exchanges in the global coastal margin: Implications for the organic and inorganic carbon cycles. In Vol. 13 of *The Sea*, edited by A. Robinson and K.H. Brink, Chap. 7. Cambridge, MA: Harvard University Press.

Mayer, L.M., R.G. Keil, S.A. Macko, S.B. Joye, K.C. Ruttenberg, and R.C. Aller. 1998. Importance of suspended particulates in riverine delivery of bioavailable nitrogen to coastal zone. *Global Biogeochemical Cycles* 12 (4):573–579.

Meybeck, M. 1982. Carbon, nitrogen and phosphorus transport by world rivers. *American Journal of Science* 282:401–450.

Middelburg, J.J., and K. Soetaert. 2004. The role of sediments in shelf ecosystem dynamics. In Vol. 13 of *The Sea*, edited by A. Robinson and K.H. Brink, 353–374. Cambridge, MA: Harvard University Press.

Milliman, J.D., and R.H. Meade. 1983. World-wide delivery of river sediment to the oceans. *The Journal of Geology* 91:1–21.

Moore, W.S. 1996. Large groundwater inputs to coastal waters revealed by ^{226}Ra enrichments. *Nature* 380:612–614.

Morel, F.M.M., and J.G. Hering. 1993. *Principles and Applications of Aquatic Chemistry*. Chichester, UK: John Wiley & Sons.

Naqvi, S.W.A., H. Naik, D.A. Jayakumar, M.S. Shailaja, and P.V. Narvekar. 2006. The Indian Ocean. In *Past and Present Water Column Anoxia*, edited by L. Neretin,

195–224. Proceedings of the NATO Advanced Research Workshop, Yalta, Crimea, Ukraine, 4–8 October 2003. Vol. 64 of NATO Science Series 4: Earth and Environmental Sciences. Berlin: Springer.

Niencheski, L.F.H., H.L. Windom, W.S. Moore, and R.A. Jahnke. 2007. Submarine groundwater discharge of nutrients to the ocean along a coastal lagoon barrier, Southern Brazil. *Marine Chemistry* 106:546–561.

Ohlendiek, U., A. Stuhr, and H. Siegmund. 2000. Nitrogen fixation by diazotrophic cyanobacteria in the Baltic Sea and transfer of the newly fixed nitrogen to picoplankton organisms. *Journal of Marine Systems* 25:213–219.

OSPAR Commission. 2000. Regional QSR II: Greater North Sea. In *Quality Status Report 2000*, 136. London: OSPAR Commission.

Pan, G., and P.S. Liss. 1998. Metastable-equilibrium adsorption theory. *Journal of Colloid Interface Science* 201:71–76.

Rabalais, N.N., R.E. Turner, Q. Dortch, D. Justic, V.J. Biermann, and W.J. Wiseman. 2002. Nutrient-enhanced productivity in the northern Gulf of Mexico: Past, present and future. *Hydrobiologia* 475/476:39–63.

Rabalais, N.N., R.E. Turner, D. Justic, Q. Dortch, W.J. Wiseman, and B.K. Sen Gupta. 2000. Gulf of Mexico biological system responses to nutrient changes in the Mississippi River. In *Estuarine Science: A synthetic approach to research and practice*, edited by E. Hobbie, 241–268. Washington, DC: Island Press.

Rabouille, C., F. Mackenzie, and L.M. Ver. 2001. Influence of the human perturbation on carbon, nitrogen and oxygen biogeochemical cycles in the global coastal ocean. *Geochimica et Cosmochimica Acta* 65:3,615–3,639.

Raymond, P.A., and J.J. Cole. 2003. Increase in the export of alkalinity from North America's largest river. *Science* 301:88–91.

Sabine, C.L., R.A. Feely, N. Gruber, R.M. Key, K. Lee, J.L. Bullister, R. Wanninkhof, C.S. Wong, D.W.R. Wallace, B. Tilbrook, F.J. Millero, T.-H. Peng, A. Kozyr, T. Ono, A.F. Rios. 2004. The oceanic sink for anthropogenic CO_2. *Science* 305:367–371.

Sañudo-Wilhelmy, S.A., I. Rivera Duarte, and A.R. Flegal. 1996. Distribution of colloidal trace metals in the San Francisco Bay estuary. *Geochimica et Cosmochimica Acta* 60:4,933–4,944.

Seitzinger, S., J.A. Harrison, E. Dumont, A.H.W. Beusen, and A.F. Bouwmann. 2005. Sources and delivery of carbon, nitrogen and phosphorus to the coastal zone: An overview of global Nutrient Export from Watersheds (NEWS) models and their application. *Global Biogeochemical Cycles* 19:GB4S01, doi: 10 1029/2005GB002606.

Smith, S.V., and J.T. Hollibaugh. 1993. Coastal metabolism and the oceanic carbon balance. *Review of Geophysics* 31:75–89.

Thomas, H., Y. Bozec, H.J.W. deBaar, K. Elkalay, M. Frankignoulle, W. Kühn, H.J. Lenhart, A. Moll, J. Pätsch, G. Radach, L.-S. Schiettecatte, and A. Borges. 2009. Carbon and nutrient budgets of the North Sea. In *Carbon and Nutrient Fluxes in Global Continental Margins*, edited by L.Atkinson, K.K. Liu, R. Quinones, and L. Talaue-McManus. Berlin: Springer.

Thomas, H., Y. Bozec, H.J.W. de Baar, K. Elkalay, M. Frankignoulle, L.-S. Schiettecatte, G. Kattner, and A.V. Borges. 2005. The carbon budget of the North Sea. *Biogeosciences* 2:87–96.

Thomas, H., Y. Bozec, K. Elkalay, and H.J.W. de Baar. 2004. Enhanced open ocean storage of CO_2 from shelf sea pumping. *Science* 304:1,005–1,008, doi: 10 1126/science 1095491.

Thomas, H., V. Ittekkot, C. Osterroht, and B. Schneider. 1999. Preferential recycling of nutrients: The ocean's way to increase new production and to pass nutrient limitation? *Limnology and Oceanography* 44:1,999–2,004.

Thomas, H., J. Pempkowiak, F. Wulff, and K. Nagel. 2003. Autotrophy, nitrogen accumulation and nitrogen limitation in the Baltic Sea: A paradox or a buffer for eutrophication? *Geophysical Research Letters* 30 (21):2,130, doi: 10 1029/2003GL017937.

Thomas, H., and B. Schneider. 1999. The seasonal cycle of carbon dioxide in Baltic Sea surface waters. *Journal of Marine Systems* 22:53–67.

Tong, L., and J. Zhang. 2007. Chinese IMBER/GLOBEC program progress: Dead zone study. *IMBER Update*, Issue No. 6, March 2007, 6–8. Integrated Marine Biogeochemistry and Ecosystem Research (IMBER). www.imber.info/newsletters.html.

Tsunogai, S., S. Watanabe, and T. Sato. 1999. Is there a "continental shelf pump" for the absorption of atmospheric CO_2? *Tellus* 51B:701–712.

Turner, R.E., and N.N. Rabalais. 1991. Changes in Mississippi River water quality this century and implications for coastal food webs. *BioScience* 41:140–147.

Valiela, I., J. Coasta, K. Foreman, J.M. Teal, B. Howes, and D. Aubrey. 1990. Transport of groundwater-borne nutrients from watersheds and their effects on coastal waters. *Biogeochemistry* 10 (3):177–197.

Vicchi, M., P. Ruardij, and J.W. Baretta. 2004. Link or sink: A modelling interpretation of the open Baltic biogeochemistry. *Biogeosciences* 1:79–100.

Voss, M., D. Bombar, N. Loick, and J.W. Dippner. 2006. Riverine influence on nitrogen fixation in the upwelling region off Vietnam, South China Sea. *Geophysical Research Letters* 33:L07604, doi: 10 1029/2005GL025569.

Wang, H., Z. Yang, Y. Saito, J.P. Liu, X. Sun, and Y. Wang. 2007. Stepwise decreases of the Huanghe (Yellow River) sediment load (1950–2005): Impacts of climate change and human activities. *Global and Planetary Change* 57:331–354.

Watson, A.J., D.C.E. Bakker, A.J. Ridgwell, P.W. Boyd, and C.S. Law. 2000. Effect of iron supply on Southern Ocean CO_2 uptake and implications for glacial atmospheric CO_2. *Nature* 407:730–733.

Wollast, R. 1993. Interactions of carbon and nitrogen cycles in the coastal zone. In *Interaction of C, N, P and S Biogeochemical Cycles*, edited by R. Wollast, F.T. Mackenzie, and L. Chou, 195–210. Berlin: Springer.

Wollast, R. 1998. Evaluation and comparison of the global carbon cycle in the coastal zone and in the open ocean. In Vol. 10 of *The Sea*, edited by A. Robinson and K.H. Brink, 213–252. New York: John Wiley & Sons.

Wollast, R., and L. Chou. 2001. The carbon cycle at the ocean margin in the northern Gulf of Biscay. *Deep-Sea Research II* 48:3,265–3,293.

Yang, Z.S., X.G. Sun, Z.R. Chen, and C.G. Pang. 1998. Sediment discharge of the Yellow River to the sea: Its past, present, future and human impact on it. In *Health of the Yellow Sea*, edited by G.H. Hong, J. Zhang, and B.K. Park 109–127. Seoul: The Earth Love Publication Association.

Zhang, J., S.M. Liu, J.L. Ren, Y. Wu, and G.L. Zhang. 2007a. Nutrient gradients from the eutrophic Changjiang (Yangtze River) Estuary to the oligotrophic Kuroshio waters and re-evaluation of budgets for the East China Sea Shelf. *Progress in Oceanography* 74:449–478.

Zhang, J., and J.L. Su. 2006. Nutrient dynamics of the Chinese seas: The Bohai, Yellow Sea,

East China Sea and South China Sea. In Vol. 14 of *The Sea*, edited by A.R. Robinson and K.H. Brink, 637–671. Cambridge, MA: Harvard University Press.

Zhang, J., Z.G. Yu, T. Raabe, S.M. Liu, A. Starke, L. Zou, H.W. Gao, and U. Brockmann. 2004. Dynamics of inorganic nutrient species in the Bohai seawaters. *Journal of Marine Systems* 44:189–212.

Zhang, G.S., J. Zhang, and S.M. Liu. 2007b. Characterization of nutrients in the atmospheric wet and dry deposition observed at the two monitoring sites over Yellow Sea and East China Sea. *Journal of Atmospheric Chemistry* 57:41–57.

10

Dynamics and Vulnerability of Marine Food Webs in Semi-Enclosed Systems

Catharina J.M. Philippart and Werner Ekau

Important features of semi-enclosed marine systems (SEMSs) are the species present, the biological productivity of the systems, the complexity of the ecosystems, and human interference through exploitation of commercially important species and pollution. These factors are linked to physical and chemical conditions in SEMSs and related to the systems' connectedness with the open ocean. This chapter describes the ecosystem structure of the SEMSs included in this book and analyzes the relationships among primary productivity, fish catch (total, and benthic versus pelagic), trophic levels, carbon transfer, and connectedness of the systems with the open ocean.

Food Webs, Trophic Level, and Transfer Efficiencies

The trophic structure of marine food webs can be described by pyramids in which primary productivity generated at the lowest trophic level (TL1) is moved up toward higher trophic levels, with a huge fraction of that productivity being used in the process for maintenance, reproduction, and other activities of the animals in the systems (Pauly and Christensen 1995). An energy pyramid is typically characterized by three aspects, namely, the width of the base (primary productivity), the height (number of trophic levels), and the transfer efficiency from one trophic level to the next (Figure 10-1) which is about 10% (Odum 1969). Pimm (1980, 2002) proposed that food chain length is limited by primary productivity and/or energy transfer efficiency.

The "productive space hypothesis" by Schoener (1989) predicts that food chain length should increase as a function of total ecosystem productivity, the product of per-unit-size productivity or resource availability (e.g., g C m^{-2} y^{-1}), and ecosystem size (area or volume). In addition, the number of trophic levels is thought to be determined by the complexity of the food web. A simple food web, such as a food chain, is characterized by only

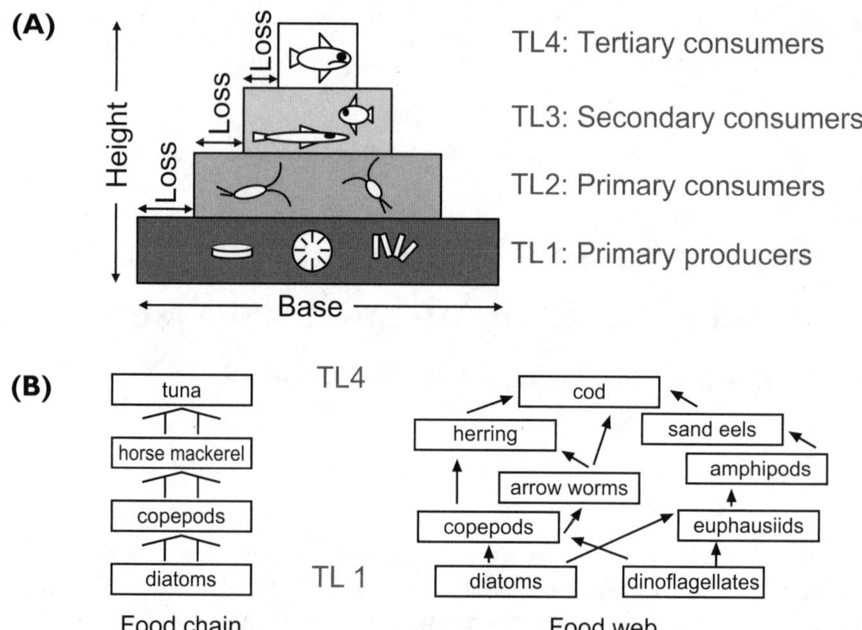

Figure 10-1. (A) Hypothetical marine trophic pyramid, showing the subsequent trophic levels (TL). (B) A food chain with a few trophic links (left panel) and more complex food web with numerous trophic links (right panel).

a few trophic links, with loss of energy during transfer (Figure 10-1B). More-complex food webs have more trophic relationships, resulting in lower transfer efficiencies between trophic levels. The history of community organization (which includes colonization and in situ evolution) is an important, but difficult to quantify, constraint on food chain length (Kitching 2001). Subsequently, the complexity of food webs may increase from the evolutionarily young coastal seas to the older open ocean (e.g., Cury et al. 2000).

A recent assemblage of published data from North Atlantic food webs suggests that seas are basically bottom-up controlled and that disturbances may cause shifts to top-down control (Frank et al. 2007). Ecosystem susceptibility to such shifts is related to both species richness and oceanic temperature conditions. For example, warmer areas with greater species richness appear to be able to withstand excessive exploitation because of higher maximum population growth rates and a potentially greater pool of compensatory species. This implies that very cold and species-poor areas are more vulnerable to shifts from bottom-up to top-down control, while warmer areas might oscillate between these controls, depending on the strength of the perturbations and, possibly, changing temperature regimes (Frank et al. 2007).

For these reasons, we hypothesize that large, undisturbed, and open seas will have more-complex food webs than small, heavily exploited, and enclosed ones, and subsequently lower transfer efficiencies. In this chapter, we will examine the current relationships between primary productivity and fish catch (as an index of ecosystem services) for a set of SEMSs (see Chapter 1 for selection criteria). In addition, we will examine regional dynamics of food webs by analyses of fifty-year time series on several features of the catch (mean trophic level and proportion of pelagic and benthic organisms of fisheries landings). Data on primary productivity, height, and mean trophic level of the fish catch are used to estimate regional transfer efficiencies.

Primary Productivity

Phytoplankton are the most important biomass producers in the ocean. These microscopic algae remove carbon dioxide from the environment and transfer carbon to other trophic levels. Obviously, any change affecting primary producers' structure and dynamics may have consequences for the entire marine food web. This is called bottom-up control. It has been observed in several seas that a shift in phytoplankton community composition and/or timing of phytoplankton blooms was followed by changes in higher trophic levels (e.g., Cushing 1975; Fromentin and Planque 1996; Møller et al. 2007; Philippart et al. 2007b). Conversely, changes in top predators (top-down control) within ecosystems can result in cascading effects through the trophic levels below, resulting in a complete reorganization of the food web (e.g., Frank et al. 2005). Strong changes in bottom-up (resource-driven) and top-down (consumer-driven) control, that eventually could lead to so-called regime shifts (see Chapter 3, this volume), may be irreversible, making recovery slow or even impossible (Hutchings and Reynolds 2004; Shelton et al. 2005).

Within the SEMSs that are considered in this chapter (Figure 10-2), primary productivity ranges from 63 mg C m^{-2} d^{-1} in the polar Kara Sea to 1,804 mg C m^{-2} d^{-1} in the temperate Baltic Sea (Table 10-1). Primary productivity is significantly related to the climate zone in which the sea is located ($n = 13$, $r^2 = 0.72$, $p = 0.008$). Highest average values were found for temperate seas (1,105 ± 118 mg C m^{-2} d^{-1}), followed by tropical (634 ± 204 mg C m^{-2} d^{-1}), subtropical (401 ± 204 mg C m^{-2} d^{-1}), and polar (208 ± 166 mg C m^{-2} d^{-1}) seas.

Although productivity in temperate seas appeared to be higher than in others, we cannot generalize these findings. On a global scale, light and nutrients have to be considered as main growth-limiting factors. Light generally decreases from the equator to the poles, while nutrient levels go up (Levitus et al. 1994). In polar seas, primary productivity is restricted to short summer blooms, while tropical productivity is more or less stable through the year (Parsons et al. 1983). Temperate seas are generally characterized by a bloom in spring and autumn (Parsons et al. 1983). Due to their relative shallowness and the additional nutrient inputs from rivers, coastal seas and continental shelf areas generally support higher primary productivity than the open ocean. Productivity of a SEMS

Table 10-1. Regional Characteristics of SEMSs

Regional Sea	*Connectivity (%)*	*Climate Zone*	*Total Area (km^2)*	*Mean Depth (m)*	*Primary Production (mgC)*	*Fish Catch, Pelagics ($mgC\ m^{-2}\ d^{-1}$)*	*Fish Catch, Demersals ($mgC\ m^{-2}\ d^{-1}$)*	*Trophic Index (-)*	*Transfer Efficiency (%)*
Baltic Sea	0.61	Temperate	412,560	52	1,804	0.58	0.03	3.35	3.35
Bay of Bengal	19.23	Tropical	3,660,127	2,600	568	0.15	0.19	3.51	5.16
Black Sea	0.06	Temperate	460,151	1,189	882	0.27	0.03	3.08	2.15
East China & Yellow Seas	23.33	Temperate	1,212,441	188	1,058	1.54	0.32	3.30	6.31
Gulf of Mexico	7.20	Subtropical	1,529,669	1,486	417	0.13	0.06	2.60	0.82
Gulf of Thailand	16.07	Tropical	386,878	45	700	0.33	0.40	3.52	6.54
Hudson Bay	4.55	Polar	841,214	100	442	0.00	0.00	4.26	0.53
Kara Sea	13.97	Polar	883,000	111	63	0.00	0.00	—	—
Laptev Sea	28.16	Polar	662,000	53	120	0.00	0.00	—	—
Mediterranean Sea	0.18	Subtropical	2,516,484	2,000	385	0.08	0.03	3.25	2.73
North Sea	7.60	Temperate	575,000	94	1,067	0.70	0.08	3.26	4.09
Sea of Okhotsk	9.09	Temperate	1,552,663	838	774	0.30	0.09	3.47	4.62
Gulf of St. Lawrence	4.67	Temperate	155,000	152	1,044	0.25	0.28	3.00	2.27

Connectivity was taken from large-scale maps as the fraction of the total coastline of the SEMS that is an opening.

Other information was compiled from various publications: Alekseev et al. 2006, Carmack and Wassmann 2006, Hirche et al. 2006, Lääne et al. 2005, Lozan 1996, Orysova et al. 2005, Qu et al. 2005, Schmid et al. 2006, Sea Around Us 2008,Teng et al. 2005, and Tsyban et al. 2005.

Transfer efficiency (E) was calculated from the relationship between primary productivity (PP), fish catch (FC), and marine trophic index (MTI) according to $E = (FC/PP)^{1/(MTI-1)}$.

Data on catch (fish catch and trophic index) are mean values for the period 1990–1999.

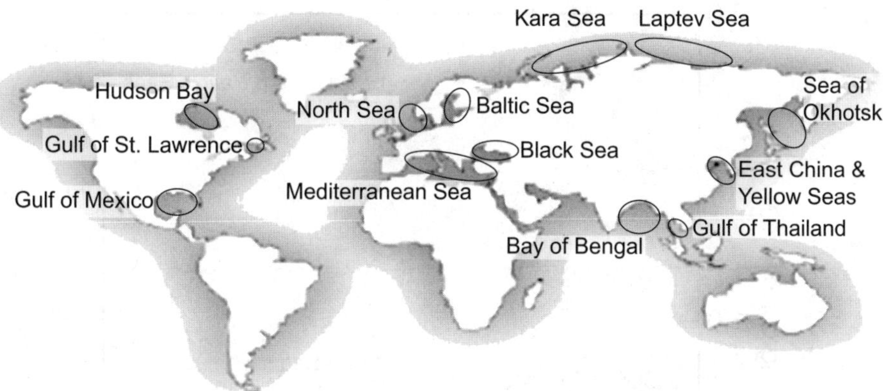

Figure 10-2. Map of the thirteen SEMSs considered in Chapter 10: Baltic Sea, Bay of Bengal, Black Sea, East China and Yellow seas, Gulf of Mexico, Gulf of Thailand, Hudson Bay, Kara Sea, Laptev Sea, Mediterranean Sea, North Sea, Sea of Okhotsk, and Gulf of St. Lawrence.

is, therefore, a mix of global and regional factors and might even change over time (e.g., as the result of eutrophication).

Fish Production

If fish-catch data represent the biomass of the species occurring in the area, they should also be a good measure to represent changes in the biomass of different species through changes in catches over time. In reality, we have to be aware that the fish-catch data we can work with are only the reported part of the whole catch. There is an unreported share that comprises landed but unreported fish (especially in remote or undeveloped areas), illegal catches, discards (caught but subsequently thrown overboard), and ghost kills (killed by lost gear). For the North Sea, for example, the total annual quantity of discards is estimated to be nearly one-third of the total weight landed and one-tenth of the estimated total biomass of fish in this area (Catchpole et al. 2005). Discarding leads to a direct loss of potential income and to a loss of potential yield when small and/or juvenile commercial fish are caught and killed, which subsequently reduces potential growth and contribution to stock replacement. In addition, the fish catch is often a strong selection of species and sizes of all marine organisms present, determined by fishing efforts, catchability, and preferences. For example, the strong increase of stocks of codfish in the North Sea during the 1960s resulted in high catches and development of new fisheries (Cushing 1984). Nevertheless, fish-catch data are the most complete data sets that can be used for this purpose and are thought to reflect species composition in the respective area sufficiently (Pauly and Watson 2005).

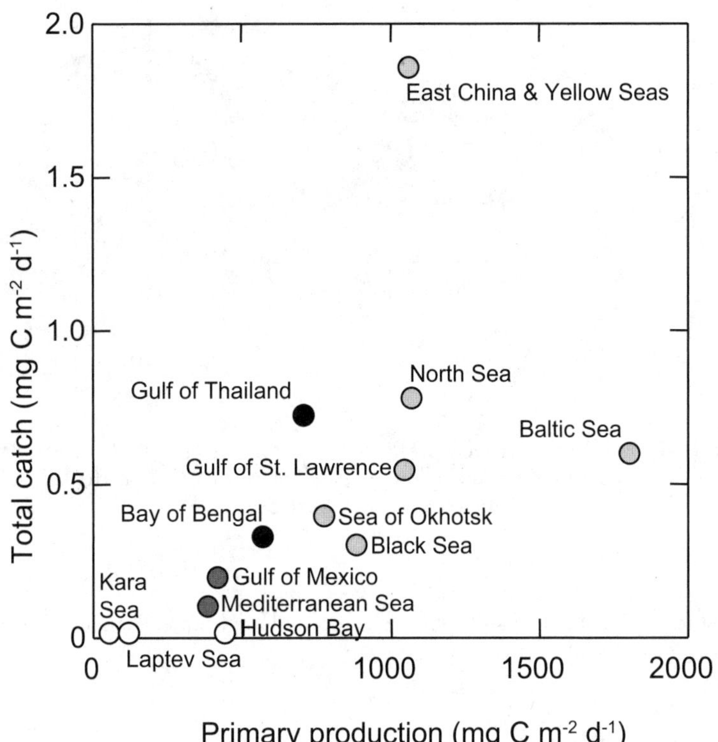

Figure 10-3. Relationship between primary production (recent value, in units of mg C m^{-2} d^{-1}) and total fish landings (average value in 1990–1999, in units of mg C m^{-2} d^{-1}). The shade of the dot indicates the climatic conditions: polar—white; temperate—light gray; subtropical—dark gray; tropical—black.

The original database distinguishes thirty functional groups, based on taxonomy, habitat preferences, feeding habits, and maximum size (Sea Around Us 2008). For direct comparison of catch data with primary productivity (mg C m^{-2} d^{-1}) we converted the total annual catch to a time- and area-related unit (mg FW m^{-2} d^{-1}; FW = fresh weight). For comparison of primary productivity and fish catch was a 9:1 ratio for the conversion of wet weight to carbon (Strathmann 1967; Pauly and Christensen 1995).

On average, most fish is caught in seas that have high primary productivity ($n = 13$, $r^2 = 0.34$, $p = 0.036$) (Figure 10-3). These results are in agreement with the positive relationship between the annual mean phytoplankton biomass (expressed as chlorophyll *a* concentration) and the long-term annual yield of resident fish observed for regions in the northeast Pacific (Ware and Thomson 2005). Although few if any fish were caught in polar seas (Hudson Bay, Kara Sea, and Laptev Sea) and highest catches (> 10 mg FW

m^{-2} d^{-1}) were taken from temperate seas (East China and Yellow seas and North Sea), fish catches were not significantly related to climate zone ($n = 13$, $r^2 = 0.44$, $p = 0.144$).

Characteristics of Fish Catch

Changes in fish communities over time were characterized by combining two approaches. First, we calculated the ratio of pelagic to demersal (benthic) fish (PDR) as a measure of changes in the system from a benthic- to a pelagic-dominated regime or vice versa. We calculated the pelagic catch as the sum of the catches of pelagic fish, benthopelagic fish, shrimp, and squid, and we calculated the demersal catch as the sum of catches of demersal fish, flatfish, reef-associated fish, shark, lobster, crab, and other demersal invertebrates. Second, we calculated the marine trophic index (MTI) as the mean trophic level of fisheries landings. This index was originally used by Pauly and colleagues (1998) to demonstrate that fisheries, since 1950, increasingly rely on the smaller, short-lived fish and on the invertebrates from the lower parts of both marine and freshwater food webs ("fishing down the foodweb"). Changes in MTI additionally reflect bottom-up processes such as the eutrophication of some areas of the Mediterranean Sea, which led to increases in the biomass and production of small herbivorous pelagic fish species, such as anchovies and sardines (Caddy 1993; Caddy et al. 1998).

For most seas, the changes in fish catch coincided with changes in MTI (Figure 10-4) and in the PDR (Figure 10-5) during the past fifty years.

- *Baltic Sea*—The increase in catch from the 1950s up to the mid-1970s coincided with an increase in herring (*Clupea harengus*: TL = 3.46, pelagic) and cod (*Gadus morhua*: TL = 4.01, benthopelagic) (Sea Around Us 2008). From 1980 to 1990, pelagics were exceptionally high in the catches (Figure 10-5A). Thereafter pelagic catches returned to the ratio of before, with sprat (*Sprattus sprattus*: TL = 3.00, pelagic) being the dominant species in the catch. The increase in fish catch, on average, coincided with a decrease in MTI (Figure 10-4A). As for the Black Sea, the PDR for the Baltic Sea is an order of magnitude higher than for other SEMSs. This may be due to the occurrence of anoxia in both seas, suppressing demersal fish.
- *Bay of Bengal*—Pelagic species were fished off in the first decade, 1950–1960, causing a tremendous drop in the PDR (Figure 10-5B). A steady increase of the fish catch from the 1960s up to the early 2000s was observed, coinciding with an overall increase of the most dominant group of species in these catches, threadfin breams (*Nemipterus* spp.: TL = 3.50, pelagic), coupled with an increase of pelagics in general (Figure 10-5B). These species contributed less than 5% of the catch in the early 1950s but up to more than 15% in the 1980s and 1990s (Sea Around Us 2008). For this area, a suite of relatively rare organisms was lumped into a group of "other taxa" with a contribution to the fish catch ranging from 60% in the 1950s to less than 40% in later years, indicating the high diversity of the fish community. Increase in fish catch, on average, coincided with a decrease in MTI at least until the mid-1990s (Figure 10-4B).

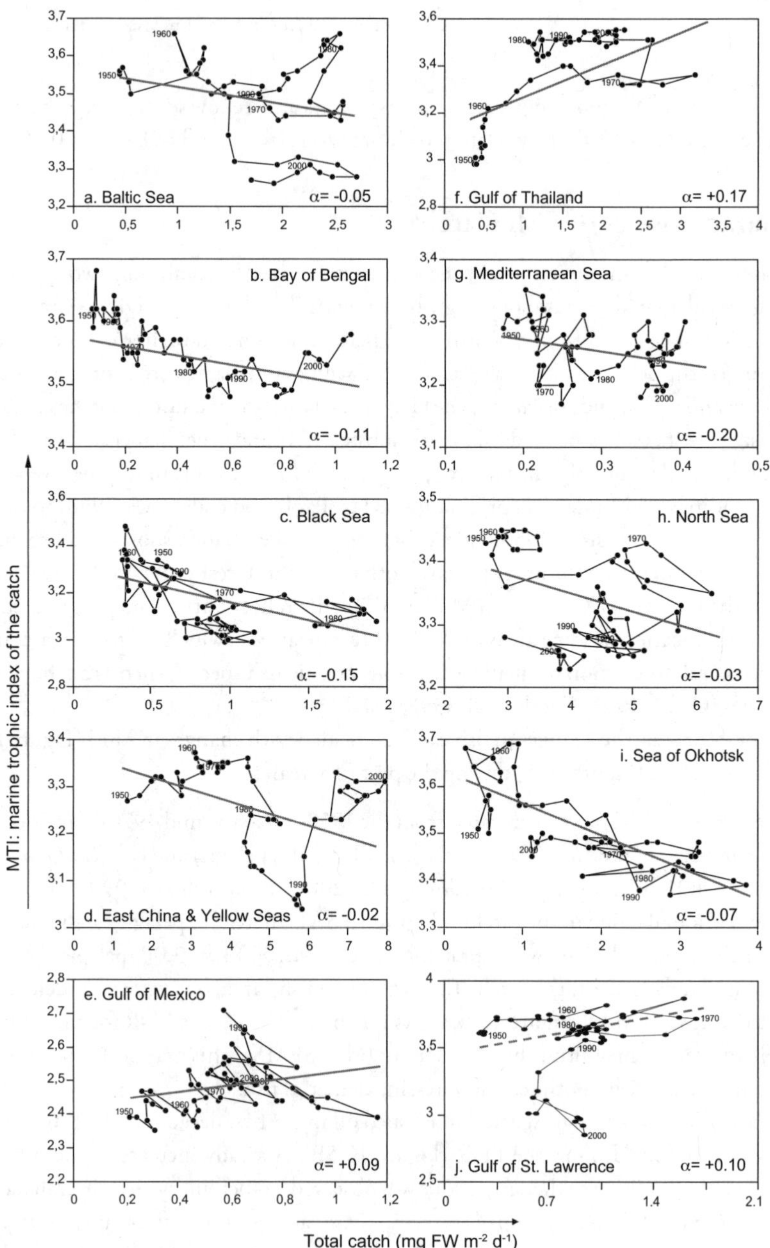

Figure 10-4. Regional relationships between fish catch (FC, in units of mg FW m^{-2} d^{-1}) and marine trophic index (MTI) between 1950 and 2003: (A) Baltic Sea, (B) Bay of Bengal, (C) Black Sea, (D) East China and Yellow seas, (E) Gulf of Mexico, (F) Gulf of Thailand, (G) Mediterranean Sea, (H) North Sea, (I) Sea of Okhotsk, (J) Gulf of St. Lawrence. The slopes of the regression line between MTI and total catch indicate the positive or negative correlation of the parameters; significance level is expressed by α. Please note that these lines do not indicate a change in the trend in MTI over time. Sufficient data for the Kara Sea, Laptev Sea, and Hudson Bay were not available, so these SEMSs are not represented here.

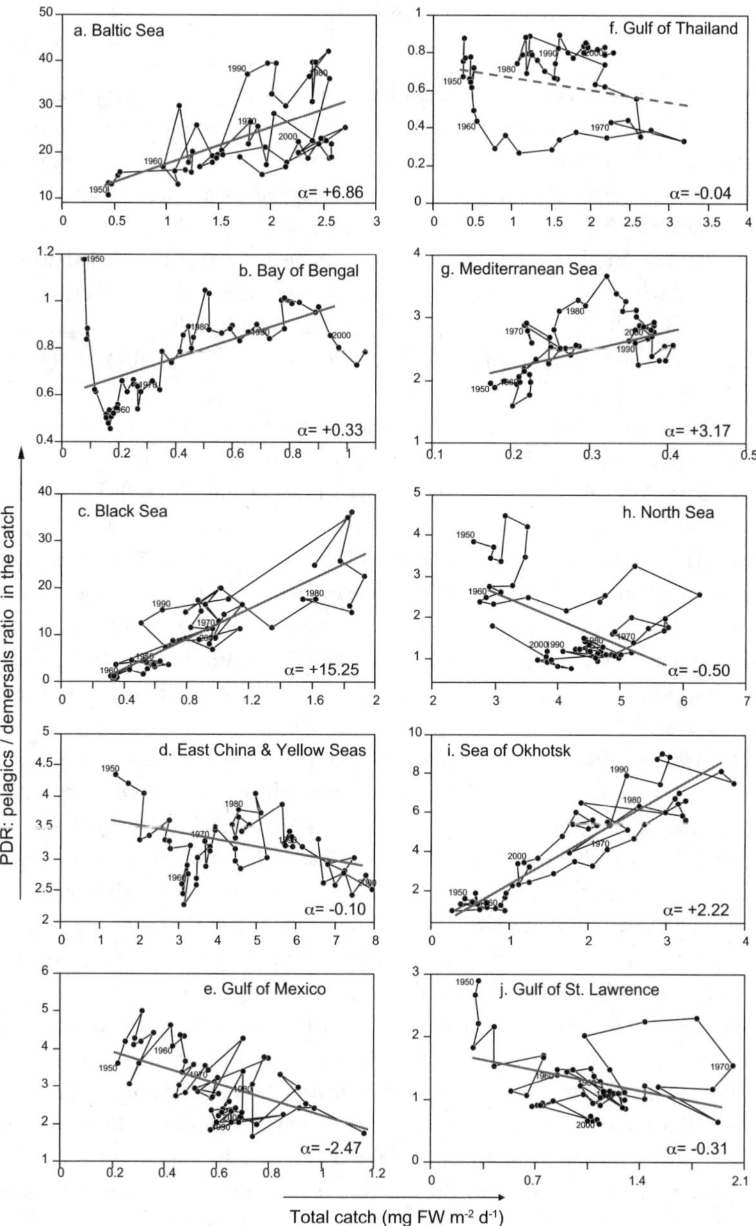

Figure 10-5. Regional relationships between fish catch (FC, in units of mg FW $m^{-2} d^{-1}$) and ratio of pelagic and demersal catch (PDR) between 1950 and 2003: (A) Baltic Sea, (B) Bay of Bengal, (C) Black Sea, (D) East China and Yellow seas, (E) Gulf of Mexico, (F) Gulf of Thailand, (G) Mediterranean Sea, (H) North Sea, (I) Sea of Okhotsk, (J) Gulf of St. Lawrence. The slopes of the regression line between PDR and total catch indicate the positive or negative correlation of the parameters; significance level is expressed by α. Please note that these lines do not indicate a change in the trend in PDR over time. Sufficient data for the Kara Sea, Laptev Sea, and Hudson Bay were not available, so these SEMSs are not represented here.

- *Black Sea*—The fish catch of the dominant anchovy (*Engraulis encrasicolus*: TL = 2.92, pelagic) strongly increased at the end of the 1960s, resulting in an increase of the fish catch (Sea Around Us 2008). The catch of this species collapsed at the end of the 1980s, most probably due to a combination of predation by an invasive comb jelly (*Mnemiopsis leidyi*) and overfishing. In 1998, another gelatinous carnivore, *Beroe ovata*, a known predator of *M. leidyi*, was introduced into this area via ballast waters and spread within a year over most parts of the sea, followed by an increase in fish stocks (Kideys 2002; Shiganova et al. 2003). As a result, fish catch could increase again but this time did not affect the MTI significantly (Figure 10-4C) nor cause a significant change in the PDR (Figure 10-5C).
- *East China and Yellow seas*—The system shows a continuous and steady increase of fish catch between the 1950s and present years, connected to a more or less steady decrease of MTI (Figure 10-4D) and pelagic species (Figure 10-5D). Two extraordinary periods can be observed, one from 1960 to 1970, with a breakdown of pelagic catches (Figure 10-5D), and a second from 1980 to 1992, with high catches of low-trophic-index species (Figure 10-4D). These events can be related to the increase of horse mackerel in the 1960s and its consequent disappearance. Sardine catches (*Sardinops sagax*: TL = 2.46, pelagic) peaked in the 1980s until the early 1990s and have also disappeared from catches, while anchovies have reached up to 15% of the total catch during the last ten years (Sea Around Us 2008).
- *Gulf of Mexico*—The increase of fish catches has mainly been due to demersal species. Increasing catches until the mid-1980s go along with decreasing PDR and an inverse development thereafter (Figure 10-5E). This was mainly due to changes in the catch of gulf menhaden (*Brevoortia patronus*: TL = 2.19, pelagic) which comprised approximately 60% of the fish catch (Sea Around Us 2008). Increase in fish catch, on average, coincided with an increase in MTI (Figure 10-4E), indicating a higher proportion of top predator species in the catches. This is one of the two examples within this chapter where MTI and total catch show a positive correlation.
- *Gulf of Thailand*—This is the second example where increasing fish catches are positively correlated with MTI (Figure 10-4F). The main increase in fishing was during the 1960s, and then the catches dropped dramatically, followed only in the 1970s by a gradual increase again during the 1980s and 1990s. Interesting is the development of PDR, which started out pelagic-dominated, reversed during the 1950s until 1961, remained rather constant during the 1960s, and then reversed again over the next ten years to remain at a pelagic-dominated value (Figure 10-5F). Changes were mainly related to changes in threadfin breams (*Nemipterus* spp.: TL = 3.50, pelagic) and jacks (*Caranx* spp; TL = 3.80, demersal) (Sea Around Us 2008). These species jointly contributed to approximately 50% of the catch.
- *Mediterranean Sea*—The development shows two distinct periods: Until 1985 pelagic species were of increasing importance (Figure 10-5G). From then onward, pelagic species declined due to overfishing of the dominating species, pilchard (*Sardina pilchardus*: TL = 3.10, pelagic) and anchovies (*Engraulis encrasicolus*: TL =

2.92, pelagic) (Sea Around Us 2008). Increase in fish catch, on average, coincided with a decrease in MTI (Figure 10-4G).

- *North Sea*—Pelagic fish were removed during the 1950s without changing the MTI significantly (Figures 10-4H and 10-5H). Then during the 1960s the two indicators, MTI and PDR, remained in the same range until the early 1970s, even if this was a time of the greatest increase in fish catch. From the mid-1970s, after the breakdown of the catch-dominating herring (*Clupea harengus*: TL = 3.46, pelagic), a steady decrease in fish production was observed, accompanied by a decrease in PDR and MTI. Herring appeared to be replaced by sand eels (*Ammodytes* spp.: TL = 3.10, pelagic) (Sea Around Us 2008).
- *Sea of Okhotsk*—Landings were dominated by the Alaska pollack (*Theragra chalcogramma*: TL = 3.45, benthopelagic), which contributed up to 80% of the fish catch. The increase of MTI in the 1950s and 1960s reflects the increase of Alaska pollack catches during that period, peaking in the late 1980s and then dropping back to the 1960s level (Figure 10-4I). Interestingly in this case, MTI is increasing again and PDR is dropping inversely, showing a functional relationship of MTI and PDR to total catch (Figures 10-4I and 10-5I).
- *Gulf of St. Lawrence*—Here the fish fauna changed already in the 1950s, with only slightly increasing fishing pressure indicated by shifts in the PDR (Figure 10-5J). Until the end of the 1960s, fishing increased and consequently PDR decreased, but MTI remained high (Figure 10-4J). From the 1970s onward, MTI decreased dramatically with decreasing catches, while PDR remained stable except during a five-year period in the early 1970s when PDR increased. Up to the end of the 1970s, the catch was dominated by both Atlantic herring (*Clupea harengus*: TL = 3.46, pelagic) and cod (*Gadus morhua*: TL = 4.01, benthopelagic) (Sea Around Us 2008). Cod strongly declined thereafter and are now seldom caught.

Transfer Efficiencies

As a measure of the efficiency of the food web in the different SEMSs, we calculated regional transfer efficiencies. The transfer efficiency (E) was estimated from the relationship between primary productivity (PP; mg C m^{-2} d^{-1}), MTI, and the fish catch (FC; mg C m^{-2} d^{-1}), according to $FC = PP\ E^{(MTI-1)}$, which can be rewritten as $E = (FC/PP)^{1/(MTI-1)}$.

Estimates of transfer efficiencies ranged from < 1% for Hudson Bay and the Gulf of Mexico to > 6% for the East China and Yellow seas and the Gulf of Thailand (Table 10-1). These values are less than the expected 10% (Odum 1969), most probably as the result of the fact that fish catch does not represent the production of the entire population. Still, these values are remarkably close to the transfer efficiencies if all fish in the sea were caught.

In contrast to the expectations, transfer efficiency of a SEMS was not related to the size of the system ($n = 11, r^2 = 0.02, p = 0.662$) nor with historical exploitation pressure

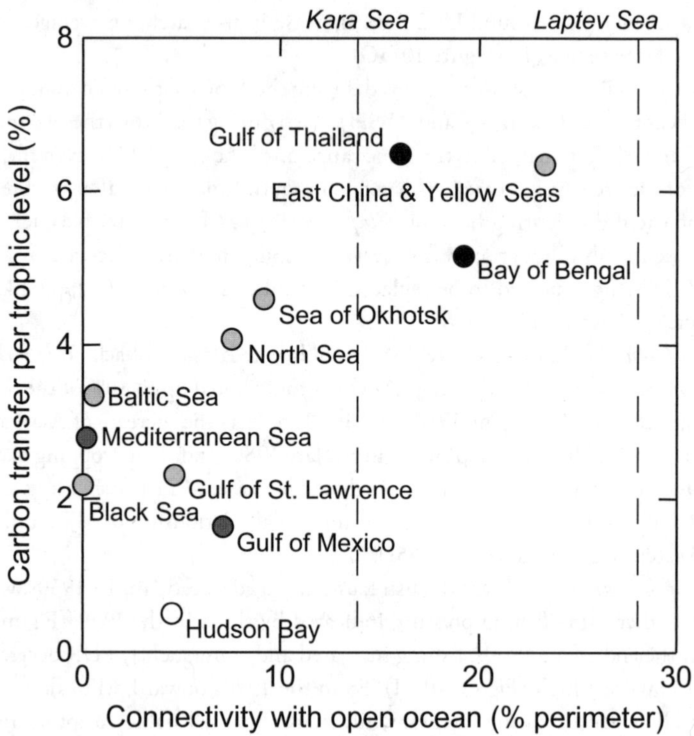

Figure 10-6. Relationship between transfer efficiency (expressed as percent carbon transfer per trophic level) and connectivity of a SEMS with the open ocean (percent of perimeter of the SEMS based on maps).

as indexed by long-term change in MTI ($n = 10, r^2 = 0.02, p = 0.713$), but it was related positively to the connectivity with the open ocean ($n = 11, r^2 = 0.57, p = 0.007$) (Figure 10-6). We cannot conclude if this relationship is causal or spurious, because we did not include all Large Marine Ecosystems in our calculations, and we do not know how much of the total fish production is actually caught and landed.

Based upon the equation to calculate transfer efficiency (E), a low value of E can be the result of a (relatively) high value of primary productivity (PP), a low value of fish catch (FC) and/or a low value of MTI. Several mechanisms can be considered when interpreting the observed relationship between connectivity and transfer efficiency, for example:

Bycatch—On average, the proportion of discards is higher for coastal systems and shelves (31%–44%) than for upwelling areas and open oceans (15%–20%) (Pauly and Christensen 1995). Lower transfer efficiencies (E) in the more enclosed seas could then result from an underestimation of the fish production, indexed by fish catch (FC), due to the higher proportion of discards in these ecosystem types.

Microbial loop—Carbon available to higher trophic levels is determined not only by primary productivity but also by the trophic efficiency of the microbial loop, a trophic pathway in aquatic environments where dissolved organic carbon (DOC) is reintroduced to the food web through incorporation into bacteria. A higher efficiency of the microbial food web may, therefore, add to the primary productivity by the "classical" food web and subsequently increase the biomass and production of fish (e.g., Pavés and González 2008). High transfer efficiencies (E) in the more open seas could then result from an underestimation of fish catch (FC) due to the higher proportion of discards in these areas.

Residence time—Regional primary productivity may be advected more strongly into the surrounding ocean for open seas with a high connectivity than for more-enclosed areas. Low transfer efficiencies (E) in the more enclosed seas may then be the result of an overestimation of the biomass available to higher trophic levels (PP). Residence time might also be higher for seas that are deep and/or have low tidal amplitude.

Anoxia—Some seas, in particular eutrophied enclosed seas, may experience periods of hypoxia and anoxia. These conditions may result in increased mortality of the (benthic) communities and subsequently result in lower biomass and production than to be expected from primary productivity levels (e.g., Kemp et al. 2005). Under such conditions, lower transfer efficiencies (E) in the more enclosed seas could then result from a hypoxia-induced lowering of the fish populations and subsequently of the fish catch (FC).

Resilience of Ecosystems to Fisheries

Resilience is the extent to which ecosystems can absorb recurrent natural and human perturbations and continue to regenerate without slowly degrading or unexpectedly flipping into alternate states (Hughes et al. 2005; Chapter 3, this volume). Investigating the impact of any disturbance, human or natural, on an ecosystem is difficult. We can measure different parameters such as temperature, oxygen, or nutrients, but they give us only an indication of changes of certain conditions in the system. Here we were looking for a more holistic indicator that allows a definition of the status of the whole ecosystem. MTI, which is mainly the average trophic level of the organisms caught by the fisheries, was the most complete biological data set available for the SEMSs.

In the Bay of Bengal, the Mediterranean Sea, and the East China and Yellow seas, fishing pressure has increased steadily. In all other systems, fisheries have drawn back since some maximum catches, mostly in the 1970s or 1980s. The only systems where we can identify some resilience are the Sea of Okhotsk and, to a lesser degree, the Black Sea. After the fisheries in these SEMSs were reduced, the MTI has been developing back on the same track as during the period of fishing increase. The Gulf of Mexico perhaps

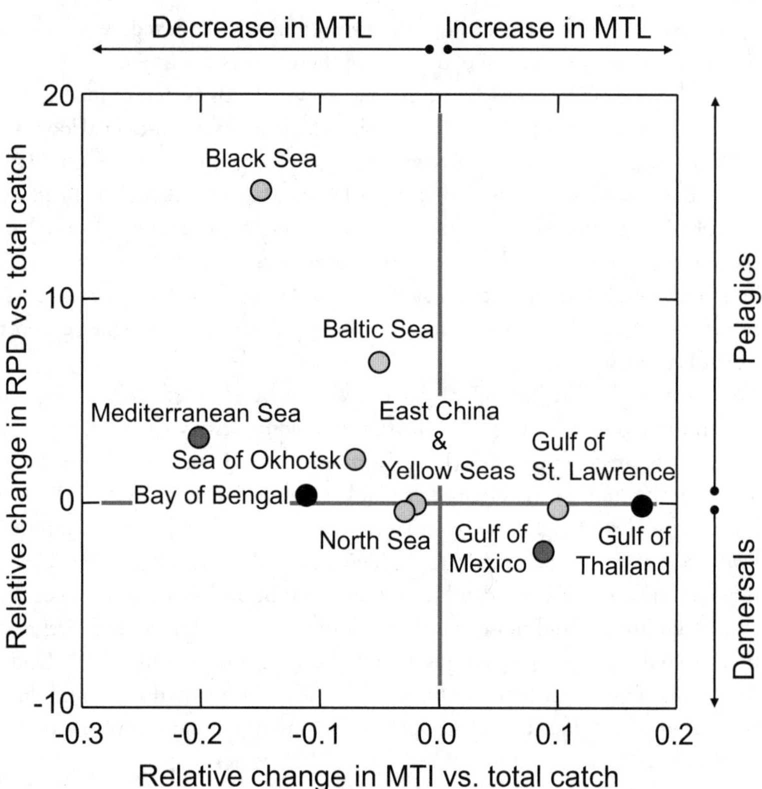

Figure 10-7. Vulnerability of SEMSs to perturbation, as indicated by the relationship between fish catch and marine trophic index (*x*-axis; α in Figure 10-4) and the relationship between fish catch and the ratio between pelagic and demersal proportions of landings (*y*-axis: α in Figure 10-5). Positive values along the *x*-axis indicate that increase in fish catch coincides with an increase in marine trophic index of that catch, while positive values along the *y*-axis indicate that increase in fish catch coincides with an increase in the fraction of pelagic fish of that catch. The shade of the dot indicates the climatic conditions: temperate—light gray; subtropical—dark gray; tropical—black.

belongs to this group also. The other four systems (North Sea, Baltic Sea, Gulf of St. Lawrence, and Gulf of Thailand) reached a completely different MTI when fisheries drew back, indicating a completely different species community.

For many seas, the maximum catch appears to have been reached one decade ago (e.g., Mediterranean Sea, Sea of Okhotsk), two decades ago (e.g., Baltic Sea, Black Sea, Gulf of Mexico), or even three decades ago (e.g., Gulf of Thailand, North Sea, Gulf of St. Lawrence) (Figures 10-4 and 10-5). These regional trends are in agreement with global fishery landings, which have been declining since the late 1980s despite (or because of) the increase in fishing efforts (Watson and Pauly 2001). In general, most pelagic fish tend to feed on lower trophic levels, while the diet of demersal species comprises a wider range. It is expected, therefore, that decreases in MTI over time are concurrent with increases in PDR, and vice versa. This general pattern appears to be true for most regional seas examined in this chapter (Figure 10-7). A coinciding decrease in MTI and increase in the proportion of pelagics in the catch was observed for the Baltic Sea, Bay of Bengal, Black Sea, Mediterranean Sea, and Sea of Okhotsk (Figures 10-4, 10-5, and 10-7). For the Gulf of Mexico and, to a lesser extent, the Gulf of Thailand and the Gulf of St. Lawrence, MTI increased and the proportion of pelagics decreased with increasing catch. Only for the East China and Yellow seas region and the North Sea did MTI and PDR jointly decrease with catch. It must be noted, however, that coinciding changes in composition of the catch (MTI and PDR; Figure 10-7) may have been both a cause and an effect of fishing effort. On the one hand, fishing often follows developments within marine ecosystems, while on the other hand, fishing efforts changed species composition within these seas.

Future Developments

Worldwide observations suggest that productivity of the seas is related to climate and species richness (e.g., Worm et al. 2006; Frank et al. 2007). Subsequently, marine biodiversity is setting the stage for ecosystem services, that is, high biodiversity increases the ocean's capacity to provide food, maintain water quality, and recover from perturbations (Worm et al. 2006). Such perturbations may come from exploitation, eutrophication (e.g., Schindler 2006), invasive species (e.g., Van der Weijden et al. 2007), and climate change (e.g., Philippart et al. 2007a). The results of our analyses corroborate earlier findings that the impacts of fisheries on ecosystems in most cases are not reversible. Fishing often has led to completely different species communities, at least on the higher trophic levels, and we assume that this is also true for lower trophic levels, such as present in the phytoplankton, zooplankton, and benthic invertebrate communities. If fisheries pressure is reduced, it is not very likely that ecosystems will return to their original states, or they will take a long time doing so.

The foreseen climate change–induced reduction of the Arctic Sea ice cover is expected to open up the relatively inaccessible polar seas such as the Kara Sea and Laptev Sea for fisheries, shipping, and oil exploitation. This will additionally increase the risk of biological invasions in these areas. Based on the high connectivity with the open oceans (14% and 28%, respectively) of these polar regions, the carbon transfer of primary productivity to exploitable fish could be rather high after sea ice reduction (Figure 10-6). However, it is unlikely that the Kara Sea and Laptev Sea ecosystems will have time to redevelop into a mature state under the ice-free conditions so that they can withstand the run on their living and nonliving resources. If climate-induced changes of marine ecosystems are not taken into account in the management of human activities in SEMSs, we are likely to observe even stronger impacts of fisheries on marine communities in the future.

Acknowledgments

We would like to thank the organizers of the PACKMEDS workshop for the facilitation and stimulation to prepare this chapter and thank three anonymous reviewers for the very helpful comments on previous manuscripts.

References

Alekseev, A.R., P. Baklanov, I. Arzamastrev, Y. Blinov, A. Fedorovskii, F. Khrapchenkov, I. Medvedeva, P. Minakir, G.L.F. Titova, A. Vlasov, B. Voronov, and H. Ishitobi. 2006. *Sea of Okhotsk: GIWA Regional Assessment 30*. Global International Waters Assessment. Kalmar, Sweden: University of Kalmar on behalf of United Nations Environment Programme.

Caddy, J. 1993. Toward a comparative evaluation of human impacts on fisheries ecosystems of enclosed and semi-enclosed seas. *Reviews in Fisheries Science* 1:57–95.

Caddy, J.F., J. Csirke, S.M. Garcia, and R.J.R. Grainger. 1998. How pervasive is "Fishing down marine foodwebs"? *Science* 282:1,383, doi: 10.1126/science.282.5393.1383a.

Carmack, E., and P. Wassmann. 2006. Food webs and physical–biological coupling on pan-Arctic shelves: Unifying concepts and comprehensive perspectives. *Progress in Oceanography* 71:446–477.

Catchpole, T.L., C.L.J. Frid, and T.S. Gray. 2005. Discards in North Sea fisheries: Causes, consequences and solutions. *Marine Policy* 29:421–430.

Cury, P., A. Bakun, R.J.M. Crawford, A. Jarre, R.A. Quiñones, L.J. Shannon, and H.M. Verheye. 2000. Small pelagics in upwelling systems: Patterns of interaction and structural changes in "wasp-waist" ecosystems. *ICES Journal of Marine Science* 57: 603–618.

Cushing, D.H. 1975. *Marine Ecology and Fisheries*. Cambridge, UK: Cambridge University Press.

Cushing, D.H. 1984. The gadoid outburst in the North Sea. *Journal du Conseil International du Exploration de la Mer* 41:159–166.

Frank, K.T., B. Petrie, J.S. Choi, and W.C. Leggett. 2005. Trophic cascades in a formerly cod-dominated ecosystem. *Science* 308:1,621–1,623.

Frank, K.T., B. Petrie, and N.L. Shackell. 2007. The ups and downs of trophic control in continental shelf systems. *Trends in Ecology and Evolution* 22:236–242.

Fromentin, J., and B. Planque. 1996. *Calanus* and the environment in the eastern North Atlantic. II. Influence of the North Atlantic Oscillation on *C. finmarchicus* and *C. helgolandicus*. *Marine Ecology Progress Series* 134:111–118.

Hirche, H.J., K.N. Kosobokova, B. Gaye-Haake, I. Harms, B. Meon, and E. Nöthig. 2006. Structure and function of contemporary food webs on Arctic shelves: A panarctic comparison. The pelagic system of the Kara Sea: Communities and components of carbon flow. *Progress in Oceanography* 71:288–313.

Hughes, T.P., D.R. Bellwood, C. Folke, R.S. Steneck, and J. Wilson. 2005. New paradigms for supporting the resilience of marine ecosystems. *Trends in Ecology & Evolution* 20:380–386.

Hutchings, J.A., and J.D. Reynolds. 2004. Marine fish population collapses: Consequences for recovery and extinction risk. *Bioscience* 54: 297–309.

Kemp, W.M., W.R. Boynton, J.E. Adolf, D.F. Boesch, W.C. Boicourt, G. Brush, J.C. Cornwell, T.R. Fisher, P.M. Glibert, J.D. Hagy, L.W. Harding, E.D. Houde, D.G. Kimmel, W.D. Miller, R.I.E. Newell, M.R. Roman, E.M. Smith, and J.C. Stevenson. 2005. Eutrophication of Chesapeake Bay: Historical trends and ecological interactions. *Marine Ecology Progress Series* 303:1–29.

Kideys, A.E. 2002. Fall and rise of the Black Sea ecosystem. *Science* 297:1,482–1,484.

Kitching, R.L. 2001. Food webs in phytotelmata: "Bottom-up"and "top-down" explanations for community structure. *Annual Review of Entomology* 46:729–760.

Lääne, A., E. Kraav, and G. Titova. 2005. *Baltic Sea: GIWA Regional Assessment 17.* Global International Waters Assessment. Kalmar, Sweden: University of Kalmar on behalf of United Nations Environment Programme.

Levitus, S., R. Burgett, and T. Boyer. 1994. Nutrients. In Vol. 3 of *World Ocean Atlas 1994.* NOAA Atlas NESDIS (National Environmental Satellite, Data, and Information Service) 3. Washington, DC: U.S. Government Printing Office.

Lozán, J. (ed.). 1996. *Warnsignale aus der Ostsee: Wissenschaftliche Fakten.* Hamburg: Parey.

Møller, A.P., E. Flensted-Jensen, and W. Mardal. 2007. Agriculture, fertilizers and life history of a coastal seabird. *Journal of Animal Ecology* 76:515–525.

Odum, E.P. 1969. The strategy of ecosystem development. *Science* 164:262–270.

Orysova, O., A. Kondakov, S. Palearl, E. Rautalahti-Miettinen, F. Stolberg, and D. Daler. 2005. *Eutrophication in the Black Sea Region: Impact Assessment and Causal Chain Analysis.* Global International Waters Assessment (GIWA). Kalmar, Sweden: University of Kalmar on behalf of United Nations Environment Programme.

Parsons, T.R., M. Takahashi, and B. Hargrave. 1983. *Biological Oceanographic Processes.* 3rd ed. Paris: Pergamon/Elsevier.

Pauly, D., and V. Christensen. 1995. Primary production required to sustain global fisheries. *Nature* 374:255–257.

Pauly, D., V. Christensen, J. Dalsgaard, R. Froese, and F.C. Torres, Jr. 1998. Fishing down marine food webs. *Science* 279:860–863.

Pauly, D., and R. Watson. 2005. Background and interpretation of the "Marine Trophic Index" as a measure of biodiversity. *Philosophical Transactions of the Royal Society B* 360:415–423.

Pavés, H.J., and H.E. González. 2008. Carbon fluxes within the pelagic food web in the coastal area off Antofagasta (23°S), Chile: The significance of the microbial versus classical food webs. *Ecological Modelling* 212:218–232.

Philippart, C.J.M., R. Ánadon, R. Danovaro, J.W. Dippner, K.F. Drinkwater, S.J. Hawkins, G. O'Sullivan, T. Oguz, and P.C. Reid. 2007a. *Impacts of Climate Change on the European Marine and Coastal Environment: Ecosystems Approach.* Marine Board Position Paper 9. Strasbourg: European Science Foundation.

Philippart, C.J.M., J.J. Beukema, G.C. Cadée, R. Dekker, P.W. Goedhart, J.M. van Iperen, M.F. Leopold, and P.M.J. Herman. 2007b. Impact of nutrient reduction on coastal communities. *Ecosystems* 10:96–119.

Pimm, S.L. 1980. Properties of food webs. *Ecology* 61:219–225.

Pimm, S.L. 2002. *Food Webs.* Chicago: The University of Chicago Press.

Qu, J., Z. Xu, Q. Long, L. Wang, X. Shen, J. Zhang, and Y. Cai. 2005. *East China Sea: GIWA Regional Assessment 36.* Global International Waters Assessment. Kalmar, Sweden: University of Kalmar on behalf of United Nations Environment Programme.

Schindler, D.W. 2006. Recent advances in the understanding and management of eutrophication. *Limnology and Oceanography* 51:356–363.

Schmid, M.K., D. Piepenburg, A.A. Golikov, K. Juterzenka, V. Petryashov, and M. Spindler. 2006. Trophic pathways and carbon flux patterns in the Laptev Sea. *Progress in Oceanography* 71:314–330.

Schoener, T.W. 1989. Food webs from the small to the large. *Ecology* 70:1,559–1,589.

Sea Around Us. 2008. www.seaaroundus.org.

Shelton, P.A., A.F. Sinclair, G.A. Chouinard, R. Mohn, and D.E. Duplise. 2005. Fishing under low productivity conditions is further delaying recovery of northwest Atlantic cod (*Gadus morhua*). *Canadian Journal of Fisheries and Aquatic Sciences* 63:235–238.

Shiganova, T.A., E.I. Musaeva, Yu. V. Bulgakova, Z.A. Mirzoyan, and M.L. Martynyuk. 2003. Invaders ctenophores *Mnemiopsis leidyi* (A. Agassiz) and *Beroe ovata* Mayer 1912, and their influence on the pelagic ecosystem of northeastern Black Sea. *Biology Bulletin* 30:180–190.

Strathmann, R.R. 1967. Estimating the organic carbon content of phytoplankton from cell volume or plasma volume. *Limnology and Oceanography* 12:411–418.

Teng, S., H. Yu, Y. Tang, L. Tong, L. Loi, D. Kang, H. Liu, Y. Chun, R. Juliano, E. Rautalahti-Miettinen, and D. Daler. 2005. *Yellow Sea: GIWA Regional Assessment 34*. Global International Waters Assessment. Kalmar, Sweden: University of Kalmar on behalf of United Nations Environment Programme.

Tsyban, A., G. Titova, S. Shchuka, V. Ranenko, and Y. Izrael. 2005. *Russian Arctic: GIWA Regional Assessment 1a*. Global International Waters Assessment. Kalmar, Sweden: University of Kalmar on behalf of United Nations Environment Programme.

Van der Weijden, W., R. Leeuwis, and P. Bol. 2007. *Biological Globalisation: Bio-invasions and Their Impacts on Nature, the Economy and Public Health*. Zeist, The Netherlands: KNNV.

Ware, D.M., and R.E. Thomson. 2005. Bottom-up ecosystem trophic dynamics determine fish production in the northeast Pacific. *Science* 308:1,280–1,284.

Watson, R., and D. Pauly. 2001. Systematic distortion in world fisheries catch trends. *Nature* 424:534–536.

Worm, B., E.B. Barbier, N. Beaumont, J.E. Duffy, C. Folke, B.S. Halpern, J.B.C. Jackson, H.K. Lotze, F. Micheli, S.R. Palumbi, E. Sala, K.A. Selkoe, J.J. Stachowicz, and R. Watson. 2006. Impacts of biodiversity loss on ocean ecosystem services. *Science* 314:787–790.

11

Distribution and Consequences of Hypoxia

Nancy N. Rabalais and Denis Gilbert

Introduction

Anoxic (no oxygen) and hypoxic (low oxygen) conditions in estuarine, coastal, and oceanic waters exist naturally, but they also occur as a result of human activities. Anoxia, as its Latin roots imply, means no oxygen. Hypoxia means that oxygen is present in low concentrations within a certain limit or threshold that is derived from the response of aquatic animals. There is no single definition of hypoxia, because the relationship between the oxygen concentration and physiological stress and related responses in invertebrate and fish fauna varies depending on the organism, its life stage, and prior exposure and on the pressure, temperature, and salinity of the ambient waters (Rabalais and Turner 2001). The threshold below which large changes occur in benthic and demersal communities, however, is mostly defined as oxygen levels lower than 2 mg dissolved O_2 L^{-1} (or 1.4 mL L^{-1} at standard temperature and pressure; Tyson and Pearson 1991). In the northern Gulf of Mexico (27–35 salinity and 21°C–31°C), this threshold is approximately 30% oxygen saturation during spring/summer.

The negative effects of hypoxia include loss of suitable and required habitat for many bottom-dwelling fishes and benthic fauna, direct mortality, increased predation, decreased food resources, altered migration, and potentially reduced fisheries, including of valuable finfishes and crustaceans (Rabalais and Turner 2001). Increasing nutrient loads that also change the nutrient ratios can affect the composition of the phytoplankton community and can shift trophic interactions. Hypoxia also alters or interrupts ecosystem functions and services such as nutrient cycling (Turner et al. 1998; Dortch et al. 2001; Childs et al. 2002, 2003; Rabalais 2004a).

Hypoxia and anoxia occur naturally in regions of reduced ventilation—deep basins, fjords, and terminal ends of the global thermohaline circulation—and are associated

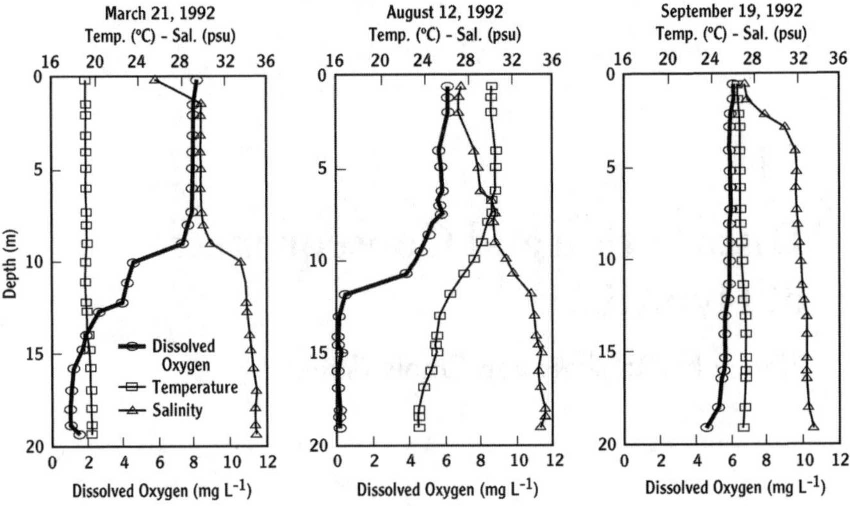

Figure 11-1. Examples of seasonal progression and variability of thermoclines, haloclines, and oxyclines at a 20 m station in the northern Gulf of Mexico west of the Mississippi River, where hypoxia occurs from March through November. Salinity differences are not always a condition of strong stratification in many coastal areas but do contribute substantially on coasts where freshwater discharge is high. Adapted from Rabalais and Turner 2001, with permission of the American Geophysical Union.

with eastern boundary upwelling systems (Kamykowski and Zentara 1990; Helly and Levin 2004). Oxygen depletion can also result when increased nutrient loads are injected into relatively stagnant and strongly stratified coastal waters. Stratification refers to density differences in the water column caused by vertical temperature and/or salinity gradients (Figure 11-1). Under calm conditions, stratification is intensified, but mixing events during fall and winter will reduce and break down the stratification, though not always. In most coastal waters hypoxia is a seasonal phenomenon.

In many areas of the ocean, hypoxia is a recent phenomenon that reflects increasing nutrient fluxes into coastal waters, enhanced productivity, and increased oxygen demand by bacteria oxidizing organic carbon (Diaz and Rosenberg 1995; Nixon 1995; Rabalais 2004a) (Figure 11-1). Increased nutrient loads are caused by the activities of increasing human populations: application of nitrogen and phosphorus fertilizers, planting of leguminous crops, atmospheric deposition of nutrients, and discharge of municipal and industrial wastewater (Vitousek et al. 1997; Bennett et al. 2001). Over the last half of the twentieth century, the impacts of nutrient enrichment, including hypoxia, increased in frequency and severity and expanded geographically (Diaz and Rosenberg 1995; Nixon 1995; Boesch 2002). With an ever-increasing human population, expansion of fertilizer use in support of crops and biofuel production, and nutrient production as by-products of burning fossil fuels, the flux of nutrients into coastal systems will continue to increase (Seitzinger et al. 2002; National Research Council 2007). Hypoxia will become more widespread and frequent (Plate 6) (Rabalais 2004a). Semi-enclosed marine systems (SEMSs)—entire basins or portions of systems—are especially vulnerable because of longer residence times of the water and stronger stratification.

Causes

Hypoxia occurs when the rate of oxygen consumption exceeds the rate of supply from the atmosphere, horizontal advection, vertical mixing, and photosynthesis. The biological and physical water column characteristics that support the development and maintenance of hypoxia include (1) the supply of organic matter to the lower water column and sediment, (2) reduced vertical mixing because of strong stratification, and (3) long residence time in the lower water column (Plate 7). Senescent and dead algae, zooplankton fecal pellets, and marine aggregates contribute significant amounts of organic detritus to the lower water column and seabed.

Oxygen is consumed in both the water column and the sediments. The relative contributions of water column and benthic respiration in the hypoxic part of the northern Gulf of Mexico have been estimated with oxygen concentration and stable isotope measurements (Quiñones-Rivera et al. 2007). The model indicates that the severe bottom water oxygen depletion in July 2001 was due predominantly to benthic respiration (73%). Subsequent work by Quiñones-Rivera and colleagues (unpubl. data) indicates that the relative proportion of water column and benthic respiration varies with year and season, consistent with Rowe and colleagues (2002), who showed that sediment oxygen consumption is dependent on bottom-water oxygen concentration.

When the oxygen concentration in the water overlying the sediments falls below 0.2 mg L^{-1} and approaches anoxia, patches of sulfur-oxidizing bacteria, such as *Beggiatoa*, *Thiovolum*, and *Thioploca*, grow at the interface between oxic and sulfidic zones and cover the mud (Jørgensen 1980; Harper et al. 1981; Rosenberg and Diaz 1993; Rumohr et al. 1996; Rabalais et al. 2001a). When the bottom water is depleted of oxygen, hydrogen sulfide builds up in the bottom waters as anaerobic bacterial metabolism reduces sulfate to H_2S (Jørgensen 1980), the sediment becomes almost uniformly black, and there are no signs of aerobic life. Hydrogen sulfide is toxic to most metazoans and contributes to the overall benthic mortality.

Anthropogenic Coastal Hypoxia in SEMSs

Diaz and Rosenberg (1995) noted that no other environmental variable of such ecological importance to estuarine and coastal marine ecosystems as dissolved oxygen has changed so drastically, in such a short period of time. They noted a consistent trend of increasing severity in duration, intensity, or extent in areas where hypoxia has a long history. They also noted that hypoxia now occurs in areas where it did not occur before, including estuarine and coastal areas. The trend is consistent with an increase in human activities. What follows are examples of hypoxic conditions in SEMSs that have resulted wholly or in part from human activities.

Hypoxic areas rim the Baltic Sea where increased nutrient loads have led to eutrophication and decreased dissolved oxygen concentrations. The aggregated coastal areas of the Baltic Sea, along with the natural deep-basin hypoxia, result in a large expanse of hypoxic waters. The Baltic Sea, a brackish ecosystem, is isolated from the North Sea by a series of shallow sills in the Danish straits, and inflows of higher-salinity oxygenated waters are limited (see Chapter 6, this volume). Few major inflows have taken place since the mid-1970s (HELCOM 2001), resulting in areas of hypoxia in the lower water column between 12,000 km^2 and 70,000 km^2, or 5% to 27% of the total bottom area (Conley et al. 2002a). Danish marine waters display all the classic symptoms associated with eutrophication, including periods of hypoxia and anoxia in bottom waters (Conley and Josefson 2001; Conley et al. 2002a, b) and the death of benthic-dwelling organisms during anoxia (Fallesen et al. 2000; Conley et al. 2002b). Hypoxia is widespread in areas of the Skagerrak and Kattegat, the coastal waters of the North Sea, and many coastal embayments and fjords. Nutrient loading from Denmark ranks among the highest in Europe per unit area (Conley et al. 2002b) and reflects the density of the population and the intensity of the agriculture.

The deep central part of the Black Sea is a permanently anoxic basin, but the northwestern margin, which receives the freshwater inflows of the Danube River and the smaller Dneiper and Dneister rivers to the north of the Danube, experienced a surge of increased nitrogen and phosphorus loads in the 1960s to 1970s, similar to many other coastal areas downstream of developed nations (Galloway and Cowling 2002; Rabalais 2004a). Hypoxia existed historically on the northwestern Black Sea shelf, but hypoxic and anoxic events became more frequent and widespread in the 1970s and 1980s, covering areas of the seafloor up to 40,000 km^2, in depths of 8 to 40 m (Tolmazin 1985; Zaitsev 1992; Mee 2001; Mee et al. 2005) (Figure 11-2). The conditions on the northwestern shelf of the Black Sea improved from 1990 to 2000 when nutrient loads from the Danube River decreased substantially. Decreasing nutrient loads were followed by a decrease in the extent of bottom-water hypoxia (Mee 2001).

Historical data from the Northern Adriatic Sea, going back to 1911, show increasing oxygen concentrations in the surface water and decreasing oxygen concentrations in the bottom water as well as decreasing water clarity associated with an increase in phytoplankton biomass. These trends are consistent with increasing nutrient loads in the Po River and development of hypoxia in the Northern Adriatic Sea (Figure 11-3) (Justic et al. 1987; Justic 1988, 1991). The loss of taxa of pelagic medusae, which depend on benthic stages as part of their life cycle, is also consistent with the expanding hypoxia (Benović et al. 1987). Paleoindicators in sediment cores from the vicinity of the Po River delta plume indicate a gradual increase in eutrophication at the end of the nineteenth

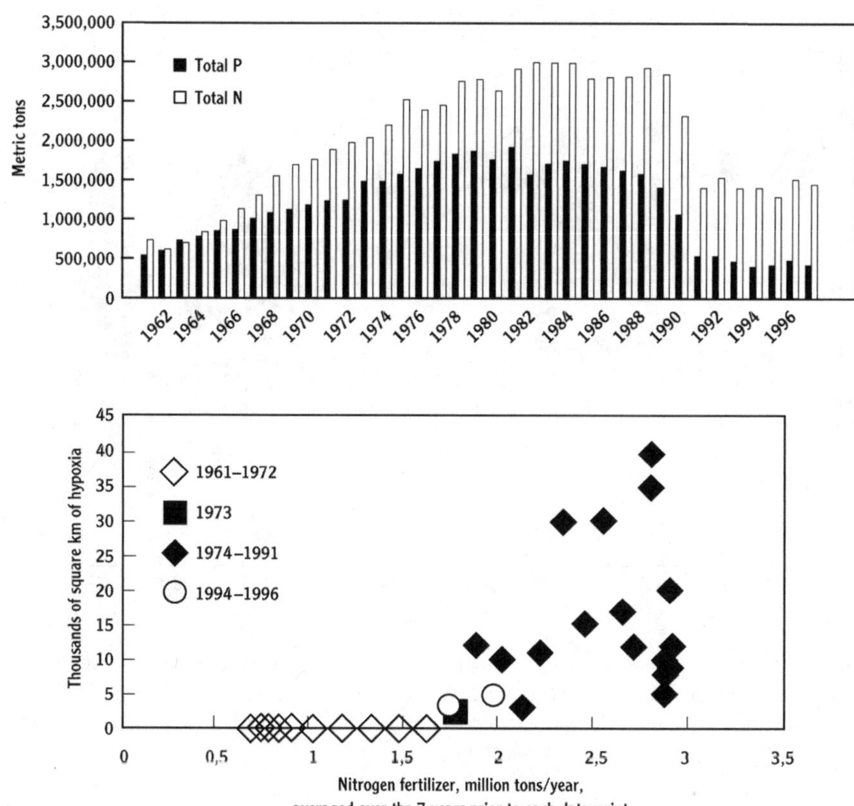

Figure 11-2. Extent of bottom-water hypoxia on the northwestern shelf of the Black Sea in relation to use of nitrogen fertilizer in the Danube River watershed in different groups of years (Mee 2001). Trends for nitrogen and phosphorus fertilizer use in the watersheds draining to the northwestern shelf of the Black Sea show that nitrogen and phosphorus loads in the SEMS corresponded to fertilizer use (Mee 2001). Used with permission from Dahlem University Press.

century, which accelerated after 1930. Seasonal hypoxia began in 1960 and became more intense and prolonged in 1980 (Barmawidjaja et al. 1995).

The hypoxic zone on the continental shelf of the northern Gulf of Mexico is one of the largest human-caused hypoxic zones in the coastal ocean (Rabalais et al. 2007a). The Mississippi River creates a strongly stratified coastal system west of the delta for much of the year and delivers more nutrients now than in the 1950s. The area of bottom covered by hypoxic water can reach 22,000 km^2, and it averaged 13,500 km^2 between 1985 and 2007 (Figure 11-4) (Rabalais et al. 2007a). The hypoxia occurs from March through November and is nearly continuous from mid-May through mid-September. In mid-

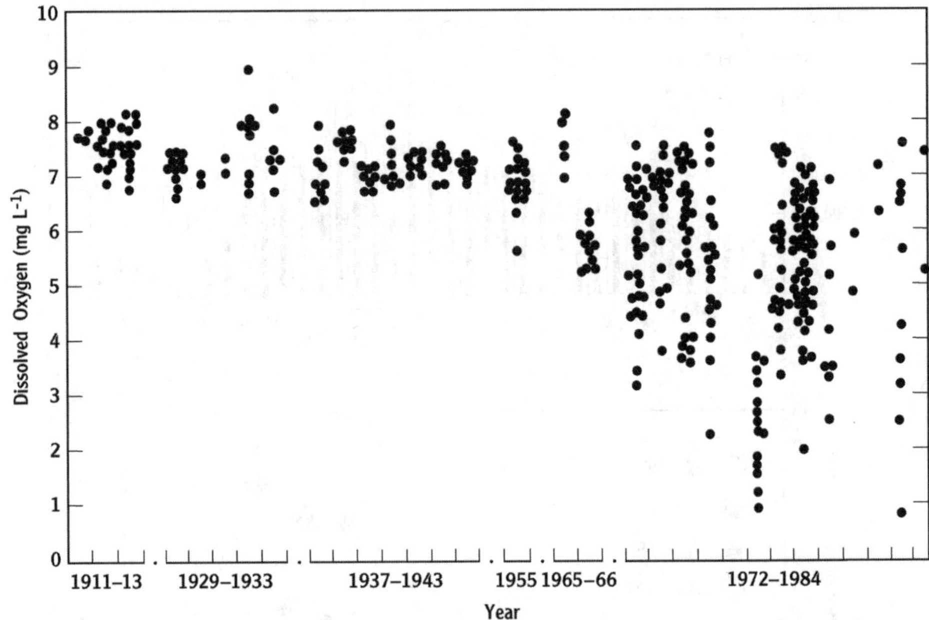

Figure 11-3. Oxygen content 2 m above the bottom during August–September in the Northern Adriatic Sea from 1911 to 1984 for the periods indicated. Similar trends were found for increasing surface water oxygen and decreased Secchi disk depth. Redrawn from Justic 1991 with permission of The Geological Society (London).

summer, the size of the hypoxic zone is most closely related to the nitrate load of the Mississippi River in the two months prior to the mapping, and the same size hypoxic area is now formed with a lower nitrate load than historically (Turner et al. 2006, 2008). Changes in the nitrate loads over time are due mostly to the change in nitrate concentration in the Mississippi River from agricultural activities (80%), and the remainder is due to increased freshwater discharge (20%) (Donner et al. 2002; Justic et al. 2002).

Paleoindicators of primary productivity and bottom water hypoxia on the continental shelf beneath the Mississippi River plume indicate increasing eutrophication in the last half of the twentieth century, with concomitant development of hypoxia. The hypoxia has been increasing in extent, duration, and intensity since the early 1970s (Rabalais et al. 2007b). Dissolved oxygen concentrations—in an area where hypoxia occurs frequently—decreased from 1978 to 1995 (Turner et al. 2005) (Figure 11-5).

Hypoxia also occurs in other coastal areas of the Gulf of Mexico. Intermittent and isolated areas of hypoxia exist east of the Mississippi River (Rabalais et al. 2007a). Low-oxygen waters have been associated with massive, noxious algal blooms along the west coast of Florida. An incipient hypoxic area forms off the Coatzacoalcos River discharge

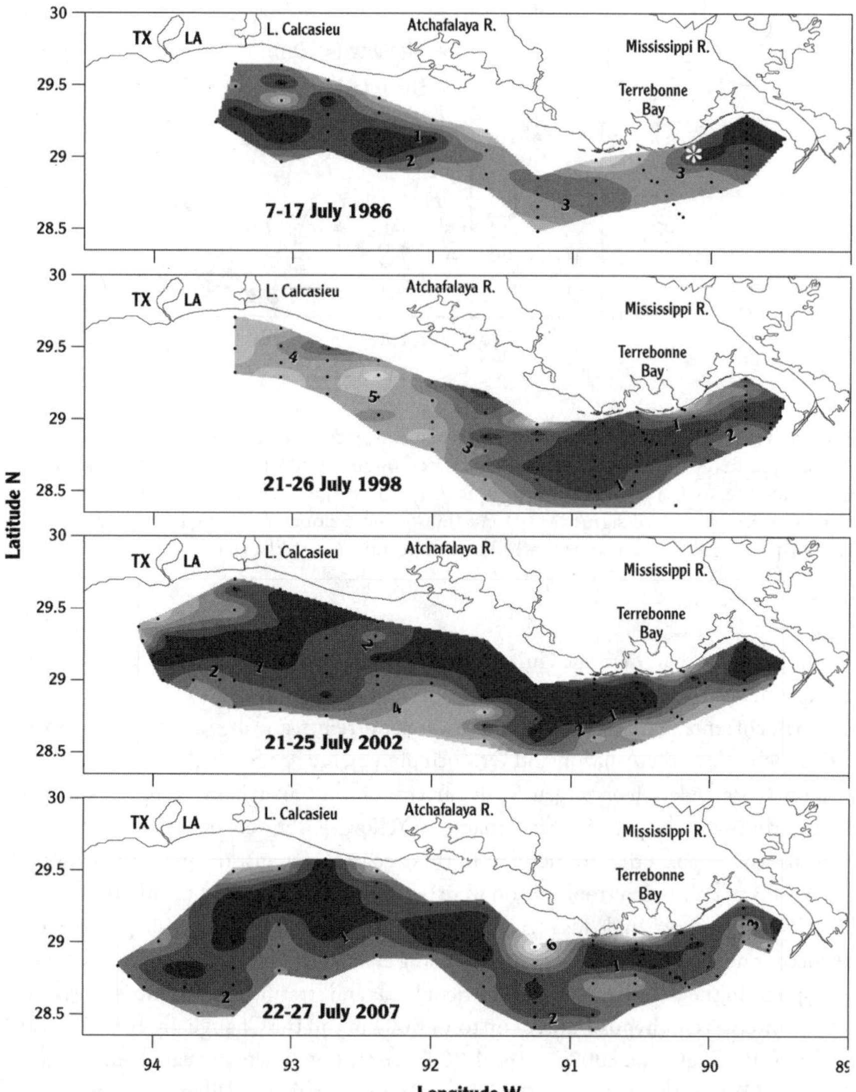

Figure 11-4. Distribution of bottom-water oxygen concentration (mg L^{-1}) on the northern Gulf of Mexico continental shelf of Louisiana and Texas for the dates indicated. Asterisk in 1986 panel refers to data in Figure 11-5. Data from N.N. Rabalais; maps from Adam Sapp, Louisiana Universities Marine Consortium.

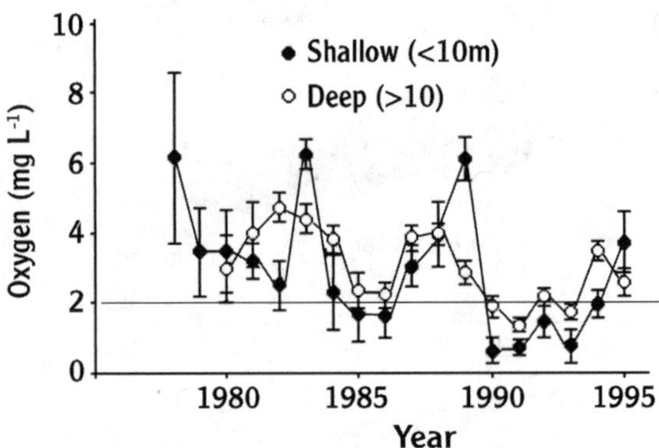

Figure 11-5. Declining bottom-water oxygen concentration at stations west of the Mississippi River delta. The location is indicated by an asterisk in Figure 11-4, 1986 panel. Values are means ± standard error for June, July, and August. The linear regression for the bottom-water data set is significant at $P = 0.005$ with a slope of -0.5 mg L^{-1} y^{-1}. Reprinted from Turner et al. 2005, copyright 2005, with permission of Elsevier.

on the southern margin of the Gulf of Mexico (Rabalais 2004b). The potential for other hypoxic areas to form is real.

Swift currents that quickly move freshwater, nutrients, and organic matter away from a river delta, as in the Amazon and Orinoco plumes, are not conducive to accumulation of biomass or depletion of oxygen. Swift currents off the Changjiang (Yangtze) River and high turbidity in the plume of the Huanghe (Yellow) River were once thought to be the reasons why hypoxia did not develop in these coastal systems. Incipient indications of symptoms of cultural eutrophication were becoming evident at the terminus of both of these systems as nutrient loads increased (Turner et al. 1990; Liu et al. 2003). The severely reduced, minimal flow of the Huanghe (Wang et al. 2006) has prevented the formation of hypoxia in the Bohai by reducing nutrient loads and stratification. There is now, however, an intermittent hypoxic area of up to 14,000 km^2 off the Changjiang estuary (Li and Daler 2004; Chen et al. 2007; Wei et al. 2007) corresponding to increased nutrient loads from the Changjiang (Liu et al. 2003). In addition, harmful algal blooms, another symptom of eutrophication, have become more frequent in the East China Sea (Zhang 1994) and correlate with high atmospheric deposition rates of nitrogen.

While the preceding examples of coastal hypoxia in SEMSs are clearly related to increased nutrient loadings and human activities, changes in ocean circulation may also have a strong effect on the supply of oxygen to the bottom waters. The deep waters in the Lower St. Lawrence Estuary are presently hypoxic, with low-oxygen waters (< 2.0 mg L^{-1}) covering a 1,300 km^2 area (Plate 8) (Gilbert et al. 2005). Historical data indicate that dissolved oxygen concentrations in the 300 to 355 m depth range have decreased by nearly 50% over the last seventy years, from 4.0 mg L^{-1} in the 1930s to an average of 2.1 mg L^{-1} for the 1984–2003 period (Gilbert et al. 2005). Two-thirds of the 1.9 mg L^{-1} oxygen decline and a concomitant 1.7°C warming of the bottom waters are attributed to a decreasing proportion of oxygen-rich Labrador Current Water in the water mass entering the Gulf of St. Lawrence from the northwest Atlantic Ocean. The remaining one-third of the change could be due to oxygen consumption resulting from increased primary production and greater carbon flux to the deep water. The organic carbon content and the accumulation rates of dinoflagellate cysts and benthic foraminifera have increased over the last four decades, and a shift in the stable carbon isotope signature of the organic carbon suggests enhanced accumulation of marine organic carbon (Thibodeau et al. 2006). As in the northern Gulf of Mexico, Long Island Sound, and Chesapeake Bay (reviewed by Rabalais et al. 2007b), there has been a shift in dominance toward benthic foraminifera that are tolerant to low dissolved-oxygen concentrations and prefer high organic content of sediments (Thibodeau et al. 2006).

Consequences

Hypoxia is but one of the symptoms of eutrophication. Eutrophied coastal ecosystems exhibit a series of symptoms, such as reduced water clarity; excessive, noxious, and sometimes harmful algal blooms; loss of critical macro-algal or sea grass habitat; and, in some cases, loss of fishery resources (Rabalais 2004a). More subtle responses of coastal ecosystems to eutrophication include shifts in phytoplankton and zooplankton communities, shifts in food webs, loss of biodiversity, changes in trophic interactions, and changes in ecosystem functions and biogeochemical processes.

Direct Effects

The obvious effects of hypoxia/anoxia are displacement of pelagic organisms, selective loss of demersal and benthic organisms, and reduced biodiversity. Environments that are impacted infrequently may recover between hypoxic events. Environments impacted seasonally may recover partially, or their ecosystem structure and function may be altered permanently (Diaz and Rosenberg 1995). Mobile animals, such as shrimp, fish, and some crabs, flee waters when the oxygen concentration falls below 2 to 3 mg L^{-1} (Renaud 1986; Rabalais and Turner 2001; Rabalais et al. 2001a). As depletion of dissolved oxygen continues, less-mobile infauna become stressed, move out of the sediments, attempt to leave the seabed, and often die. Between 0.5 mg L^{-1} oxygen and anoxia, the decrease in benthic infaunal diversity, abundance, and biomass is fairly linear (Rabalais et al. 2001b). Mass mortality events in the Northern Adriatic Sea and the Gulf of Trieste associated with complex bioherms of sponges, cnidarians, and ascidians were recorded by Stachowitsch (1984, 1991). The initial response of motile organisms such as brittle stars and hermit crabs was to move high up in the water column on the available epifaunal bioherms.

Eventually, the supporting biogenic structure and the organisms associated with it died. The recovery of these biogenically structured substrates took several years.

Losses of entire higher taxa are features of the depauperate benthic fauna in severely stressed seasonal hypoxic/anoxic zones (Rabalais 2005). Larger, longer-lived burrowing infauna are replaced by short-lived, smaller surface-deposit-feeding polychaetes, and certain typical marine invertebrates are absent from the fauna, for example, pericaridean crustaceans, bivalves, gastropods, and ophiuroids on the northern Gulf of Mexico impacted area (Rabalais et al. 2001b). Long-term trends for the Skagerrak coast of western Sweden in semi-enclosed fjordic areas experiencing increased oxygen stress showed declines in the total abundance of macroinfauna, mollusks, and suspension feeders and carnivores (Rosenberg 1990). These changes in benthic communities result in an impoverished diet for bottom-feeding fish and crustaceans and contribute, along with low dissolved oxygen, to altered sediment biogeochemical cycles. In waters of Scandinavia and the Baltic, there was a 3 million ton reduction in benthic macrofaunal biomass during the worst years of hypoxia (Karlson et al. 2002). This loss, however, may have been partly compensated by the biomass increase that occurred in well-flushed organically enriched coastal areas.

Fishery Effects

An increase in nutrient availability results in an increase of fisheries yield to a maximal point; then there are declines in various compartments of the fishery as further increases in nutrients lead to seasonal hypoxia and permanent anoxia in semi-enclosed seas (Figure 11-6) (Caddy 1993). Documenting loss of fisheries is complicated by poor fisheries data, inadequate economic indicators, increase in overharvesting at the time that habitat degradation progressed, natural variability of fish populations, shifts in harvestable populations, and climatic variability (Caddy 2000; Boesch et al. 2001; Jackson et al. 2001; Rabalais and Turner 2001).

Eutrophication often leads to the loss of habitat (rooted vegetation or macro-algae) or low dissolved oxygen, both of which may lead to loss of fisheries production. In the deepest bottom waters of the Baltic Sea proper, animals have long been scarce or absent because of low oxygen availability. Above the halocline in areas not influenced by local pollution, benthic biomass has increased, due mostly to an increase in mollusks (Cederwall and Elmgren 1990). On the other hand, many reports document instances where local pollution resulting in severely depressed oxygen levels has greatly impoverished or even annihilated the soft-bottom macrofauna (Cederwall and Elmgren 1990).

Eutrophication of surface waters accompanied by oxygen-deficient bottom waters can lead to a shift in dominance from demersal fishes to pelagic fishes (de Leiva Moreno et al. 2000). In the Baltic Sea and Kattegat, Elmgren and Larsson (2001) showed that changes in fish stocks have been both positive, due to increased food supply (e.g., pike

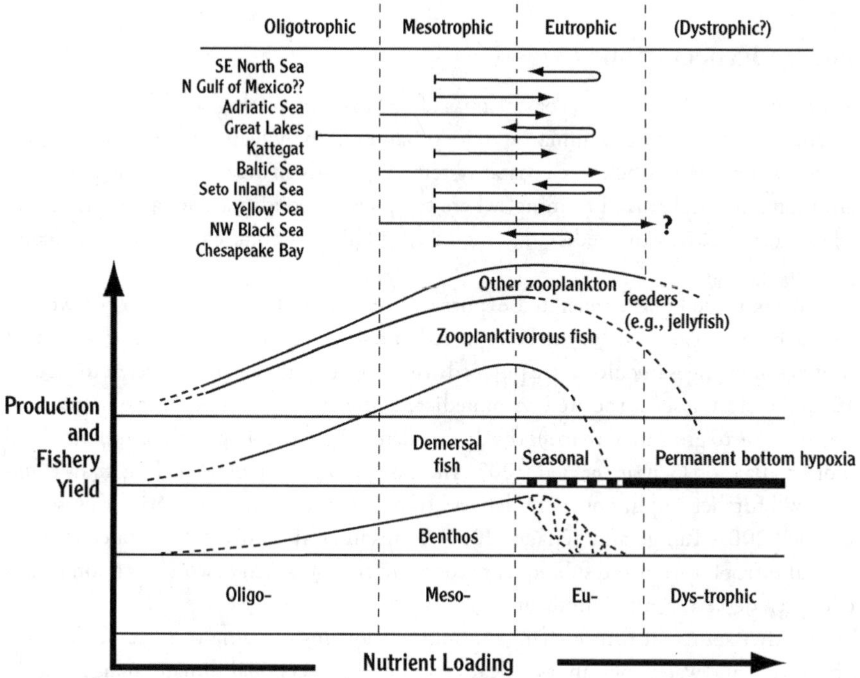

Figure 11-6. The generalized relationship between production/fishery yield and nutrient loading, with varying effects of eutrophication expressed as seasonal and permanent bottom-water anoxia for a spectrum of enclosed seas. Modified from Caddy 1993 and reproduced with permission from Taylor & Francis.

perch in Baltic archipelagoes), and negative (e.g., oxygen deficiency reducing Baltic cod recruitment and eventual harvest). Similar shifts are inferred with limited data on the Mississippi River–influenced shelf, with the increase in two pelagic species in bycatch from shrimp trawls and a decrease in some demersal species (Chesney and Baltz 2001). Commercial fisheries in the Black Sea declined as eutrophication led to the loss of macroalgal habitat and oxygen deficiency, coincident with intense fishing pressure and strong climate variability (Mee 2001; Oguz and Gilbert 2007). After the mid-1970s, benthic fish populations (e.g., turbot) collapsed, and pelagic fish populations (small pelagic fish such as anchovy and sprat) started to increase. The commercial fisheries diversity declined from about twenty-five fished species to about five in twenty years (1960s to 1980s), while anchovy stocks and fisheries increased rapidly (Mee 2001). The point on the continuum of increasing nutrients versus fishery yields remains vague as to where benefits are subsumed by environmental problems that lead to decreased landings or reduced quality of production and biomass.

Future Expectations

The occurrence of hypoxia in coastal areas of SEMSs is increasing, and the trend is consistent with the increase in human activities that result in increased fluxes of nutrients to coastal waters. More and more coastal systems, especially in areas of increased industrialization and mechanized farming, where the physical conditions are appropriate and where nutrient loads are predicted to increase, will likely become eutrophic with accompanying hypoxia.

The continued and accelerated export of nitrogen and phosphorus to the world's coastal ocean is the trajectory to be expected unless societal intervention in the form of controls or changes in culture are pursued. The largest increases are predicted for southern and eastern Asia, associated with predicted large increases in population, increased fertilizer use to grow food to meet the dietary demands of that population, and increased industrialization (Seitzinger et al. 2002). Increased production of biofuels in many countries will further amplify nutrient delivery from the land to the sea (National Research Council 2007; Runge and Senauer 2007; Simpson et al. 2007). The implications for coastal eutrophication and subsequent ecosystem changes, such as worsening conditions of oxygen depletion, are significant.

Another source of future change in nutrient loadings leading to increased organic production and water column stratification is the climate. Global climate changes within the range predicted to occur in the twenty-first century could have profound consequences to hypoxia in the northern Gulf of Mexico (Justic et al. 2003b, 2005). A modeling study that examined the impacts of global warming on the annual discharge of the thirty-three largest rivers of the world (Miller and Russell 1992) suggested that the average annual discharge of the Mississippi River would increase 20% if the concentration of atmospheric CO_2 doubled. If discharge increased this much, nutrient loads would increase, stratification would strengthen, and hypoxia would intensify and expand on the Louisiana continental shelf (Justic et al. 2003a, b). Other studies have shown that the runoff estimates for the Mississippi River discharge would decrease by 30% by the year 2099 (Wolock and McCabe 1999). Increases in surface water temperature would strengthen the summer pycnocline and perhaps worsen hypoxia (Justic et al. 2003a, b). On the other hand, warmer Atlantic Ocean temperatures could also increase tropical storm activity (Mann and Emanuel 2006) and severity (Trenberth 2005; Kerr 2006) resulting in more mixing and re-aeration events. Whichever occurs, the increase or decrease in flow, flux of nutrients, and water temperature are likely to have important, but as yet not clearly identifiable, influences on hypoxia.

Coastal water quality in SEMSs with regard to hypoxia is currently on the decline, and the future, based on the continued increase in the global occurrence of hypoxia and current and projected increased loads of nutrients, is trending to more hypoxia. The likelihood of strengthened stratification alone, from increased surface water temperature as the global climate warms, is sufficient to worsen hypoxia where it currently exists and facilitate its formation in additional coastal waters. The interplay of increased nutrients and stratification may be offset by the potential for more frequent and/or severe tropical storm systems, but the tropical storm projections with increased global temperatures remain under debate. The overall forecast, however, is for hypoxia to worsen, with increased occurrence, frequency, intensity, and duration. The need remains for water and resource managers to reduce nutrient loads even, if at a minimum, to maintain the current status of global hypoxia and not allow further degradation.

References

Barmawidjaja, D.M., G.J. van der Zwaan, F.J. Jorissen, and S. Puskaric. 1995. 150 years of eutrophication in the northern Adriatic Sea: Evidence from a benthic foraminiferal record. *Marine Geology* 122:367–384.

Bennett, E.M., S.R. Carpenter, and N.F. Caraco. 2001. Human impact on erodable phosphorus and eutrophication: A global perspective. *BioScience* 51:227–234.

Benović, A., D. Justic, and A. Bender. 1987. Enigmatic changes in the hydromedusan fauna of the Northern Adriatic Sea. *Nature* 326:597–600.

Boesch, D.F. 2002. Challenges and opportunities for science in reducing nutrient over-enrichment of coastal ecosystems. *Estuaries* 25:744–758.

Boesch, D.F., E. Burreson, W. Dennison, E. Houde, M. Kemp, V. Kennedy, R. Newell, K. Paynter, R. Orth, and R. Ulanowicz. 2001. Factors in the decline of coastal ecosystems. *Science* 293:1,589–1,590.

Caddy, J.F. 1993. Toward a comparative evaluation of human impacts on fishery ecosystems of enclosed and semi-enclosed seas. *Reviews in Fisheries Science* 1:57–95.

Caddy, J.F. 2000. Marine catchment basin effects versus impacts of fisheries on semi-enclosed seas. *ICES Journal of Marine Science* 57:628–640.

Cederwall, H., and R. Elmgren. 1990. Biological effects of eutrophication in the Baltic Sea, particularly the coastal zone. *Ambio* 19:109–112.

Chen, C.-C., G.-C. Gong, and F.-K. Shiah. 2007. Hypoxia in the East China Sea: One of the largest coastal low-oxygen areas in the world. *Marine Environmental Research* 64:399–408.

Chesney, E.J., and D.M. Baltz. 2001. The effects of hypoxia on the northern Gulf of Mexico coastal ecosystem: A fisheries perspective. In *Coastal Hypoxia: Consequences for Living Resources and Ecosystems,* edited by N.N. Rabalais and R.E. Turner, 321–354. Coastal and Estuarine Studies Vol. 58. Washington, DC: American Geophysical Union.

Childs, C.R., N.N. Rabalais, R.E. Turner, and L.M. Proctor. 2002. Sediment denitrification in the Gulf of Mexico zone of hypoxia. *Marine Ecology Progress Series* 240:285–290.

Childs, C.R., N.N. Rabalais, R.E. Turner, and L.M. Proctor. 2003. Erratum. *Marine Ecology Progress Series* 247:310.

Conley, D.J., C. Humborg, L. Rahm, O.P. Savchuk, and F. Wulff. 2002a. Hypoxia in the Baltic Sea and basin-scale changes in phosphorus biogeochemistry. *Environmental Science & Technology* 36:5,315–5,320.

Conley, D.J., and A.B. Josefson. 2001. Hypoxia, nutrient management and restoration in Danish waters. In *Coastal Hypoxia: Consequences for Living Resources and Ecosystems,* edited by N.N. Rabalais and R.E. Turner, 425–434. Coastal and Estuarine Studies Vol. 58. Washington, DC: American Geophysical Union.

Conley, D.J., S. Markager, J. Andersen, T. Ellermann, and L.M. Svendsen. 2002b. Coastal eutrophication and the Danish National Aquatic Monitoring and Assessment Program. *Estuaries* 25:848–861.

de Leiva Moreno, J.I., V.N. Agostini, J.F. Caddy, and F. Carocci. 2000. Is the pelagic-demersal ratio from fishery landings a useful proxy for nutrient availability? A preliminary data exploration for the semi-enclosed seas around Europe. *ICES Journal of Marine Science* 57:1,091–1,102.

Diaz, R.J., and R. Rosenberg. 1995. Marine benthic hypoxia: A review of its ecological effects and the behavioural responses of benthic macrofauna. *Oceanography and Marine Biology Annual Review* 33:245–303.

Donner, S.D., M.T. Coe, J.D. Lenters, T.E. Twine, and J.A. Foley. 2002. Modeling the impact of hydrological changes on nitrate transport in the Mississippi River Basin from 1955 to 1994. *Global Biogeochemical Cycles* 16 (3), doi: 10.1029/2001GB001396.

Dortch, Q., N.N. Rabalais, R.E. Turner, and N.A. Qureshi. 2001. Impacts of changing Si/N ratios and phytoplankton species composition. In *Coastal Hypoxia: Consequences for Living Resources and Ecosystems*, edited by N.N. Rabalais and R.E. Turner, 37–48. Coastal and Estuarine Studies Vol. 58. Washington, DC: American Geophysical Union.

Elmgren, R., and U. Larsson. 2001. Eutrophication in the Baltic Sea area: Integrated coastal management issues. In *Science and Integrated Coastal Management*, edited by B. von Bounce and R.K. Turner, 15–35. Berlin: Dahlem University Press.

Fallesen, G., F. Andersen, and B. Larsen. 2000. Life, death and revival of the hypertrophic Mariager Fjord, Denmark. *Journal of Marine Systems* 25:313–321.

Galloway, J.N., and E.B. Cowling. 2002. Reactive nitrogen and the world: Two hundred years of change. *Ambio* 31:64–71.

Gilbert, D., D. Chabot, P. Archambault, B. Rondeau, and S. Hébert. 2007. Appauvrissement en oxygène dans les eaux profondes du Saint-Laurent marin: Causes possibles et impacts écologiques. *Le Naturaliste Canadien* 131:67–75.

Gilbert, D., B. Sundby, C. Gobeil, A. Mucci, and G.-H. Tremblay. 2005. A seventy-two year record of diminishing deep-water oxygen in the St. Lawrence estuary: The northwest Atlantic connection. *Limnology and Oceanography* 50 (5):1,654–1,666.

Harper, D.E., Jr., L.D. McKinney, R.R. Salzer, and R.J. Case. 1981. The occurrence of hypoxic bottom water off the upper Texas coast and its effects on the benthic biota. *Contributions in Marine Science* 24:53–79.

HELCOM. 2001. Fourth Periodic Assessment of the State of the Marine Environment of the Baltic Sea Area 1994–1998. Baltic Sea Environment Proceedings No. 82. Helsinki: Helsinki Commission.

Helly, J.J., and L.A. Levin. 2004. Global distribution of naturally occurring marine hypoxia on continental margins. *Deep-Sea Research I* 51:1,159–1,168.

Jackson, J.B.C., M.X. Kirby, W.H. Berger, K.A. Bjorndal, L.W. Botsford, B.J. Bourque, R.H. Bradbury, R. Cooke, J. Erlandson, J.A. Estes, T.P. Hughes, S. Kidwell, C.B. Lange, and R.R. Warner. 2001. Historical overfishing and the recent collapse of coastal ecosystems. *Science* 293:629–638.

Jørgensen, B.B. 1980. Seasonal oxygen depletion in the bottom waters of a Danish fjord and its effect on the benthic community. *Oikos* 34:68–76.

Justic, D. 1988. Trend in the transparency of the Northern Adriatic Sea 1911–1982. *Marine Pollution Bulletin* 19:32–35.

Justic, D. 1991. Hypoxic conditions in the Northern Adriatic Sea: Historical development and ecological significance. In *Modern and Ancient Continental Shelf Anoxia*, edited by R.V. Tyson and T.H. Pearson, 95–105. Special Publication 58. London: Geological Society.

Justic, D., T. Legović, and L. Rottini-Sandrini. 1987. Trends in oxygen content 1911–1984 and occurrence of benthic mortality in the Northern Adriatic Sea. *Estuarine and Coastal Shelf Science* 25:435–445.

Justic, D., N.N. Rabalais, and R.E. Turner. 2002. Modeling the impacts of decadal changes in riverine nutrient fluxes on coastal eutrophication near the Mississippi River Delta. *Ecological Modeling* 152:33–46.

Justic, D., N.N. Rabalais, and R.E. Turner. 2003a. Climatic influences on riverine nitrate flux: Implications for coastal marine eutrophication and hypoxia. *Estuaries* 26:1–11.

Justic, D., N.N. Rabalais, and R.E. Turner. 2003b. Simulated responses of the Gulf of Mexico hypoxia to variations in climate and anthropogenic nutrient loading. *Journal of Marine Systems* 42:115–126.

Justic, D., N.N. Rabalais, and R.E. Turner. 2005. Coupling between climate variability and marine coastal eutrophication: Historical evidence and future outlook. *Journal of Sea Research* 54:25–35.

Kamykowski, D., and S.J. Zentara. 1990. Hypoxia in the world ocean as recorded in the historical data set. *Deep-Sea Research* 37:1,861–1,874.

Karlson, K., R. Rosenberg, and E. Bonsdorff. 2002. Temporal and spatial large-scale effects of eutrophication and oxygen deficiency on benthic fauna in Scandinavian and Baltic waters: A review. *Oceanography and Marine Biology Annual Review* 40:427–489.

Kerr, R. 2006. Global warming may be homing in on Atlantic hurricanes. *Science* 314:910–911.

Li, D., and D. Daler. 2004. Ocean pollution from land-based sources: East China Sea, China. *Ambio* 33:107–113.

Liu, S.M., J. Zhang, H.T. Chen, Y. Wu, H. Xiong, and Z.F. Zhang. 2003. Nutrients in the Changjiang and its tributaries. *Biogeochemistry* 62:1–18.

Mann, M.E., and K.A. Emanuel. 2006. Atlantic hurricane trends linked to climate change. *Eos, Transactions of the American Geophysical Union* 87:233–244.

Mee, L.D. 2001. Eutrophication in the Black Sea and a basin-wide approach to its control. In *Science and Integrated Coastal Management*, edited by B. von Bodungen and R.K. Turner, 71–91. Berlin: Dahlem University Press.

Mee, L.D., J. Friedrich, and M.T. Gomoiu. 2005. Restoring the Black Sea in times of uncertainty. *Oceanography* 18:100–111.

Miller, J.R., and G.L. Russell. 1992. The impact of global warming on river runoff. *Journal of Geophysical Research* 97:2,757–2,764.

National Research Council. 2007. *Water Implications of Biofuels Production in the United States*. Washington, DC: National Academies Press.

Nixon, S.W. 1995. Coastal marine eutrophication: A definition, social causes, and future concerns. *Ophelia* 41:199–219.

Oguz, T., and D. Gilbert. 2007. Abrupt transitions of the top-down controlled Black Sea pelagic ecosystem during 1960–2000: Evidence for regime-shifts under strong fishery exploitation and nutrient enrichment modulated by climate-induced variations. *Deep-Sea Research I* 54:220–242.

Quiñones-Rivera, Z.J., B. Wissel, D. Justic, and B. Fry. 2007. Partitioning oxygen sources and sinks in a stratified, eutrophic coastal ecosystem using stable oxygen isotopes. *Marine Ecology Progress Series* 342:60–83.

Rabalais, N.N. 2004a. Eutrophication. In Vol. 13 of *The Sea*, edited by A.R. Robinson, J. McCarthy, and B.J. Rothschild, 819–865. Cambridge, MA: Harvard University Press.

Rabalais, N.N. 2004b. Hipoxia en el Golfo de México. In *Diagnóstico Ambiental del Golfo de México*, Vol. II., M. Caso, I. Pisanty y E. Excurra (compiladores). Mexico, DF: Instituto Nacional de Ecología.

Rabalais, N.N. 2005. The potential for nutrient overenrichment to diminish marine biodiversity. In *Marine Conservation Biology: The Science of Maintaining the Sea's Biodiversity*, edited by E.A. Norse and L.B. Crowder, 109–122. Washington, DC: Island Press.

Rabalais, N.N., D.E. Harper, Jr., and R.E. Turner. 2001a. Responses of nekton and demersal and benthic fauna to decreasing oxygen concentrations. In *Coastal Hypoxia: Consequences for Living Resources and Ecosystems*, edited by N.N. Rabalais and R.E. Turner, 115–128. Coastal and Estuarine Studies Vol. 58. Washington, DC: American Geophysical Union.

Rabalais, N.N., L.E. Smith, D.E. Harper, Jr., and D. Justic. 2001b. Effects of seasonal hypoxia on continental shelf benthos. In *Coastal Hypoxia: Consequences for Living Resources and Ecosystems*, edited by N.N. Rabalais and R.E. Turner, 211–240. Coastal and Estuarine Studies Vol. 58. Washington, DC: American Geophysical Union.

Rabalais, N.N., and R.E. Turner (eds.). 2001. *Coastal Hypoxia: Consequences for Living Resources and Ecosystems*. Coastal and Estuarine Studies Vol. 58. Washington, DC: American Geophysical Union.

Rabalais, N.N., R.E. Turner, B.K. Sen Gupta, D.F. Boesch, P. Chapman, and M.C. Murrell. 2007a. Characterization and long-term trends of hypoxia in the northern Gulf of Mexico: Does the science support the Action Plan? *Estuaries and Coasts* 30:753–772.

Rabalais, N.N., R.E. Turner, B.K. Sen Gupta, E. Platon, and M.L. Parsons. 2007b. Sediments tell the history of eutrophication and hypoxia in the northern Gulf of Mexico. *Ecological Applications* 17:S129–S143.

Renaud, M. 1986. Hypoxia in Louisiana coastal waters during 1983: Implications for fisheries. *Fishery Bulletin* 84:19–26.

Rosenberg, R. 1990. Negative oxygen trends in Swedish coastal bottom waters. *Marine Pollution Bulletin* 21:335–339.

Rosenberg, R., and R.J. Diaz. 1993. Sulfur bacteria (*Beggiatoa* spp.) mats indicate hypoxic conditions in the inner Stockholm Archipelago. *Ambio* 22:32–36.

Rowe, G.T., M.E. Cruz Kaegi, J.W. Morse, G.S. Boland, and E.G. Escobar-Briones. 2002. Sediment community metabolism associated with continental shelf hypoxia, northern Gulf of Mexico. *Estuaries* 25:1,097–1,106.

Rumohr, H., E. Bonsdorff, and T.H. Pearson. 1996. Zoobenthic succession in Baltic sedimentary habitats. *Archive of Fishery and Marine Research* 44:170–214.

Runge, C.F., and B. Senauer. 2007. How biofuels could starve the poor. *Foreign Affairs* May/June 2007. http://www.foreignaffairs.org/20070501faessay86305/c-ford-runge-benjamin-senauer/how-biofuels-could-starve-the-poor.html.

Seitzinger, S.P., C. Kroeze, A.F. Bouwman, N. Caraco, F. Dentener, and R.V. Styles. 2002. Global patterns of dissolved inorganic and particulate nitrogen inputs to coastal systems: Recent conditions and future projections. *Estuaries* 25:640–655.

Selman, M., S. Greenhalgh, R. Diaz, and Z. Sugg. 2008. *Eutrophication and Hypoxia in Coastal Areas: A Global Assessment of the State of Knowledge*. Washington, DC: World Resources Institute.

Simpson, T., J. Pease, B. McGee, M. Smith, and R. Korcak. 2007. *Biofuels and Water*

Quality: Meeting the Challenge and Protecting the Environment. Mid-Atlantic Regional Water Program Report 07-04. University of Maryland. http://www.mawaterquality.org/Publications/pdfs/Biofuels_and_Water_Quality.pdf.

Stachowitsch, M. 1984. Mass mortality in the Gulf of Trieste: The course of community destruction. *Marine Ecology* 5:243–264.

Stachowitsch, M. 1991. Anoxia in the Northern Adriatic Sea. Rapid death, slow recovery. In *Modern and Ancient Continental Shelf Anoxia*, edited by R.V. Tyson and T.H. Pearson, 119–129. Special Publication 58. London: Geological Society.

Thibodeau, B., A. de Vernal, and A. Mucci. 2006. Recent eutrophication and consequent hypoxia in the bottom water of the Lower St. Lawrence Estuary: Micropaleontological and geochemical evidence. *Marine Geology* 231:37–50.

Tolmazin, R. 1985. Changing coastal oceanography of the Black Sea. I. Northwestern shelf. *Progress in Oceanography* 15:227–276.

Trenberth, K. 2005. Uncertainty in hurricanes and global warming. *Science* 308:1,753–1,754.

Turner, R.E., N. Qureshi, N.N. Rabalais, Q. Dortch, D. Justic, R.F. Shaw, and J. Cope. 1998. Fluctuating silicate:nitrate ratios and coastal plankton food webs. *Proceedings of the National Academy of Science USA* 95:13,048–13,051.

Turner, R.E., N.N. Rabalais, and D. Justic. 2006. Predicting summer hypoxia in the northern Gulf of Mexico: Riverine N, P, and Si loading. *Marine Pollution Bulletin* 52:139–148.

Turner, R.E., N.N. Rabalais, and D. Justic. 2008. Gulf of Mexico hypoxia: Alternate states and a legacy. *Environmental Science & Technology* 42:2,323–2,327. doi: 10.1021/es071617k.

Turner, R.E., N.N. Rabalais, E.M. Swenson, M. Kasprzak, and T. Romaire. 2005. Summer hypoxia in the northern Gulf of Mexico and its prediction from 1978 to 1995. *Marine Environmental Research* 59:65–77.

Turner, R.E., N.N. Rabalais, and Z.-N. Zhang. 1990. Phytoplankton biomass, production and growth limitations on the Huanghe (Yellow River) continental shelf. *Continental Shelf Research* 10:545–571.

Tyson, R.V., and T.H. Pearson. 1991. Modern and ancient continental shelf anoxia: An overview. In *Modern and Ancient Continental Shelf Anoxia*, edited by R.V. Tyson and T.H. Pearson, 49–64. Special Publication 58. London: Geological Society.

Vitousek, P.M., J.D. Abler, R.W. Howarth, G.E. Likens, P.A. Matson, D.W. Schindler, W.H. Schlesinger, and D.G. Tilman. 1997. Human alterations of the global nitrogen cycle: Sources and consequences. *Ecological Applications* 7:737–750.

Wang, H., Z. Yang, Y. Saito, J.P. Liu, and X. Sun. 2006. Interannual and seasonal variation of the Huanghe (Yellow River) water discharge over the past 50 years: Connections to impacts from ENSO events and dams. *Global and Planetary Change* 50:212–225.

Wei, H., Y. He, Q. Li, Z. Liu, and H. Wang. 2007. Summer hypoxia adjacent to the Changjiang Estuary. *Journal of Marine Systems* 67:292–303.

Wolock, D.M., and G.J. McCabe. 1999. Estimates of runoff using water-balance and atmospheric general circulation models. *Journal of the American Water Resources Association* 35:1,341–1,350.

Zaitsev, Y.P. 1992. Recent changes in the trophic structure of the Black Sea. *Fisheries Oceanography* 1 (2):180–189.

Zhang, J. 1994. Atmospheric wet depositions of nutrient elements: Correlations with harmful biological blooms in the northwest Pacific coastal zones. *Ambio* 23:464–468.

12

Ecosystem Services of Semi-Enclosed Marine Systems

Heike K. Lotze and Marion Glaser

Introduction

Semi-enclosed marine systems (SEMSs) are important ecological and social systems that have sustained human and marine life throughout history. They are linked to watersheds and rivers on the land side, linked to harbor estuaries and lagoons in the coastal zone, and have connections to the open ocean. The multiple influences from land and ocean, fresh- and seawater create highly dynamic, productive, and diverse marine ecosystems. Generally, SEMSs provide high primary productivity to sustain marine life, including species important for human food, as well as protected habitats for human settlement and for breeding, nursing, and foraging animals, and they link human transport and animal migration routes among ocean, estuaries, and rivers. Therefore, high concentrations of human populations and intense human–nature interactions (Figure 12-1) are characteristic for SEMSs.

Today, many ecosystem services are depleted or degraded as a result of increasing human pressures on SEMSs, and this is eroding human well-being (Figure 12-1; Vitousek et al. 1997; Millennium Ecosystem Assessment 2005; Lotze et al. 2006; Worm et al. 2006).

Ecosystem Services

Ecosystem services are the benefits people obtain from natural ecosystems (Daily 1997; de Groot et al. 2002; Millennium Ecosystem Assessment 2005). All humans have similar needs in relation to natural ecosystems, including food, water, fiber, and shelter. However, there are further direct and indirect ecosystem goods and services that are not immediately visible or regionally specific, such as aesthetic and cultural functions. Ecosystem services have been divided into four major categories:

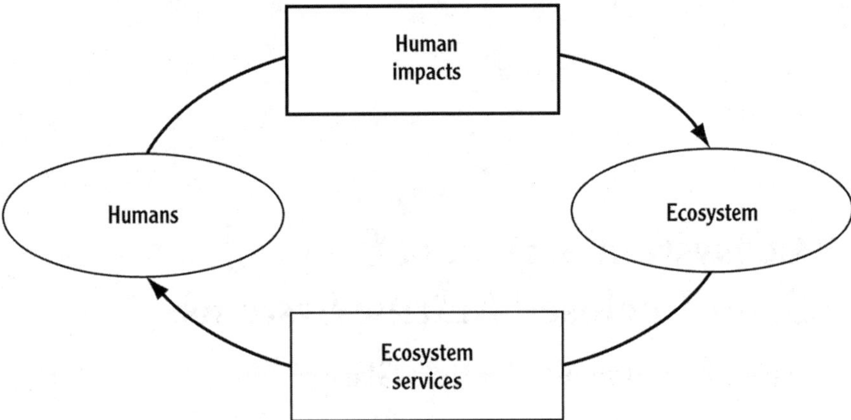

Figure 12-1. Conceptual diagram of the interactions between humans (social system) and ecosystems (ecological system) in SEMSs via human impacts and ecosystem services.

- *provisioning services*—products or goods obtained from ecosystems, including food, fiber, and medicines
- *regulating services*—benefits obtained from regulation of ecosystem processes, including climate regulation, water purification, and erosion control
- *cultural services*—including nonmaterial benefits such as spiritual enrichment and recreation and aesthetic values
- *supporting services*—necessary for the production of all other ecosystem services, which differ from provisioning, regulating, and cultural services because their benefits to people are indirect or occur over a very long time, whereas changes in the other services have relatively direct and short-term effects on people

Some services provided by SEMSs are important for social and economic systems but do not directly depend on ecosystem health. These include shipping, energy, and mineral resources (Table 12-1). There are cases, however, where land-driven sedimentation leads to channel filling and expensive long-term dredging of shipping canals.

Ecosystem services can be assessed in various ways and monitored over time. In general, provisioning services occur in measurable quantities and have a market value. Regulating and supporting services can sometimes be measured directly, such as the amount of carbon that is absorbed by the ocean, the amount of nutrients that are cycled in wetlands, or the amount of habitat that is provided. In other cases, an indirect assessment is possible, such as a comparison between the costs of building a water purification system and the costs of protecting a natural wetland that would perform the same regulating services. Cultural services are probably the most difficult to assess. However, the particular value that, for instance, aesthetic or recreational services have to an individual person can be measured and compared.

Table 12-1. Ecosystem Services of SEMSs

Provisioning or Production Services	
Food	Fisheries, mariculture, low-trophic-level harvesting, hunting
Fiber	Furs, skins, feathers, wood
Genetic resources	Species and genetic diversity
Biochemical resources	Medicines, pharmaceuticals, alginates
Ornamental resources	Shells, bones, skins, feathers
Energy resources	Wind, water, tidal, wave
Transport	Shipping, trading
Mineral resources	Mining, oil and gas extraction
Regulating Services	
Air quality regulation	Exchange of molecules with atmosphere
Climate regulation	Land cover (temperature, precipitation), carbon storage
Water regulation	Land cover (runoff, flooding, aquifer recharge, water storage)
Water purification and waste treatment	Filtration, denitrification, decomposition, detoxification
Erosion regulation	Vegetative cover (soil retention, prevention of landslides)
Disease regulation	Pathogen abundance, disease vectors
Natural hazard regulation	Coastal protection (mangroves, coral reefs)
Cultural Services	
Spiritual and religious values	Sacred species (e.g., sharks, whales, dolphins), symbols, places
Cultural values	Fishing societies, nomadic herding, agricultural societies
Knowledge and education	Traditional, formal
Inspiration	Art, folklore, national symbols, architecture
Aesthetic values	Landscape, seascape, diversity
Recreation and ecotourism	Swimming, boating, nature watching
Supporting Services	
Photosynthesis	Oxygen production
Primary production	Assimilation of energy and nutrients into organic matter
Nutrient cycling	About 20 essential nutrients (e.g., N, P), cycling, storage, transformation
Water cycling	Water cycling, storing, transformation
Habitat provision	Nursery, spawning, breeding, foraging, refuge

Adapted from de Groot et al. 2002, Millenium Ecosystem Assessment 2005.

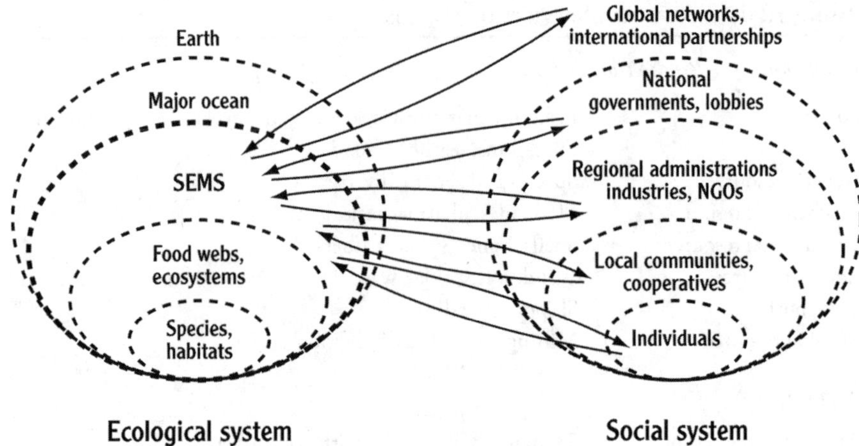

Figure 12-2. Generic model of the linkages of SEMSs to different levels of the ecological and social (including economic) systems. Circles indicate elements that are embedded in the next higher level; arrows indicate the interactions between SEMSs and different levels of the social system.

The provision of ecosystem services depends on underlying physical, chemical, and biological processes (see Chapters 6–11, this volume) and social environments. The combination and interplay of these environments determine the abundance, distribution, and diversity of species and habitats and ultimately the structure and functioning of food webs, ecosystems, and the services they provide. Therefore, all factors that influence single or multiple components of social and ecological systems result in changes in the availability of ecosystem services (Figure 12-2). Beyond this, any factors that alter species abundance or environmental conditions at larger scales, such as in the open ocean or at global levels, also impact ecosystem services at local or regional scales (Figure 12-2).

Social–Ecological Interactions

The ecosystem services provided by SEMSs flow to different levels of the associated social systems (Figure 12-2). Individual people benefit from food resources, income, inspiration, and recreation. Local communities are built around opportunities for fishing, transport, recreation, and protected land for houses and farms. Regional administrations benefit from tax income; urban centers and industries benefit from economic opportunities, transport, and waste removal; and nongovernmental organizations (NGOs) benefit from the necessities of biodiversity conservation and habitat protection. Often, strong cultural values associated with regional land- and seascapes determine whether regions are developed for industrial or municipal uses or are preserved as natural ecosystems. National governments and industrial or environmental lobbies can play crucial roles in such regional development decisions. At the global scale, supporting services such as climate regulation and oxygen production benefit all humans. Throughout history, goods and services essential for human survival, options for economic development, and cultural identification have been provided by SEMSs, mostly at local and regional levels. However, in the wake of globalization, global flows of ecosystem services have changed predominantly regional sectors, such as fisheries, into globally connected sectors. Most major population centers have developed along coastlines, often adjacent to SEMSs (Limburg 1999; Millennium Ecosystem Assessment 2005). At the end of the twentieth century, about half of the world's major cities with more than 500,000 inhabitants were found within 50 km of the coasts, and human well-being is about four times greater along the coasts compared with inland communities (Millennium Ecosystem Assessment 2005).

Humanity not only benefits from, but also influences, SEMSs. At all levels of society, multiple stakeholders and users have positive and negative impacts on the ecosystem services provided by SEMSs. Negative impacts include overexploitation of resources, habitat destruction, water pollution, and waste dumping by individuals, communities, and industries. Positive impacts comprise successful stewardship and governance, including harvest regulations, pollution control, and protection and restoration of species and habitats by management agencies and NGOs. Regional and national administrations play a crucial role in regulating and managing different uses and impacts. International agreements and treaties integrate and streamline efforts across countries linked through shorelines and drainage basins within SEMSs. However, national and international economic and political pressures can overturn local and regional efforts. Thus, the impacts of and interplay among different levels of society strongly influence the status of SEMSs and the sustainability of ecosystem services and human well-being.

Regional Differences

Around the globe, ecological and social systems differ from region to region. Table 12-2 provides an overview of selected ecological indicators for the fourteen SEMSs that are the focus of this book (we consider the East China Sea and Yellow Sea separately in this chapter). Ecologically, most of them are considered Large Marine Ecosystems (LMEs), except for the Gulf of St. Lawrence, which is part of the Scotian Shelf LME, and for the Adriatic Sea, embedded in the Mediterranean LME (Table 12-2). Most importantly, the fourteen systems are located in different climate zones, which strongly affect primary productivity and species richness. Generally, primary productivity is higher in cold-temperate waters and species richness in warm-temperate and tropical zones (Table 12-2). Exceptions are the Yellow Sea (with very high) and Black Sea (with very low) fish species richness in cold- versus warm-temperate waters, respectively. Species richness of cephalopods is high in polar systems and the Black Sea, while marine mammal richness is highest in the East China Sea (Table 12-2). In terms of ecosystem services, species richness is an important indicator of the variety of genetic, biodiversity, and food resources and ocean wildlife, while primary productivity translates into oxygen production, fisheries productivity, and climate regulation (Table 12-1).

Fisheries productivity is perhaps the most prominent marine ecosystem service. In terms of average fisheries landings per unit area and per unit primary productivity from 1950 to 2003, the North Sea, Sea of Okhotsk, East China Sea, Yellow Sea, and Bay of Bengal are the most productive SEMSs (Table 12-2; Figure 12-3). The Kara and Laptev seas, Hudson Bay, and the Bay of Bengal experienced the strongest increases in fisheries landings from 1950 to their maximum, while the North, Black, and Mediterranean seas showed only small increases. This may reflect the long history of fishing in European systems, whereas polar and tropical regions became important fishing areas more recently. However, the polar systems in particular, but also the Scotian Shelf and North Sea, showed the strongest catch declines, with < 50% of their maximum catch remaining in 2003 (Table 12-2; Figure 12-3). In polar systems, this could reflect the boom-and-bust cycle of exploratory fisheries, whereas long-term trends in the Scotian Shelf and North Sea indicate that declining fish abundance is undermining fisheries services (Christensen et al. 2003; Jennings and Blanchard 2004; Rosenberg et al. 2005). In turn, some tropical and warm-temperate systems—including the East China and Yellow seas, the Bay of Bengal, and the Mediterranean Sea—showed steady increases in fisheries landings from 1950 to 2003 (Figure 12-3).

Other ecosystem services that the fourteen SEMSs provide include habitat provision, water filtration, and coastal protection. These are based on different vegetation and suspension-feeding organisms, depending on climate conditions and geological features. Tropical systems and some warm-temperate systems host mangroves, sea grass beds, and coral reefs, while warm- and cold-temperate systems host sea grass and rockweed beds and some kelp forests (Table 12-2; Lüning 1990; Steneck et al. 2002; Orth et al. 2006). Salt marshes and oyster reefs are also found in many estuarine and coastal ecosystems (Adam 2002; Lotze et al. 2006). All these habitat-building species provide spawning, breeding, nursery, and foraging grounds for numerous species; act as carbon sinks and nutrient filters; and reduce the severity of floods and erosion (Millennium Ecosystem Assessment 2005). Other important habitats in SEMSs include seamounts and sea ice (Table 12-2; Horner et al. 1992; Rogers 1994).

Most SEMSs are bordered by human settlements, industries, agriculture, aquaculture, and other enterprises and are considered strongly impacted coastal zones in a recent global survey (Table 12-2; Kennish 2002). Exceptions are the sparsely populated polar systems; however, the Kara and Laptev seas are highly impacted by oil spills and toxic wastes (Tsyban et al. 2005). A long-term consequence of land-based human activities is increased nutrient loading into the coastal ocean. Since prehistoric times, nutrient loading has increased most in the North, Baltic, Black, Mediterranean, and Yellow seas, followed by the Gulf of Mexico, the East China Sea, and the Bay of Bengal (Table 12-2;

Table 12-2. Overview of Selected Ecological Indicators for Fourteen SEMSs

System (LME)[a]	Climate[b]	Size[a] (km^2)	PP[a] ($mg\ C\ m^{-2}\ d^{-1}$)	Species Richness[a]			Fish Landings 1950–2003[a]			Increase to Max. (X-fold)	Decrease from Max (% Left)
				Fish	Ceph.	Mam.	($mt\ y^{-1}$)	Max. mt	Max. y		
Kara Sea	Polar	797,171	410	19	58	17	64	176	1981	176.0	27.3
Laptev Sea	Polar	499,039	479	43	57	13	1	23	1968	23.0	0.0
Hudson Bay	Polar	841,214	442	18	0	11	39	249	1968	24.9	18.5
Sea of Okhotsk	Cold-temp	1,552,663	774	375	52	32	2,921,371	6,058,118	1985	8.9	32.7
Gulf of St Lawrence (Scotian Shelf)	Cold-temp.	282,953	1,044	197	8	26	421,016	817,139	1970	6.4	33.4
Baltic Sea	Cold-temp.	390,077	1,804	156	23	21	699,204	1,056,290	1997	6.1	62.9
North Sea	Cold-temp.	693,840	1,067	189	22	25	2,997,628	4,393,569	1968	2.4	48.0
Yellow Sea	Cold-temp.	437,376	1,643	1,904	48	31	2,390,106	4,411,596	2000	5.4	77.8
Black Sea	Warm-temp.	460,151	882	150	62	5	409,133	890,334	1984	3.4	51.0
Adriatic Sea (Mediterranean)	Warm-temp.	2,516,484	385	698	26	15	776,453	1,083,968	1994	2.4	85.1
Gulf of Mexico	Warm-temp.	1,529,669	417	977	5	31	917,088	1,801,383	1984	5.3	53.6
East China Sea	Warm-temp.	775,065	728	1,036	47	46	3,393,800	5,661,966	1999	5.7	84.1
Bay of Bengal	Tropical	3,660,127	568	693	34	31	1,667,066	4,005,393	2003	13.8	100.0
Gulf of Thailand	Tropical	386,878	700	613	35	19	600,922	1,249,252	1969	8.4	73.6

(continued)

Table 12-2. Overview of Selected Ecological Indicators for Fourteen SEMSs *(continued)*

	Sea Mounts[a] *(% Global)*	*Coral Reefs*[a] *(% Global)*	*Mangroves*[c]	*Sea Grasses*[d]	*Rockweeds*[e]	*Kelp Forests*[f]	*Sea Ice*[g]	*Strongly Impacted*[h]	*N-flux Increase*[i]	*Dead Zones*[j]	*HAB*[k]
Kara Sea	0	0	-	-	x	(x)	x	-	1–50	-	-
Laptev Sea	0	0	-	-	x	(x)	x	-	1–50	-	-
Hudson Bay	0	0	-	(x)	x	(x)	x	-	50–100	-	-
Sea of Okhotsk	0.04	0	-	(x)	x	x	x	-	1–50	-	-
Gulf of St. Lawrence (Scotian Shelf)	0	0	-	x	x	x	x	(x)	50–100	1	x
Baltic Sea	0	0	-	x	x	(x)	x	x	400–500	34	x
North Sea	0	0	-	x	x	(x)	-	x	> 500	12	x
Yellow Sea	0	0	-	x	x	(x)	-	x	> 500	-	-
Black Sea	0	0	-	x	x	-	-	x	400–500	2	-
Adriatic Sea (Mediterranean)	0.41	0	-	x	x	-	-	x	400–500	6	x
Gulf of Mexico	0.02	0.49	x	x	x	-	-	x	200–400	12	x
East China Sea	0.02	0.34	x	x	x	(x)	-	x	200–400	6	-
Bay of Bengal	0.12	3.63	x	x	x	-	-	x	200–400	-	x
Gulf of Thailand	0	0.46	x	x	x	-	-	x	1–50	-	x

x = present; (x) = present but of limited importance.

[a] Source: LME database, Sea Around Us 2008; PP = primary productivity, Ceph. = cephalopods, Mam. = marine mammals.

[b] Source: Lüning 1990; climate zone for vegetation based on summer/winter isotherms for polar (10/0°C), cold-temperate (15/10°C), warm-temperate (25/20°C), and tropical (> 25/> 20°C) zones.

[c] Source: Alongi 2002.

[d] Source: Orth et al. 2006; x = present, (x) = species present but not forming extended habitats.

[e] Source: Lüning 1990; Fucales (e.g., *Fucus, Ascophyllum, Cystoseria, Sargassum*).

[f] Source: Steneck et al. 2002, Lüning 1990; Laminariales (e.g., *Laminaria, Chorda*).

[g] Source: Lüning 1990.

[h] Source: Kennish 2002; strongly impacts coastal zones.

[i] Source: Millennium Ecosystem Assessment 2005; increase in nitrogen flux in rivers to the coastal zone.

[j] Source: Diaz et al. 2004; number of oxygen-depleted dead zones.

[k] Source: Sellner et al. 2003; occurrence of harmful algal blooms (HAB).

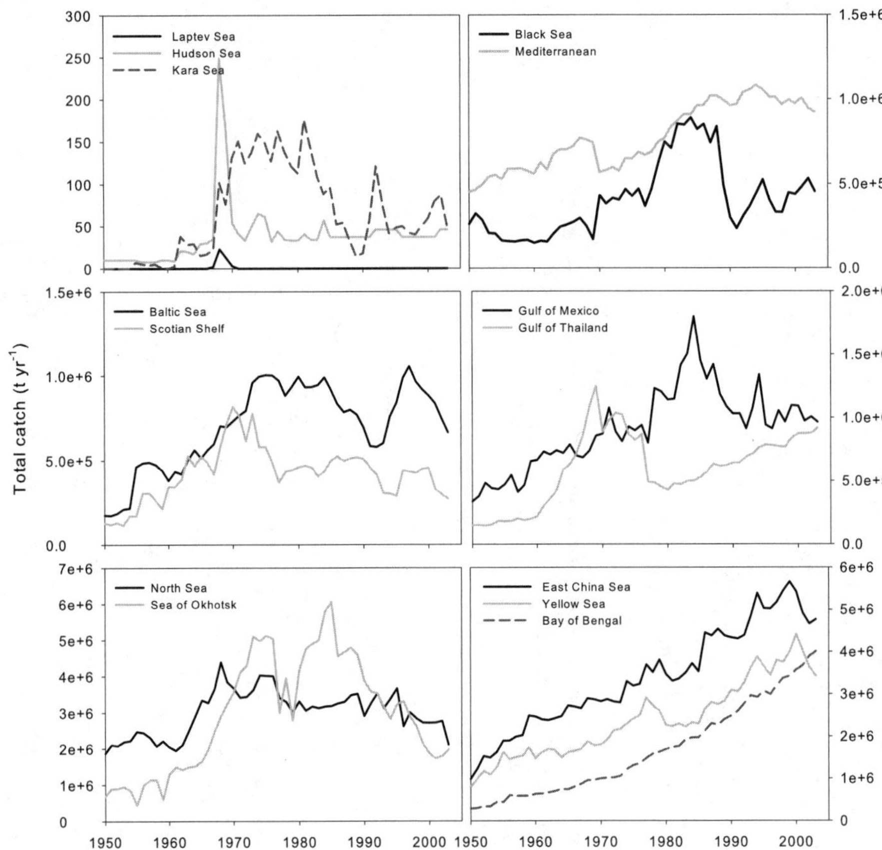

Figure 12-3. Regional differences in fisheries landings across fourteen SEMSs or their respective Large Marine Ecosystems (LMEs) from 1950 to 2003. Note the different scales on the different *y*-axes. From LME database, Sea Around Us 2008.

Millennium Ecosystem Assessment 2005). These are also the systems showing oxygen-depleted "dead zones" and the occurrence of harmful algal blooms (Table 12-2; Sellner et al. 2003; Diaz et al. 2004). While both dead zones and harmful algal blooms can occur naturally in some regions, they tend to increase with eutrophication, becoming more frequent, intense, and widespread in many coastal systems around the world (see Chapter 11, this volume).

Effective management and governance of SEMSs, especially regarding nutrient and sediment inputs from land, should include all countries of the drainage basins. For example, deforestation and river-use practices in Nepal, Bhutan, and China potentially drive sedimentation and nutrient input to the Bay of Bengal. Here, however, we focus on comparing only coastal bordering countries because these are the main beneficiaries of direct marine ecosystem services (Table 12-3). Social and economic data related to SEMSs are available on the basis of administrative units. While human communities and activities in coastal zones interact directly with the ecological system, market demands and human pressures are rooted in wider national socioeconomic systems, for which data are available. Table 12-3 thus pulls together some central social and economic indicators to characterize our fourteen systems in terms of variables important for ecosystem services. Included are indicators describing the "system to be governed," for example, human population size, gross domestic product (GDP), human development index (HDI), and economic value of ecosystem services. Also included are indicators describing the "governing system," such as the perceived corruption index. Further governance indicators are yet to be developed (Hockings et al. 2000, 2005; Ehler 2003). Our example SEMSs are then categorized according to the socioeconomic development status of the bordering countries as rich, intermediate, or poor, based on total and per capita GDP and the HDI.

In general, rich social systems display greater inequality in family income (Gini index) and a higher perceived corruption (PCI) than intermediate or poor systems (Table 12-3). Fisheries contribute much less to GDP in rich and intermediate than poor systems. This indicates that poor countries rely much more on productive and resilient ecosystems that sustain viable fisheries than do rich countries. Poor and intermediate countries also rely on the poverty alleviation function of constituent ecosystems (e.g., mangroves, coral reefs, sea grass beds, estuaries) for local subsistence. Data on the poverty alleviation function are extremely scarce because subsistence uses (fishing, harvesting) do not pass through markets and price-based evaluations. However, available data on subsistence resource uses of local populations in mangroves (Maneschy 1994; Glaser 2003) and on indigenous people in arctic regions (Tsyban et al. 2005) demonstrate the central functions of protein provision, famine protection, and emergency buffer that SEMSs provide for rural populations in isolated areas, and also for poorer populations in commercially more connected regions all over the world. In terms of governance, some SEMSs (such as the Hudson Bay and the Kara Sea) are bordered by only one nation, which simplifies management issues. Other systems are bordered by multiple countries with diverse social systems, forms of governance, and economic power. International organizations and agreements play an essential role in managing these multicountry ecosystems, for example, the Helsinki Commission (HELCOM) in the Baltic Sea and incipient attempts such as the Bay of Bengal Large Marine Ecosystem (BOBLME) and Bay of Bengal Initiative for Multi-Sectoral Technical and Economic Cooperation (BIMSTEC) in the Bay of Bengal.

Comparing social and ecological indicators, there is no strong association between socioeconomic development status and nutrient loading (Tables 12-2 and 12-3). Whether rich, intermediate, or poor, most SEMSs outside the polar regions experience nutrient loading, dead zones, and harmful algal blooms. This may reflect the combined effects of two major indirect drivers of nutrient loading: economic development and population size (Lotze et al. 2006). On the other hand, rich countries encountered greater declines in their fisheries (percent of maximum catch) from 1950 to 2003 than intermediate or

Table 12-3. Overview of Selected Social and Economic Indicators for Fourteen SEMSs

SEMS	*Development Status*	*Bordering Countries*	*Population*[a] *(# 10^3)*	*Population Density*[b] *(# km^{-2})*	*GDP*[c] *($US 10^9)*	*GDP*[d] *($US p.c.)*	*HDI*[e]	*Gini*[f]	*PCI*[g]	*Fisheries*[h] *% GDP*
North Sea	Rich	Norway	4,611	12	262	47,800	0.963	25.8	8.9	0.70
		Denmark	25,451	125	256	37,000	0.928	23.2	9.5	n/a
		UK	60,776	246	2,341	31,400	0.939	36.8	8.6	0.21
		Germany	82,422	231	2,890	31,400	0.940	28.3	8.2	0.01
		Netherlands	16,491	391	613	31,700	0.943	30.9	8.6	0.10
		Belgium	10,379	340	368	31,800	0.933	25.0	7.4	0.01
		France	62,752	108	2,154	30,100	0.938	26.7	7.5	0.00
		Total	**262,883**		**8,884**					
		Average		**208**		**34,457**	**0.941**	**28.1**	**8.4**	**0.17**
Gulf of St. Lawrence	Rich	Canada	**33,099**	**3**	**1,089**	**35,200**	**0.949**	**33.1**	**8.4**	**0.24**
Hudson Bay	Rich	Canada	**33,099**	**3**	**1,089**	**35,200**	**0.949**	**33.1**	**8.4**	**0.24**
Baltic Sea	Predominantly rich	Germany	**82,422**	**231**	**2890**	**31,400**	**0.925**	**28.3**	**8.2**	**0.01**
		Poland	38,537	122	337	14,100	0.858	34.1	3.4	0.03
		Russia	142,894	8	733	12,100	0.795	40.5	2.4	0.70
		Lithuania	3,586	53	30	15,100	0.852	32.5	4.8	2.02
		Latvia	2,275	36	16	15,400	0.863	35.0	4.2	1.15
		Estonia	1,324	30	14	19,600	0.853	33.0	6.4	n/a
		Finland	2,275	15	196	32,800	0.942	26.9	9.6	0.05
		Sweden	5,231	20	372	31,600	0.949	25.0	9.2	0.20
		Denmark	5,451	125	256	37,000	0.928	23.2	9.5	n/a
		Total	**283,994**		**4,844**					
		Average		**71**		**23,233**	**0.885**	**30.9**	**6.4**	**0.59**

Gulf of Mexico	Predominantly rich	USA	298,444	30	13,220	43,500	0.944	45.0	7.6	0.29
		Mexico	107,450	53	742	10,600	0.814	54.6	3.5	0.80
		Cuba	11,383	101	40	3,900	n/a	n/a	3.8	0.72
		Total	**417,277**		**14,002**					
		Average		**61**		**19,333**	**0.879**	**49.8**	**5.0**	**0.60**
Adriatic Sea	Intermediate (except Italy)	Italy	58,134	191	1,780	29,700	0.934	36.0	5.0	0.33
		Slovenia	2,010	99	38	23,400	0.904	28.4	n/a	n/a
		Croatia	4,495	79	37	13,200	n/a	29.0	3.4	n/a
		Bosnia-Herz.	4,499	75	9	5,500	0.786	26.2	2.9	n/a
		Montenegro	631	116	2	3,800	n/a	n/a	2.8	n/a
		Albania	3,582	108	9	5,600	0.780	26.7	2.4	0.38
		Total	**73,350**		**1,876**					
		Average		**111**		**13,533**	**0.851**	**29.3**	**3.3**	**0.36**
Black Sea	Intermediate	Turkey	70,414	90	358	8,900	0.750	42.0	3.5	0.45
		Russia	142,894	8	733	12,100	0.795	40.5	2.4	0.70
		Ukraine	46,711	79	82	7,600	n/a	29.0	2.6	0.56
		Romania	22,304	91	79	8,800	0.792	28.8	3.0	0.07
		Bulgaria	7,385	71	28	10,400	0.780	29.2	4.0	n/a
		Georgia	4,661	64	5	3,800	0.732	38.0	2.3	1.10
		Total	**294,369**		**1,285**					
		Average		**67**		**8,600**	**0.770**	**34.6**	**3.0**	**0.58**

(continued)

Table 12-3. Overview of Selected Social and Economic Indicators for Fourteen SEMSs *(continued)*

SEMS	*Development Status*	*Bordering Countries*	*Population[a] (# 10^3)*	*Population Density[b] (# km^{-2})*	*GDP[c] ($US 10^9)*	*GDP[d] ($US p.c.)*	*HDI[e]*	*Gini[f]*	*PCI[g]*	*Fisheries[h] % GDP*
East China Sea	Intermediate	Taiwan	22,859	613	354	29,000	n/a	n/a	5.9	n/a
		China	1,321,852	626	2,512	7,600	0.755	44.0	3.2	2.00
		Total	**1,344,711**		**2,866**					
		Average		**620**		**18,300**	**0.755**	**44.0**	**4.6**	**2.00**
Yellow Sea	Intermediate	China	1,321,852	626	2,512	7,600	0.755	44.0	3.2	2.00
		North Korea	23,302	n/a	n/a	1,800	n/a	n/a	n/a	n/a
		South Korea	49,045	481	897	24,200	0.901	35.8	5.1	1.00
		Total	**1,394,198**		**3,409**					
		Average		**554**		**11,200**	**0.828**	**39.9**	**4.2**	**1.50**
Kara Sea	Intermediate	Russia	**141,378**	**8**	**733**	**12,100**	**0.795**	**40.5**	**2.4**	**0.70**
Laptev Sea	Intermediate	Russia	141,378	**8**	**733**	**12,100**	**0.795**	**40.5**	**2.4**	**0.70**
Sea of Okhotsk	Intermediate	Russia	141,378	**8**	**733**	**12,100**	**0.795**	**40.5**	**2.4**	**0.70**
Bay of Bengal	Poor	Bangladesh	150,448	**1090**	**69**	**2,200**	**0.520**	**31.8**	**1.7**	**3.12**
		India	1,129,866	325	796	3,700	0.602	32.5	2.9	1.47
		Sri Lanka	20,926	184	27	4,600	0.751	50.0	3.2	2.00
		Myanmar	47,374	80	10	1,800	0.581	n/a	1.9	8.94
		Total	**1,348,615**		**902**					
		Average		**420**		**3,075**	**0.614**	**38.10**	**2.43**	**3.88**

Gulf of Thailand	Poor	Cambodia	13,996	74	7	2,600	n/a	40.0	2.3	12.00
		Thailand	65,068	126	197	9,100	0.778	51.1	3.8	2.09
		Viet Nam	85,262	243	48	3,100	0.704	36.1	2.6	4.00
		Total	**164,326**		**251**					
		Average		**148**		**4,933**	**0.741**	**42.40**	**2.90**	**6.03**

[a] Source: World Factbook 2008; total population in bordering countries.

[b] Source: UN Statistics Division 2008; population density in bordering countries.

[c] Source: World Factbook 2008; gross domestic product (GDP), official exchange rate with $U.S.

[d] Source: World Factbook 2008; gross domestic product per capita.

[e] Source: UN Development Programme 2008; the Human Development Index (HDI) is an arithmetic average of mortality, education, and income availability to describe human development levels.

[f] Source: World Factbook 2008; the Gini index measures equality of family income; the closer to 1, the higher the inequality.

[g] Source: Transparency International 2008; the Perceived Corruption Index (PCI) ranks 163 countries in terms of perceived levels of corruption among public officials and politicians; a higher index signifies greater perceived corruption.

[h] Source: FAO 2008; Fisheries contribution to GDP.

poor countries. Combined with the greater and growing strength of other parts of the economies in question, this loss of fisheries services in rich countries indicates why the current contribution of fisheries to GDP in rich countries is generally very low (Tables 12-2 and 12-3).

Long-Term Changes

Over the past fifty years, human activities have changed ecosystem structure and functioning on land and in the sea more rapidly and extensively than in any other period in history (Vitousek et al. 1997; Millennium Ecosystem Assessment 2005), and human agency has had stronger effects than any other factor of change (Crutzen 2002). This has brought the degradation of approximately 60% of global ecosystem services (Millennium Ecosystem Assessment 2005). Many changes observed in estuarine and coastal zones have a much longer history though, dating back centuries to millennia (Jackson et al. 2001; Pandolfi et al. 2003; Lotze et al. 2006). Depletion and degradation of estuarine and marine ecosystems began during Roman times (2,500 years ago) in the Mediterranean Sea, in medieval times (1,000 years ago) in the North and Baltic seas, and with European colonization (300–150 years ago) in North America and Australia (Lotze et al. 2006). Relative abundance of economically or ecologically important species rapidly declined over the past 150 to 300 years (Figure 12-4A), mostly driven by overexploitation, habitat loss, and pollution (Lotze et al. 2006). In the twentieth century, conservation efforts slowed some declines and enabled recovery in some marine mammals and birds (Lotze et al. 2006). With the decline in species, the percentage of depleted (< 50% of former abundance), rare (< 10%), and extirpated (0%) fisheries strongly increased over time (Figure 12-4B). Destructive fishing practices, habitat conversion, and water pollution caused strong declines in habitat-providing and suspension-feeding species that create oyster reefs, sea grass beds, and wetlands (Figure 12-4C; Lotze et al. 2006). In tropical systems, about 20% of coral reefs have been lost and 20% degraded over the past fifty years, and about 35% of mangrove area has been lost over the past two decades (Millennium Ecosystem Assessment 2005). Moreover, increased nutrient and sediment loading have exponentially degraded water quality and enhanced eutrophication in coastal waters (Figure 12-4D).

The long-term degradation of estuarine and coastal ecosystems has reduced essential ecosystem services, particularly food resources, habitat provision, and water purification. Such losses in ecosystem services influence all components of human well-being, including survival, health, wealth, and social relations (Millennium Ecosystem Assessment 2005). For example, Worm and colleagues (2006) documented that ecosystem degradation in twelve estuaries and coastal seas strongly increased health risks and costs to society associated with increases in harmful algal blooms, oxygen depletion, beach closures, shellfish closures, fish kills, coastal flooding, and invasive species. A eutrophication survey in 138 U.S. estuaries (Bricker et al. 1999) found deteriorating trends in > 50% of

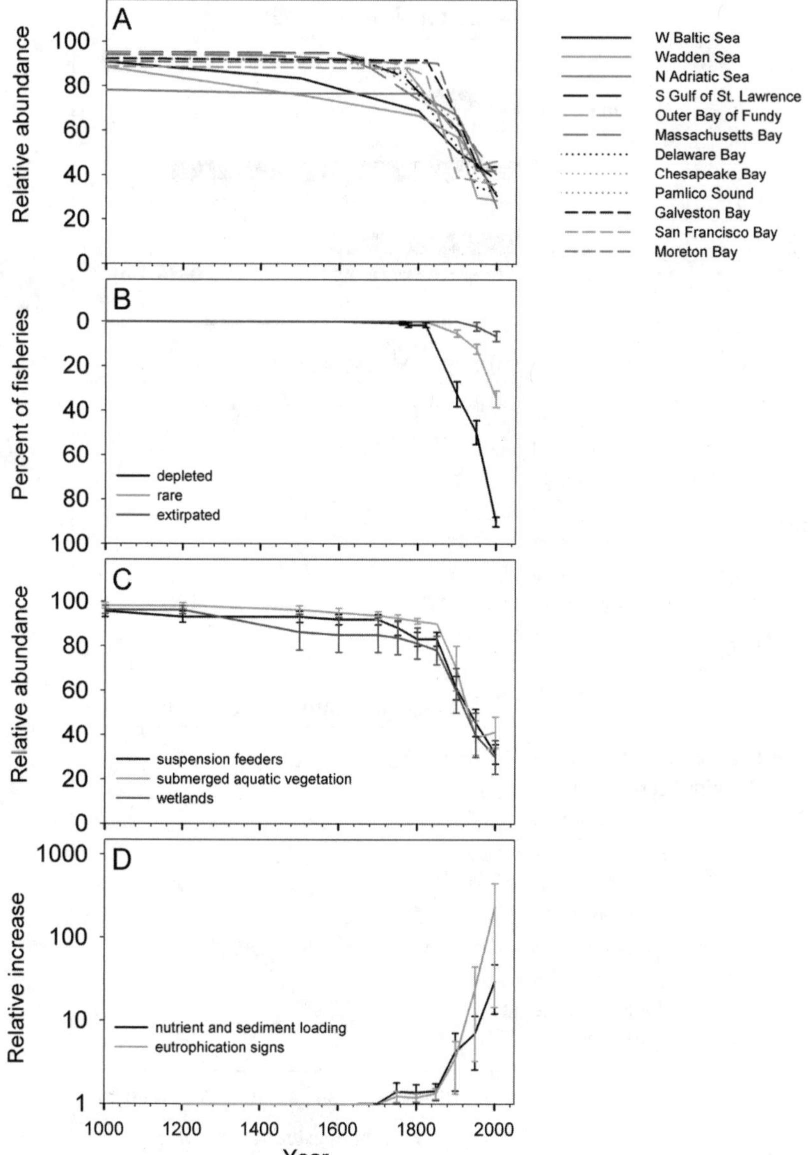

Figure 12-4. Long-term changes in estuarine and coastal ecosystems: (A) average relative abundance of 30–80 species in twelve study systems; (B) percent of fisheries that were depleted (> 50% decline), rare (> 90% decline), or extirpated (100% decline); (C) average relative abundance of filtering and habitat-providing species; (D) relative changes in water quality. Data in Panels B–D represent means and standard errors across twelve study systems except for water quality across eight systems. Data adapted from Lotze et al. 2006 and Worm et al. 2006.

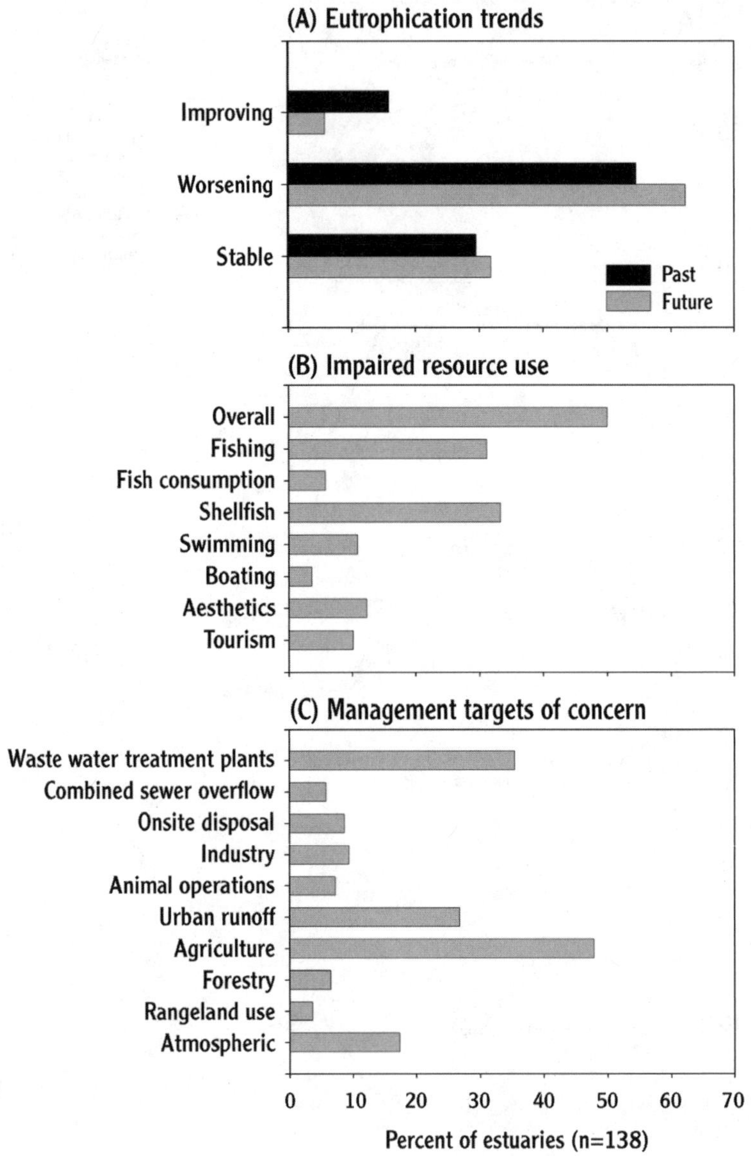

Figure 12-5. Trends and consequences of eutrophication in 138 U.S. investigated estuaries. (A) Percent in which eutrophication trends were improving, worsening, or stable in the past (1970–1995; $n = 88$), and projected future trends (toward 2020; $n = 138$). (B) Percent in which eutrophication impaired overall and selected resource uses. (C) Percent in which different management targets were identified as of concern with respect to eutrophication problems. Data adapted from Bricker et al. 1999.

estuaries from 1970 to 1995, which are projected to increase toward 2020 (Figure 12-5A). Eutrophication—mostly driven by sewage and industrial discharges and agricultural and urban runoff (Figure 12-5C)—has already significantly impaired ecosystem services such as fishing, shellfish harvesting, and tourism (Figure 12-5B).

Alterations in marine ecosystems and the services they provide to humanity are driven by combined social and economic factors such as demographic, economic, and technological change and by sociopolitical and cultural structures and dynamics (Millennium Ecosystem Assessment 2005). Over the past two to three centuries, resource depletion and ecosystem degradation were propelled by commercialization and industrialization, human population growth, and increasing demand for ecosystem resources and services (Figure 12-4; Lotze et al. 2006). Many countries experienced long-term changes in their economic structures during the nineteenth and twentieth centuries. Economic importance of living natural resources such as forest and agricultural products and fish declined, while the service sector grew. At the same time, mining, manufacturing, and construction peaked in the second half of the twentieth century (Millennium Ecosystem Assessment 2005). Some management and conservation efforts in the twentieth century stabilized and even reversed some depletion trends (Figure 12-4; Lotze et al. 2006).

Future Resilience and Sustainability

Between 1960 and 2000, world population doubled from 3 to 6 billion people, and the global economy increased more than sixfold (Millennium Ecosystem Assessment 2005). Over the next fifty years, human populations and economies will continue to increase their demands on marine ecosystems while the systems' capacities to provide services will shrink (WRI 2007). How do we govern and manage SEMSs to secure the continued provision of essential ecosystem services?

Considering the ecosystem to be governed, one way to secure the provision of ecosystem services is to strengthen the system's resilience and productivity. Resilience is the capacity of a system to experience disturbance while retaining the same structure, functions, and feedbacks.

The more resilient a system, the larger the disturbance it can absorb without shifting into an alternate regime (Walker et al. 2006). By enhancing resilience, we can reduce the risk of nonlinear changes, including accelerating, abrupt, and potentially irreversible changes that have already been observed in many strongly impacted ecosystems (Adger et al. 2005; Millennium Ecosystem Assessment 2005). In ecological systems, resilience and productivity can be strengthened by maintaining biodiversity. Experiments have shown that the more species in a marine system, the higher the primary and secondary productivity, the efficiency of resource use, and the system's stability (Worm et al. 2006). An analysis of global fisheries data in sixty-four Large Marine Ecosystems (LMEs) showed that higher species richness increased the productivity and stability of fisheries, reduced the number of fishery collapses, and enhanced recovery after a collapse (Worm et al. 2006). Conservation efforts have shown that recovery of biodiversity and restoration of degraded ecosystems is possible. In a comparative analysis of forty-eight marine protected areas and reserves, for example, species richness increased and enhanced productivity and stability (Worm et al. 2006). Protection and restoration of marine biodiversity is thus a key to maintaining ecosystem services.

In social systems, increased resilience is needed for desired social functions and structures (Glaser 2006). Social system resilience rises with societal abilities to learn and adapt. In some circumstances, the transformability of a social system, rather than its resilience, is the object of analysis and the aim of governance and management (Olsson et al. 2006).

To achieve socioeconomic resilience and/or transformability, which include maintaining the productivity of natural ecosystems, sustainable governance is also needed (Costanza et al. 1998; Millennium Ecosystem Assessment 2005). A major challenge is to reverse ecosystem degradation while meeting increasing societal demands for ecosystem services. The Millennium Ecosystem Assessment explores different future scenarios. Most require substantial changes in social institutions, practices, and policies to enhance investments in environmentally sound technology, education, and health; to reduce socioeconomic disparities and poverty; and to expand social capacities for adaptive management (Millennium Ecosystem Assessment 2005). A recent report by the World Resources Institute (WRI 2007) suggested that in order to sustain ecosystem services, people's understanding of the connection between healthy ecosystems and the attainment of social and economic goals needs to be enhanced, the rights of local people to use and manage the ecosystems they depend on for their livelihoods and well-being need to be strengthened, ecosystem services need to be managed across multiple levels and time frames, the accountability of governments and business for decisions that affect ecosystem services needs to be increased, and economic and financial incentives need to reward ecosystem stewardship. The future provision of ecosystem services by SEMSs will largely depend on the achievement of such changes in the social and governance systems.

Acknowledgments

We thank Milena Arias-Schreiber and Ursula Mendoza for assembling data on social indicators, Mohammed Mozumder for providing supportive background information, and Boris Worm for comments and discussions.

References

Adam, P. 2002. Saltmarshes in a time of change. *Environmental Conservation* 29:39–61.

Adger, W.N., T.P. Hughes, C. Folke, S.R. Carpenter, and J. Rockstrom. 2005. Social-ecological resilience to coastal disasters. *Science* 309:1,036–1,039.

Bricker, S.B., C.G. Clement, D.E. Pirhalla, S.P. Orlando, and D.R.G. Farrow. 1999. *National Estuarine Eutrophication Assessment: Effects of Nutrient Enrichment in the Nation's Estuaries.* Silver Spring, MD: National Oceanic and Atmospheric Administration (NOAA), National Ocean Service, Special Projects Office and the National Center for Coastal Ocean Science.

Christensen, V., S. Guenette, J.J. Heymans, C.J. Walters, R. Watson, D. Zeller, and D. Pauly. 2003. Hundred-year decline of North Atlantic predatory fishes. *Fish and Fisheries* 4:1–24.

Costanza, R., F. Andrade, P. Antunes, M. van den Belt, D. Boersma, D.F. Boesch, F. Catarino, S. Hanna, K. Limburg, B. Low, M. Molitor, J.G. Pereira, S. Rayner, R. Santos, J. Wilson, and M. Young. 1998. Principles for sustainable governance of the oceans. *Science* 281:198–199.

Crutzen, P. 2002. Geology of mankind. *Nature* 415:23.

Daily, G.C. 1997. *Nature's Services: Societal Dependence on Natural Ecosystems.* Washington, DC: Island Press.

de Groot, R.S., M.A. Wilson, and R.M.J. Boumans. 2002. A typology for the classification, description and valuation of ecosystem functions, goods and services. *Ecological Economics* 41:393–408.

Diaz, R.J., J. Nestlerode, and M.L. Diaz. 2004. A global perspective on the effects of eutrophication and hypoxia on aquatic biota. In *Fish Physiology, Toxicology and Water Quality.* Proceedings of the Seventh International Symposium,Tallinn, Estonia 12–15 May 2003. Washington, DC: U.S. Environmental Protection Agency. http://www.epa.gov/msbasin/taskforce/2006symposia/148fish.pdf.

Ehler, C. 2003. Indicators to measure governance performance in integrated coastal management. *Ocean and Coastal Management* 46:335–345.

FAO. 2008. Food and Agricultural Organization of the United Nations. Fisheries and Aquaculture Department. http://www.fao.org/fi/website/FISearch.do?dom=country.

Glaser, M. 2003. Ecosystem, local economy and social sustainability: A case study of Caeté Estuary, North Brazil. *Wetlands Ecology and Management* 11 (4):265–272.

Glaser, M. 2006. Conceptualising and operationalising the social dimension in ecosystem management. In *The "social" in ecosystem management: Theoretical and empirical dimensions,* Chap. 4.2. Habilitation, Faculty of Agriculture and Horticulture. Berlin: Humboldt University.

Hockings, M., S. Stoll-Kleemann, S. Bender, A. Berghöfer, M. Bertsky, S. Bhatt, F. Gatzweiler, M. Glaser, L. Hiwasaki, H. Ross, and H. Wittmer. 2005. *Social, Economic and Cultural Criteria and Indicators for Protected Area Management.* Report of 19–22 June 2005 Workshop at Humboldt University of Berlin, Germany.

Hockings, M., S. Stolton, and N. Dudley. 2000. *Evaluating Effectiveness: A Framework for Assessing the Management of Protected Areas.* IUCN Best Practice Protected Area Guidelines Series, No. 6. Cambridge, UK: IUCN (World Conservation Union).

Horner, R., S.F. Ackley, G.S. Dieckmann, B. Gulliksen, T. Hoshiai, L. Legendre, I.A. Melnikov, W.S. Reeburgh, M. Spindler, and C.W. Sullivan. 1992. Ecology of sea ice biota. I: Habitat, terminology, and methodology. *Polar Biology* 12:417–427.

Jackson, J.B.C., M.X. Kirby, W.H. Berger, K.A. Bjorndal, L.W. Botsford, B.J. Bourque, R.H. Bradbury, R. Cooke, J. Erlandson, J.A. Estes, T.P. Hughes, S. Kidwell, C.B. Lange, H.S. Lenihan, J.M. Pandolfi, C.H. Peterson, R.S. Steneck, M.J. Tegner, and R.R.

Warner. 2001. Historical overfishing and the recent collapse of coastal ecosystems. *Science* 293:629–638.

Jennings, S., and J.L. Blanchard. 2004. Fish abundance with no fishing: Predictions based on macroecological theory. *Journal of Animal Ecology* 73:632–642.

Kennish, M.J. 2002. Environmental threats and environmental futures of estuaries. *Environmental Conservation* 29:78–107.

Limburg, K.E. 1999. Estuaries, ecology, and economic decisions: An example of perceptual barriers and challenges to understanding. *Ecological Economics* 30:185–188.

Lotze, H.K., H.S. Lenihan, B.J. Bourque, R.H. Bradbury, R.G. Cooke, M.C. Kay, S.M. Kidwell, M.X. Kirby, C.H. Peterson, and J.B.C. Jackson. 2006. Depletion, degradation, and recovery potential of estuaries and coastal seas. *Science* 312:1,806–1,809.

Lüning, K. 1990. *Seaweeds: Their Environment, Biogeography and Ecophysiology*. Chichester, UK: John Wiley & Sons.

Maneschy, C. 1994. *Ajuruteua, uma comunidade pesqueira ameaçada*. Belém, Brazil: Universidade Federal do Pará.

Millennium Ecosystem Assessment. 2005. *Ecosystems and Human Well-Being: Synthesis*. Washington, DC: Island Press.

Olsson, P., L.H. Gunderson, S.R. Carpenter, P. Ryan, L. Lebel, C. Folke, and C.S. Holling. 2006. Shooting the rapids: Navigating transitions to adaptive governance of social-ecological systems. *Ecology and Society* 11 (1):18. http://www.ecologyandsociety.org/vol11/iss1/art18/.

Orth, R.J., T.J.B. Carruthers, W.C. Dennison, C.M. Duarte, J.W. Fourqurean, J. Kenneth, L. Heck, A.R. Hughes, G.A. Kendrick, W.J. Kenworthy, S. Olyarnik, F.T. Short, M. Waycott, and S.L. Williams. 2006. A global crisis for seagrass ecosystems. *BioScience* 56:987–996.

Pandolfi, J.M., R.H. Bradbury, E. Sala, T.P. Hughes, K.A. Bjorndal, R.G. Cooke, D. McArdle, L. McClenachan, M.J.H. Newman, G. Paredes, R.R. Warner, and J.B.C. Jackson. 2003. Global trajectories of the long-term decline of coral reef ecosystems. *Science* 301:955–958.

Rogers, A.D. 1994. The biology of seamounts. *Advances in Marine Biology* 30:305–350.

Rosenberg, A.A., W.J. Bolster, K.E. Alexander, W.B. Leavenworth, A.B. Cooper, and M.G. McKenzie. 2005. The history of ocean resources: Modeling cod biomass using historical records. *Frontiers in Ecology and the Environment* 3:84–90.

Sea Around Us. 2008. http://www.seaaroundus.org/lme/lme.aspx.

Sellner, K.G., G.J. Doucette, and G.J. Kirkpatrick. 2003. Harmful algal blooms: Causes, impacts and detection. *Journal for Industrial Microbiology and Biotechnology* 30:383–406.

Steneck, R.S., M.H. Graham, B.J. Bourque, D. Corbett, J.M. Erlandson, J.A. Estes, and M.J. Tegener. 2002. Kelp forest ecosystems: Biodiversity, stability, resilience and future. *Environmental Conservation* 29:436–459.

Transparency International. 2008. http://www.transparency.org/policy_research/surveys_indices/cpi.

Tsyban, A., G. Titova, S. Shchuka, V. Ranenko, and Y. Izrael. 2005. *Russian Arctic: GIWA Regional Assessment 1a*. Global International Waters Assessment. Kalmar, Sweden: University of Kalmar on behalf of United Nations Environment Programme.

UN Development Programme. 2008. *Beyond Scarcity: Power, Poverty and the Global Water*

Crisis. Human Development Report 2006. http://hdr.undp.org/en/reports/global/hdr2006/.

UN Statistics Division. 2008. http://unstats.un.org/unsd/demographic/default.htm.

Vitousek, P.M., H.A. Mooney, J. Lubchenco, and J.M. Melillo. 1997. Human domination of Earth's ecosystems. *Science* 277:494–499.

Walker, B.H., J.M. Anderies, A.P. Kinzig, and P. Ryan. 2006. Exploring resilience in social-ecological systems through comparative studies and theory development: Introduction to the special issue. *Ecology and Society* 11 (1):12. http://www.ecologyandsociety.org/vol11/iss1/art12/.

World Factbook. 2008. U.S. Central Intelligence Agency. https://www.cia.gov/library/publications/the-world-factbook/index.html.

Worm, B., E.B. Barbier, N. Beaumont, J.E. Duffy, C. Folke, B.S. Halpern, J.B.C. Jackson, H.K. Lotze, F. Micheli, S.R. Palumbi, E. Sala, K.A. Selkoe, J.J. Stachowicz, and R. Watson. 2006. Impacts of biodiversity loss on ocean ecosystem services. *Science* 314:787–790.

WRI. 2007. *Restoring Nature's Capital. An Action Agenda to Sustain Ecosystem Services.* Washington, DC: World Resources Institute (WRI).

Appendix: Workshop Participants and Other Contributors

Hans H. DÜRR
Utrecht University
The Netherlands
h.durr@geo.uu.nl

Werner EKAU
University of Bremen
Germany
wekau@zmt.uni-bremen.de

Elva ESCOBAR-BRIONES
Universidad Nacional Autonoma de Mexico
Mexico
escobri@mar.icmyl.unam.mx

Wolfgang FENNEL
Baltic Sea Research Institute
Germany
Wolfgang.fennel@io-warnemuende.de

Michael FLITNER
University of Bremen
Germany
mflitner@uni-bremen.de

Denis GILBERT
Fisheries and Oceans Canada / Pêches et Océans Canada
Canada
gilbertd@dfo-mpo.gc.ca

Marion GLASER
University of Bremen
Germany
mglaser@zmt-bremen.de

Susan GREENWOOD ETIENNE
Scientific Committee on Problems of the Environment
France
sgreenwood@icsu-scope.org

Akira HARASHIMA
National Institute for Environmental Studies
Japan
harashim@nies.go.jp

John HARRISON
Washington State University
USA
harrisoj@vancouver.wsu.edu

Venugopalan ITTEKKOT
University of Bremen
Germany
ittekkot@zmt-bremen.de

Johan van de KOPPEL
Netherlands Institute of Ecology
The Netherlands
J.vandekoppel@nioo.knaw.nl

Gennady KOROTAEV
Marine Hydrophysical Institute
Ukraine
korotaevgren@mail.ru

Carolien KROEZE
Wageningen University
The Netherlands
carolien.kroeze@wur.nl

Rik LEEMANS
Wageningen University
The Netherlands
rik.leemans@wur.nl

Kon-Kee LIU
National Central University
Taiwan
kkliu@ncu.edu.tw

Heike K. LOTZE
Dalhousie University
Canada
hlotze@dal.ca

Michael MacCRACKEN
Climate Institute
USA
mmaccrac@comcast.net

Paola MALANOTTE-RIZZOLI
Massachusetts Institute of Technology
USA
Rizzoli@mit.edu

Emilio MAYORGA
Rutgers University
USA
mayorga@marine.rutgers.edu

Jerry M. MELILLO
Marine Biological Laboratory
USA
jmelillo@mbl.edu

Katja MERTIN
Free University of Berlin
Germany
mertin@graduateschool.jfki.de

Michel MEYBECK
Université de Paris 6
France
Michel.Meybeck@ccr.jussieu.fr

Jack MIDDELBURG
Netherlands Institute of Ecology
The Netherlands
j.middelburg@nioo.knaw.nl

Wajih NAQVI
National Institute of Oceanography
India
naqvi@nio.org

Temel OGUZ
Middle East Technical University
Turkey
oguz@ims.metu.edu.tr

Gerardo M.E. PERILLO
CONICET—Instituto Argentino de Oceanografia
Argentina
perillo@criba.edu.ar

Catharina J.M. PHILIPPART
Royal Netherlands Institute for Sea Research
The Netherlands
Katja@nioz.nl

Véronique PLOCQ FICHELET
Scientific Committee on Problems of the Environment
France
vpf@icsu-scope.org

Nancy N. RABALAIS
Louisiana Universities Marine Consortium
USA
nrabalais@lumcon.edu

Maurizio RIBERA D'ALCALÀ
Stazione Zoologica 'A. Dohrn'
Italy
Maurizio@szn.it

Tim RIXEN
University of Bremen
Germany
trixen@uni-bremen.de

Sybil SEITZINGER
Rutgers University—The State University of New Jersey
USA
sybil@marine.rutgers.edu

Elizabeth H. SHADWICK
Dalhousie University
Canada
shadwick@phys.ocean.dal.ca

Paul V. R. SNELGROVE
Memorial University of Newfoundland
Canada
psnelgro@mun.ca

Penjai SOMPONGCHAIYAKUL
Prince of Songkla University
Thailand
penjai.s@psu.ac.th

Emil STANEV
University of Ulster
Northern Ireland, UK
ev.stanev@ulster.ac.uk

Jilan SU
Second Institute of Oceanography
China
sjl@sio.org.cn; su_jilan@yahoo.com.cn

Bjørn SUNDBY
Université de Québec and McGill University
Canada
Bjorn.Sundby@mcgill.ca

Paul TETT
Napier University
Scotland, UK
p.tett@ichrachan.u-net.com

Helmuth THOMAS
Dalhousie University
Canada
Helmuth.thomas@dal.ca

Daniela UNGER
University of Bremen
Germany
daniela.unger@zmt-bremen.de

Edward R. URBAN, Jr.
Scientific Committee on Oceanic Research
USA
Ed.Urban@scor-int.org

Edy YUWONO
Universitas Jenderal Soedirman (UNSOED)
Indonesia
Edy_ywn@yahoo.com

Jing ZHANG
East China Normal University
China
jzhang@sklec.ecnu.edu.cn

About the Editors

EDWARD R. URBAN, JR., is SCOR Executive Director.

BJØRN SUNDBY is a professor at the Université du Québec à Rimouski and in the Department of Earth and Planetary Sciences at McGill University and is the SCOR President.

PAOLA MALANOTTE-RIZZOLI is a professor of Physical Oceanography in the Department of Earth, Atmospheric, and Planetary Sciences at the Massachusetts Institute of Technology and is past president of IAPSO.

JERRY M. MELILLO is past president of SCOPE and codirector of The Ecosystems Center of the Marine Biological Laboratory in Woods Hole, Massachusetts.

SCOPE Series List

SCOPE 1–59 are now out of print. Selected titles from this series can be downloaded free of charge from the SCOPE Web site (http://www.icsu-scope.org).

SCOPE 1: *Global Environment Monitoring,* 1971, 68 pp
SCOPE 2: *Man-made Lakes as Modified Ecosystems,* 1972, 76 pp
SCOPE 3: *Global Environmental Monitoring Systems (GEMS): Action Plan for Phase I,* 1973, 132 pp
SCOPE 4: *Environmental Sciences in Developing Countries,* 1974, 72 pp
SCOPE 5: *Environmental Impact Assessment: Principles and Procedures,* Second Edition, 1979, 208 pp
SCOPE 6: *Environmental Pollutants: Selected Analytical Methods,* 1975, 277 pp
SCOPE 7: *Nitrogen, Phosphorus and Sulphur: Global Cycles,* 1975, 129 pp
SCOPE 8: *Risk Assessment of Environmental Hazard,* 1978, 132 pp
SCOPE 9: *Simulation Modelling of Environmental Problems,* 1978, 128 pp
SCOPE 10: *Environmental Issues,* 1977, 242 pp
SCOPE 11: *Shelter Provision in Developing Countries,* 1978, 112 pp
SCOPE 12: *Principles of Ecotoxicology,* 1978, 372 pp
SCOPE 13: *The Global Carbon Cycle,* 1979, 491 pp
SCOPE 14: *Saharan Dust: Mobilization, Transport, Deposition,* 1979, 320 pp
SCOPE 15: *Environmental Risk Assessment,* 1980, 176 pp
SCOPE 16: *Carbon Cycle Modelling,* 1981, 404 pp
SCOPE 17: *Some Perspectives of the Major Biogeochemical Cycles,* 1981, 175 pp
SCOPE 18: *The Role of Fire in Northern Circumpolar Ecosystems,* 1983, 344 pp
SCOPE 19: *The Global Biogeochemical Sulphur Cycle,* 1983, 495 pp
SCOPE 20: *Methods for Assessing the Effects of Chemicals on Reproductive Functions, SGOMSEC 1,* 1983, 568 pp
SCOPE 21: *The Major Biogeochemical Cycles and Their Interactions,* 1983, 554 pp
SCOPE 22: *Effects of Pollutants at the Ecosystem Level,* 1984, 460 pp

SCOPE 23: *The Role of Terrestrial Vegetation in the Global Carbon Cycle: Measurement by Remote Sensing,* 1984, 272 pp
SCOPE 24: *Noise Pollution,* 1986, 466 pp
SCOPE 25: *Appraisal of Tests to Predict the Environmental Behaviour of Chemicals,* 1985, 400 pp
SCOPE 26: *Methods for Estimating Risks of Chemical Injury: Human and Non-Human Biota and Ecosystems, SGOMSEC 2,* 1985, 712 pp
SCOPE 27: *Climate Impact Assessment: Studies of the Interaction of Climate and Society,* 1985, 650 pp
SCOPE 28: *Environmental Consequences of Nuclear War, Volume I: Physical and Atmospheric Effects,* 1986, 400 pp, *Volume II: Ecological and Agricultural Effects,* 1985, 563 pp
SCOPE 29: *The Greenhouse Effect, Climatic Change and Ecosystems,* 1986, 574 pp
SCOPE 30: *Methods for Assessing the Effects of Mixtures of Chemicals, SGOMSEC 3,* 1987, 928 pp
SCOPE 31: *Lead, Mercury, Cadmium and Arsenic in the Environment,* 1987, 384 pp
SCOPE 32: *Land Transformation in Agriculture,* 1987, 552 pp
SCOPE 33: *Nitrogen Cycling in Coastal Marine Environments,* 1988, 478 pp
SCOPE 34: *Practitioner's Handbook on the Modelling of Dynamic Change in Ecosystems,* 1988, 196 pp
SCOPE 35: *Scales and Global Change: Spatial and Temporal Variability in Biospheric and Geospheric Processes,* 1988, 376 pp
SCOPE 36: *Acidification in Tropical Countries,* 1988, 424 pp
SCOPE 37: *Biological Invasions: A Global Perspective,* 1989, 528 pp
SCOPE 38: *Ecotoxicology and Climate with Special References to Hot and Cold Climates,* 1989, 432 pp
SCOPE 39: *Evolution of the Global Biogeochemical Sulphur Cycle,* 1989, 224 pp
SCOPE 40: *Methods for Assessing and Reducing Injury from Chemical Accidents, SGOMSEC 6,* 1989, 320 pp
SCOPE 41: *Short-Term Toxicity Tests for Non-genotoxic Effects, SGOMSEC 4,* 1990, 353 pp
SCOPE 42: *Biogeochemistry of Major World Rivers,* 1991, 356 pp
SCOPE 43: *Stable Isotopes: Natural and Anthropogenic Sulphur in the Environment,* 1991, 472 pp
SCOPE 44: *Introduction of Genetically Modified Organisms into the Environment,* 1990, 224 pp
SCOPE 45: *Ecosystem Experiments,* 1991, 296 pp
SCOPE 46: *Methods for Assessing Exposure of Human and Non-human Biota SGOMSEC 5,* 1991, 448 pp

SCOPE 47: *Long-Term Ecological Research. An International Perspective,* 1991, 312 pp
SCOPE 48: *Sulphur Cycling on the Continents: Wetlands, Terrestrial Ecosystems and Associated Water Bodies,* 1992, 345 pp
SCOPE 49: *Methods to Assess Adverse Effects of Pesticides on Non-target Organisms, SGOMSEC 7,* 1992, 264 pp
SCOPE 50: *Radioecology after Chernobyl,* 1993, 367 pp
SCOPE 51: *Biogeochemistry of Small Catchments: A Tool for Environmental Research,* 1993, 432 pp
SCOPE 52: *Methods to Assess DNA Damage and Repair: Interspecies Comparisons, SGOMSEC 8,* 1994, 257 pp
SCOPE 53: *Methods to Assess the Effects of Chemicals on Ecosystems, SGOMSEC 10,* 1995, 440 pp
SCOPE 54: *Phosphorus in the Global Environment: Transfers, Cycles and Management,* 1995, 480 pp
SCOPE 55: *Functional Roles of Biodiversity: A Global Perspective,* 1996, 496 pp
SCOPE 56: *Global Change, Effects on Coniferous Forests and Grasslands,* 1996, 480 pp
SCOPE 57: *Particle Flux in the Ocean,* 1996, 396 pp
SCOPE 58: *Sustainability Indicators: A Report on the Project on Indicators of Sustainable Development,* 1997, 440 pp
SCOPE 59: *Nuclear Test Explosions: Environmental and Human Impacts,* 1999, 304 pp
SCOPE 60: *Resilience and the Behavior of Large-Scale Systems,* 2002, 287 pp
SCOPE 61: *Interactions of the Major Biogeochemical Cycles: Global Change and Human Impacts,* 2003, 384 pp
SCOPE 62: *The Global Carbon Cycle: Integrating Humans, Climate, and the Natural World,* 2004, 526 pp
SCOPE 63: *Alien Invasive Species: A New Synthesis,* 2004, 352 pp.
SCOPE 64: *Sustaining Biodiversity and Ecosystem Services in Soils and Sediments,* 2003, 308 pp
SCOPE 65: *Agriculture and the Nitrogen Cycle,* 2004, 320 pp
SCOPE 66: *The Silicon Cycle: Human Perturbations and Impacts on Aquatic Systems,* 2006, 296 pp
SCOPE 67: *Sustainability Indicators: A Scientific Assessment,* 2007, 448 pp
SCOPE 68: *Communicating Global Change Science to Society: An Assessment and Case Studies,* 2007, 240 pp
SCOPE 69: *Biodiversity Change and Human Health: From Ecosystem Services to Spread of Disease,* 2008, in press
SCOPE 70: *Watersheds, Bays, and Bounded Seas: The Science and Management of Semi-Enclosed Marine Systems,* 2008, in press

SCOPE Executive Committee 2005–2008

President
Prof. O. E. Sala (Argentina)

Vice President
Prof. Wang Rusong (China-CAST)

Past President
Dr. J. M. Melillo (USA)

Treasurer
Prof. I. Douglas (UK)

Secretary-General
Prof. M. C. Scholes (South Africa)

Members
Prof. W. Ogana (Kenya-IGBP)
Prof. Annelies Pierrot-Bults (The Netherlands-IUBS)
Prof. V. P. Sharma (India)
Prof. H. Tiessen (Germany)
Prof. R. Victoria (Brazil)

Index